D1037143

PROCESS ENGINEERING and DESIGN for AIR POLLUTION CONTROL

Jaime Benítez
Professor
Chemical Engineering Department
University of Puerto Rico at Mayagüez
Mayagüez, Puerto Rico

PTR Prentice Hall
Englewood Cliffs, New Jersey 07632

Library of Congress Cataloging-in-Publication Data

Benitez, Jaime
Process engineering and design for air pollution control/ Jaime Benitez.
p. cm.
Includes bibliographical references and index.
ISBN 0-13-723214-4
1. Air--Purification--Equipment and supplies--Design and construction. 2. Air--Pollution--Cost effectiveness. I. Title.
TD889.B46 1993 92-20982
628.5'3--dc20 CIP

Editorial/production supervision/interior design: *Jean Lapidus*/
Page layout on the Macintosh IIci: *Jaime Benitez*
Tables/artwork: *Jaime Benitez*
Cover design: *Lundgren Graphics*
Manufacturing buyer: *Mary E. McCartney*
Acquisitions editor: *Mike Hays*
Editorial assistant: *Dana Mercure*

The publisher offers discounts on this book when ordered in bulk quantities. For more information, contact:

Corporate Sales Department
PTR Prentice Hall
113 Sylvan Avenue
Englewood Cliffs, N. J. 07632

Phone: 201-592-2863
Fax: 201-592-2249

Printed in the United States of America

10 9 8 7 6 5 4 3 2 1

ISBN 0-13-723214-4

Prentice-Hall International (UK) Limited, *London*
Prentice-Hall of Australia Pty. Limited, *Sydney*
Prentice-Hall Canada Inc., *Toronto*
Prentice-Hall Hispanoamericana, S. A., *Mexico*
Prentice-Hall of India Private Limited, *New Delhi*
Prentice-Hall of Japan, Inc., *Tokyo*
Simon & Schuster Asia Pte. Ltd., *Singapore*
Editora Prentice-Hall do Brasil, Ltda., *Rio de Janeiro*

Dan Taylor—of Parthenon, Arkansas—deserves to be included as co-author of this book. Dan's contribution to this work spans more than two decades. Twenty-five years ago, he introduced me to the art and science of problem-solving with such a refined and masterly approach that I became deeply and forever attached to the teachings and the teacher. More recently, he read most of this manuscript, improved it with many suggestions, rewrote portions of it, and even suggested the title for the book. When I told him that he would be its co-author, he flatly refused claiming his contribution was not significant enough to justify it. But I know better.

Jaime Benítez

A mis padres, esposa e hijos; quienes me hacen pensar en el pasado, el presente y el futuro que puede ser.

Contents

Preface

Air pollution control in a cost-effective manner is one of the big engineering challenges of our times. Those technologies that constitute the very pillars of modern society—electric power generation, industrial activity, and the automobile—are also serious threats to our atmospheric environment. The governments of many industrialized countries recognize the urgency of this predicament and have developed legislation aimed at reducing emissions of air pollutants while maintaining a healthy economy. Such a task is by no means trivial. It requires a professional trained in such disciplines as: material and energy balances; thermodynamics; fluid dynamics; heat- and mass-transfer, reaction kinetics; process design and economics; and applied mathematics. The chemical engineer, with a formal training in all these disciplines, has a distinct advantage.

My objective in writing this book is to provide a means for senior chemical engineering students and practicing engineers to exploit fully their potential for the solution of air pollution control engineering problems. The book emphasizes equipment design and cost estimation. The idea for it was born out of my experiences during the last 15 years teaching air pollution control courses at the University of Puerto Rico.

The book provides a rigorous treatment of equipment design and cost estimation, including optimization techniques. Chapters are divided into sections with clearly stated objectives at the beginning. Numerous detailed examples follow each brief section of text. Abundant end-of-chapter problems are included, and problem degree of difficulty is clearly labeled for each. Most of these problems are accompanied by answers. Computer solution is emphasized, both in the examples and in the end-of-chapter problems. Major design problems are included in some chapters. Problems on occupational health and safety (an area strongly emphasized by the Accreditation Board for Engineering and Technology [ABET]) that are relevant to air pollution control are also included.

The book uses SI units exclusively, which virtually eliminates the tedious task of unit conversions and makes it "readable" to the international scientific and technical community. Although some resistance to complete adoption of SI units in the United States still exist, the transition is well under way.

FORTRAN subroutines have been included for the reader's use in the solution of computationally complex problems. Documentation of the computer code has been kept to a minimum to encourage the interested reader to explore on his or her own the intimate workings of the proposed algorithms. Although all of the subroutines have been successfully tested, computer programming is not among my main skills; therefore, there is probably ample room for improvement in that area.

The book is intended for use at the senior level in chemical and environmental engineering curricula. The material treated can serve as the subject of either a full-year or a one-term course, depending on the choice of topics covered.

Chapter 1 presents a general introduction to the air pollution problem. It identifies those substances considered air pollutants and discusses in some detail their harmful effects. It summarizes the federal legislation created in the United States to deal with air pollution, with special emphasis on the impact of the regulations on control equipment design. Examples of the application of chemical engineering skills to the solution of air pollution control problems are included.

Chapter 2 introduces the cost estimation methodology that is the backbone of the rest of the book. It presents the fundamentals of engineering economics and decision making. The methodology presented is consistent with the guidelines established by the U.S. Environmental Protection Agency (EPA) for regulatory purposes.

Chapters 3 to 6 cover some aspects of the area of control of gaseous pollutants emissions. Chapters 3 and 4 present different approaches for dealing with volatile organic compounds (VOCs) emissions: incineration and carbon adsorption. Chapter 5 covers flue gas desulfurization (FGD) systems, emphasizing the difference between throwaway and regenerative processes. Chapter 6 deals with the control of nitrogen oxides emissions, a difficult and controversial problem.

Chapters 7 to 10 cover some aspects of the area of control of particulate matter emissions. Chapter 7 is an introduction to the subject of particles dynamics in fluids. Concepts of particle size distribution functions and collection efficiency are also presented. Chapter 8 covers cyclonic devices, a simple and effective way to control emissions of relatively big particles, frequently used as precleaners for more sophisticated devices. Chapter 9 discusses electrostatic precipitators, the workhorse of fine particle control devices. Chapter 10 covers the design of fabric filters, a technology that is gaining widespread acceptance because of its simplicity and good performance removing fine particles.

It must be emphasized that this book does not intend to be a comprehensive review of all air pollution control engineering processes presently available. Instead, it aims at presenting a general methodology for process engineering and design, which can be applied to any specific air pollution control problem. The choice of specific control technologies presented in detail in the book was based on my subjective perception of their importance and on pertinent information availability.

I wish to acknowledge gratefully the contribution of the University of Puerto Rico at Mayagüez to this project. My special gratitude goes to Rafael Muñoz, an extraordinary human being and example of professionalism. Alberto Molini deserves a special mention, because he was the first to suggest the idea that I should write this book.

Abraham Rodríguez provided warm friendship and support, and access to his excellent periodicals collection. David Rivera made it all possible because he was my first contact with Prentice Hall. My students in the courses In Qu 5015 and In Qu 5018 reviewed the material of the book and gave excellent suggestions on ways to improve it.

My special gratitude goes to my wife and my children—who were always around lifting my spirits during the long, arduous hours of work—and to Elsa Castro, my source of inspiration. They made it all worthwhile.

Jaime Benítez
Mayagüez, Puerto Rico

1

Overview of the Air Pollution Problem

1.1 INTRODUCTION

Air pollution is inextricably woven into the fabric of modern life. The major cause of all air pollution is combustion, and combustion has been essential to man since the discovery of fire. The widespread use of electricity, and the ubiquity of the automobile are the two most characteristic traits of the 20th century. They both depend on combustion and account for most air pollutants.

Air pollution on a local scale is not a recent phenomenon. As early as 1272, King Edward I of England tried to clear the heavily polluted sky over London by banning the use of "sea coal". The British Parliament ordered the torturing and hanging of a man who sold and burned the outlawed coal (a far cry from the punishments incorporated into modern environmental legislation!).

Conversely, air pollution on a regional and global scale is a recent phenomenon. Never in recorded history have there been anthropogenic emissions of atmospheric pollutants capable of producing such disturbing effects as widespread acid rain, disruption of global weather patterns, and significant damage to the life-preserving ozone layer. It has become imperative to deal effectively with air pollution problems on all local, regional and global scales.

There are two strategies to reduce air pollution: prevention, and remediation. Prevention implies moving forward to the energy consumption levels of last century. Because such a shift in attitudes is unlikely, remediation becomes imperative.

Remediation is the essence of air pollution control engineering. If economies continue burning fossil fuels to generate electricity and power automobiles, and if industrial processes keep producing hazardous emissions, engineers must effectively and economically remove pollutants before they reach the atmosphere. Such a task is by no means trivial. It requires a professional trained in material and energy balances, thermodynamics, fluid dynamics, heat and mass transfer, reaction kinetics, process design and economics, and applied mathematics. The chemical engineer, skilled in all these disciplines, has a distinct advantage.

This chapter identifies those substances considered to be air pollutants, details their most important anthropogenic sources, and describes their deleterious effects. A brief review of pertinent legislation follows. Some examples illustrate the application of classical concepts of chemical engineering to problems in air pollution.

1.2 AIR POLLUTANTS

Your objectives in studying this section are to

1. Identify those substances considered air pollutants.
2. Identify major anthropogenic sources of air pollutants.

The main focus here is on major, combustion-generated compounds: sulfur dioxide, carbon monoxide and dioxide, nitrogen oxides, VOCs, and particulate matter. Countless other pollutants from specific processes pose hazards and require similar treatment.

1.2.1 Sulfur Oxides

Sulfur dioxide (SO_2) is formed from the oxidation of sulfur contained in fossil fuels and from industrial processes that treat and produce sulfur-containing compounds. By far, the greatest anthropogenic source of SO_2 in the United States is coal-generated electricity, which accounted in 1978 for more than 60% of total emissions (USEPA 1982). Primary metal smelters (7.5%), industrial burning of coal (7.2%), and oil-generated electricity (6.6%) follow distantly.

A small fraction of sulfur oxides appears in the form of primary sulfates, gaseous sulfur trioxide (SO_3), and sulfuric acid (H_2SO_4). It is estimated that, by volume, more than 90% of the total U.S. sulfur oxides emissions are in the form of SO_2, with primary sulfates accounting for most of the other 10%.

1.2.2 Carbon Monoxide (CO)

CO is a product of the incomplete combustion of carbonaceous fuels. The greatest anthropogenic source of this pollutant is internal combustion engines used for transportation. Automobiles accounted in 1981 for 77% of total emissions (USEPA, 1982b). Industrial processes (7%), and residential heating (6%) follow.

Power plants and other large furnaces usually operate to insure virtually complete combustion. The very nature of the internal combustion engine requires operation with an oxygen deficiency which causes substantial formation of carbon monoxide.

1.2.3 Carbon Dioxide (CO_2)

CO_2 is a product of the complete combustion of carbonaceous fuels. This gas has not been considered an atmospheric pollutant. However, its global background concentration has been increasing steadily leading to serious concerns about its possible effect on global climate. Combustion of fossil fuels for transportation and electric power generation released worldwide an estimated 6 billion metric tons in 1990 (Fulkerson et al. 1990)!

1.2.4 Nitrogen Oxides

Nitric oxide (NO) and nitrogen dioxide (NO_2) are the two most important nitrogen oxides from the air pollution viewpoint. They are frequently lumped together under the designation NO_x. NO is a principal by-product of combustion processes, arising from the high temperature reaction between N_2 and O_2 in the combustion air. The proposed mechanism for this oxidation, known as Zeldovich mechanism, is

$$\begin{aligned} N_2 + O &\rightarrow NO + N \\ N + O_2 &\rightarrow NO + O \end{aligned} \tag{1.1}$$

The term for nitric oxide formed this way is *thermal-*NO_x. The second major source for NO in combustion is by the oxidation of organically bound nitrogen in the fuel. Nitric oxide formed in this manner is referred to as *fuel-*NO_x.

Automobiles and electric power generation account for approximately 34%, each, of total anthropogenic emissions of NO. Industrial furnaces account for 15% (USEPA 1982b).

1.2.5 Volatile Organic Compounds (VOCs)

VOCs include all organic compounds with appreciable vapor pressure. Some are hydrocarbons, but others may be aldehydes, ketones, chlorinated solvents, refrigerants, and so on. Table 1.1 lists some of the VOCs of interest in air pollution control applications.

The major anthropogenic sources of VOCs are industrial processes (46%) and automobiles (30%). VOCs in the exhaust gases from motor vehicles consist of unburned or partially burned gasoline. Gasoline is the 313- to 537-K fraction from petroleum distillation and contains approximately 2,000 compounds. These include paraffins, olefins, and aromatics. Typical compositions vary from 4% olefins and 48% aromatics to 22% olefins and 20% aromatics. Unleaded fuel has a higher aromatic content than leaded fuel.

Table 1.1 Some VOCs Identified in Ambient Air

Family	Compound	Chemical Formula
Hydrocarbons	Methane	CH_4
	Ethane	C_2H_6
	Propane	C_3H_8
	Butane	C_4H_{10}
	Pentane	C_5H_{12}
	Hexane	C_6H_{14}
	Benzene	C_6H_6
	Toluene	C_7H_8
	Ethylene	$CH_2{=}CH_2$
	2-Butene	$CH_3CH{=}CHCH_3$
Halomethanes	Methyl chloride	CH_3Cl
	Chloroform	$CHCl_3$
	Carbon tetrachloride	CCl_4
Haloethanes	1,2-Dichloroethane	CH_2ClCH_2Cl
Halopropanes	1,2-Dichloropropane	$CH_2ClCHClCH_3$
Chloroalkenes	Trichloroethylene	$CHCl{=}CCl_2$
	Allyl chloride	$ClCH_2CH{=}CH_2$
Chloroaromatics	Monochlorobenzene	C_6H_5Cl
	Dichlorobenzene	$C_6H_4Cl_2$
Oxygenated and nitrogenated species	Formaldehyde	HCHO
	Peroxyacetyl nitrate (PAN)	$CH_3COOONO_2$
	Acrylonitrile	CHCN
Chlorofluorocarbons (CFCs)	CFC-11	$CFCl_3$
	CFC-12	CF_2Cl_2

1.2.6 Particulate Matter

Particulate matter refers to everything emitted to the atmosphere in the form of a condensed (liquid or solid) phase. Industrial processes (43%), automobiles (13%), and electric power generation (12%) are the major anthropogenic sources.

In industrial use and electric power generation, coal and, to a lesser extent, oil combustion contribute most of the particulate emissions. Coal is a slow-burning fuel with a relatively high ash content. In contrast, oil is a fast-burning, low-ash fuel. The low ash content results in formation of less particulate matter.

1.3 CONCENTRATION UNITS FOR ATMOSPHERIC POLLUTANTS

Your objectives in studying this section are to

1. Define *parts per million and micrograms per cubic meter.*
2. Establish a relation between them for gaseous pollutants.

Because concentrations of atmospheric pollutants are usually quite small, special concentration units simplify work in air pollution. The first is used for gaseous pollutants exclusively: parts per million by volume (ppm). It is defined by

$$\text{ppm} = \frac{V_{pol}}{V_{air}} \times 10^6 \qquad (1.2)$$

where

V_{pol} = partial volume occupied by the pollutant in the mixture at total pressure P and temperature T

V_{air} = total volume occupied by the mixture at the same T and P

The concentration of either gaseous pollutants or particulate matter can be expressed in terms of *micrograms of pollutant per cubic meter of air.* Symbolically,

$$C_{\mu} = \frac{\text{micrograms}}{\text{cubic meter}} = \frac{\mu\text{g}}{\text{m}^3} \qquad (1.3)$$

The relationship between these two concentration units for gaseous pollutants as derived from the ideal gas law applied to the mixture of air and pollutant (see Problem 1.1) is

$$\text{ppm} = \frac{8.314\, T\, C_{\mu}}{P\, M_i} \qquad (1.4)$$

where M_i is the molecular weight of the pollutant, T is in K, and P in Pa.

Example 1.1 Parts per Million and Micrograms per Cubic Meter

The SO_2 concentration over an urban area was found to be 0.10 ppm. A bright, young chemical engineer claims he has developed a process to recover some 50% of this compound from atmospheric air and convert it to sulfuric acid. Calculate the yield in kilograms of acid per cubic kilometer of air. Atmospheric conditions are 300 K and 1 bar.

Solution

The solution converts SO_2 concentration in ppm to micrograms per cubic meter. Yield is then calculated considering fractional recovery and stoichiometry.

Basis: 1 cubic kilometer of air at 300 K and 1 bar.

Rearrange Eq. (1.4), and substitute $M_i = 64$, $T = 300$ K, $P = 10^5$ Pa

$$C_\mu = \frac{PM_i}{8.314T}\ \text{ppm} = \frac{10^5 \times 64 \times 0.10}{8.314 \times 300} = 257\ \mu\text{g/m}^3 \tag{1.5}$$

Next, calculate the yield.

$$\text{Yield} = \frac{257\ \mu\text{g}}{\text{m}^3}\,\frac{(10^3\ \text{m})^3}{1\ \text{km}^3}\,\frac{1\ \text{kg SO}_2}{10^9\ \mu\text{g SO}_2}\,\frac{98\ \text{kg H}_2\text{SO}_4}{64\ \text{kg SO}_2}\,(0.5)$$

$$= 197\ \text{kg H}_2\text{SO}_4/\text{km}^3\ \text{air}$$

Comments

Although it is unlikely that such a process could ever be technically feasible, it corresponds precisely with what happens spontaneously in the atmosphere. Unfortunately, instead of recovering the sulfuric acid in the form of a useful product, it ends up damaging our bodies of water. The next section covers the details of this process.

1.4 SOME DELETERIOUS EFFECTS OF AIR POLLUTANTS

Your objectives in studying this section are to

1. List the most important deleterious effects of air pollutants.
2. Understand the mechanisms through which these effects operate.

A popular definition of air pollution is (Cooper and Alley 1986, p. 2):

> Air pollution is the presence in the outdoor atmosphere...of substances or pollutants in quantities which are or may be harmful or injurious to human health or welfare, animal or plant life, or property, or unreasonably interfere with the enjoyment of life or property, including outdoor recreation.

Any substance is a "pollutant" if it has some kind of deleterious effect. The identity of the substance is irrelevant as long as its concentration in ambient air is high enough to produce a measurable, undesirable effect. Under this definition, CO_2 (a substance that was considered totally harmless until recently) is now considered a very important air pollutant, as we shall soon discuss.

1.4.1 Particulate Matter

Particles can damage human and animal health, and retard plant growth. They also reduce visibility, soil buildings and other materials, damage materials, and alter local weather patterns.

An object is visible because it contrasts with its background. Particulates decrease visibility by scattering and absorbing some of the light emitted or reflected by the body reducing the contrast. Scattering of light by particles is a strong function of particle size. The most effective scatterers are particles about the same size as the wavelength of visible light, that is, in the range between 0.1 and 1.0 μm (Friedlander 1977). High humidity increases scattering because tiny particles tend to act as condensation nuclei to form droplets that increase the total number of effective particles.

Several authors have found correlations between the light scattering coefficient and the mass concentration of total suspended particulates in the air (National Research Council 1979). Higher correlations can be expected when fine particulates dominate the mass concentration of the total distribution.

Airborne particles affect human health in many ways. Of special interest are the physical effects of particles on the normal functioning of the respiratory system. To cause lung damage particles must penetrate the human respiratory system (see Figure 1.1). Particles larger than about 2 μm generally do not penetrate beyond the nasal cavity or the trachea. Nasal hairs intercept them or they settle unto the mucous membrane lining the nasal passages and the trachea. Once captured on these membranes, insoluble particles rapidly move to the larynx through the normal combined action of ciliated and mucus-secreting cells. From there particles can be swallowed or expectorated.

Very small particles (less than 0.1 μm) tend to deposit in the tracheobronchial tree by Brownian diffusion (resulting from collisions with gas molecules). These are removed in the same manner as large particles. Particles in the size range from 0.1 to 3 μm penetrate deep into the lungs where they settle in the respiratory bronchioles or the alveolar sacs. It has been found (Davison et al. 1974) that sulfur dioxide and toxic elements, such as arsenic, selenium, and cadmium, adsorb and concentrate on the surface of submicron particles during the combustion of fossil fuels. These substances are then transported and deposited deep in the lungs.

Ultimate effects of particulate on human health are difficult to quantify. Epidemiological studies show a direct relationship between particulate concentrations and hospital visits for bronchitis, asthma emphysema, pneumonia, and cardiac disease (National Research Council 1979).

1.4.2 SO_2

The early observations of SO_2 pollution associated it with extensive plant damage. One of the major effects on green plants is chlorosis, the loss of chlorophyll. Another is plasmolysis, tissue collapse of many of the leaf cells (Cooper and Alley 1986). Both effects occur with either short exposures to high concentrations or long exposures to lower concentrations.

SO_2 effects on human health depend on concentration. At concentrations above 1 ppm, some bronchoconstriction occurs; above 10 ppm, eye, nose, and throat irritation is observed. It also stimulates mucus secretion, a characteristic of chronic bronchitis. Particulates intensify the effects of SO_2. Inert particles with adsorbed SO_2 penetrate deep into the lungs and induce severe effects.

An important effect of SO_2 is its contribution to acid rain. Water droplets formed by condensation in the atmosphere normally should have a pH close to 7. However, the dissolution of atmospheric CO_2 in rainwater tends to lower the pH by forming carbonic acid. This acid is very weak, and at equilibrium the pH of rainwater would lie close to 5.65 (Wark and Warner 1981). By the early 1970s, however, pH values of rainwater from 2 to 6 were measured in different parts of the world. Thus acid concentrations from 10 to 10,000 times greater than expected from natural sources were detected. The high acid content is due to transformation of SO_2 and NO into acids in the atmosphere, followed by absorption in cloud water and raindrops. A typical reaction might be

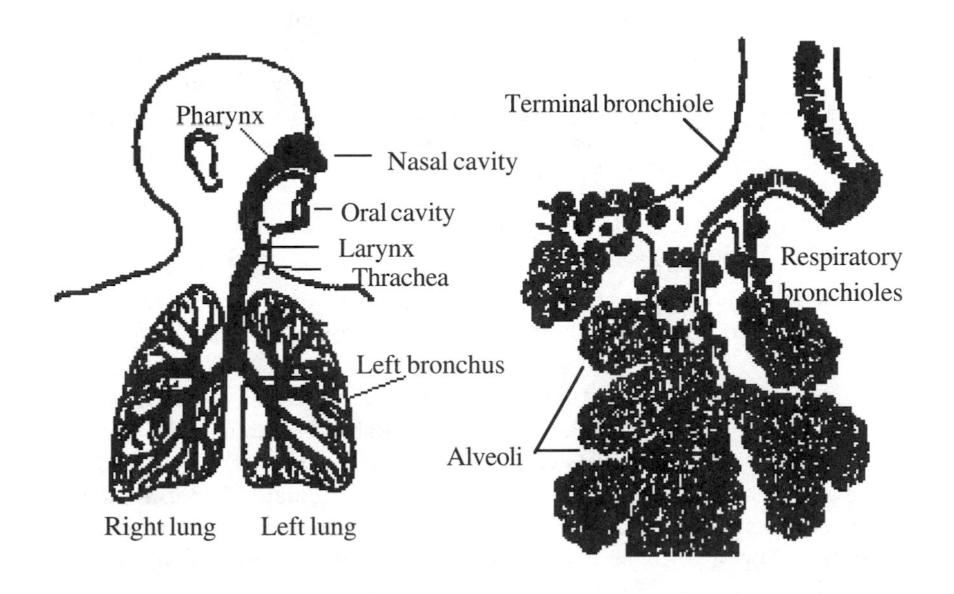

Figure 1.1 Human Respiratory System

$$2SO_2 + O_2 \rightarrow 2SO_3$$

$$SO_3 + H_2O \rightarrow 2H^+ + SO_4^-$$

These acid aerosols then deposit in significant quantities on the surface of land and water masses. This phenomenon is *acid rain.*

In clean air, SO_2 slowly oxidizes to SO_3 by homogeneous reaction. In the atmosphere the rate of SO_2 oxidation is 10 to 100 times the clean-air rate (Seinfeld 1975). Such a rapid rate of oxidation is similar to that expected of oxidation in solution in the presence of a catalyst.

SO_2 dissolves readily in water droplets and rapidly oxidizes to sulfuric acid by dissolved oxygen in the presence of metal salts such as iron or manganese. These metal salts usually exist in air as suspended particulate matter. At high humidities these particles act as condensation nuclei or undergo hydration to become solution droplets. The oxidation process then proceeds by absorption of both SO_2 and O_2 by the liquid aerosol with subsequent reaction in the liquid phase.

Although the detailed mechanism of SO_2 oxidation in solution is not well understood, most data show that it can be represented as a first-order reaction based on the gas phase concentration of SO_2. The rate constant depends on the catalyst type and the relative humidity. Cheng et al. (1971) determined a first-order rate expression for liquid-phase oxidation in the presence of $MnSO_4$:

$$R_{SO_2} = 0.0067\left[SO_2\right] \tag{1.6}$$

where R_{SO_2} is the rate of SO_2 oxidation, expressed in μg SO_2/min-mg of $MnSO_4$, $[SO_2]$ is the gas phase concentration of SO_2 in μg/m^3, and the constant factor is for drops containing 500 ppm of $MnSO_4$.

Another mechanism for the formation of sulfuric acid in the atmosphere is photochemical oxidation of SO_2. When exposed to solar radiation, SO_2 absorbs photons of wavelengths of 2940 Å and 3840 Å leading to excited states of the molecule. These excited species follow a several-step reaction sequence to form sulfur trioxide and sulfuric acid. Experimental studies (Seinfeld 1975) have shown a 0.1% to 0.2% conversion per hour through this mechanism.

The role of hydrogen peroxide in the formation of acid rain has received considerable attention recently (Gaffney et al. 1987) because it reacts quite readily with SO_2 in the aqueous phase to form acidic sulfate.

$$H_2O_2 + SO_2 \rightarrow H_2SO_4 \tag{1.7}$$

This rapid, liquid-phase reaction is acid catalyzed and is believed to be responsible for a large fraction of the conversion of SO_2 to sulfuric acid. Hydrogen peroxide is present in tropospheric air in parts per billion (ppb) levels.

There are several effects of acid rain that are deeply disturbing. There is an acidification of natural water sources that devastates aquatic life. Trout and salmon are particularly sensitive to a low pH. A leaching of nutrients occurs in the soil. This demineralization reduces the productivity of all plants and possibly changes the natural vegetation. The severity of damage to soils and bodies of water is partially determined by the minerals in the soil in a given area. Those areas that contain rock such as calcium carbonate or similar minerals are buffered against the perils of acid rain. The SO_2 in acid rain usually falls tens, or even hundreds of kilometers from the source. Thus SO_2 emission control expands to an international scope, because emissions from one country affect the nature of the rain beyond its borders.

Example 1.2 Atmospheric Oxidation of SO_2 Catalyzed by MnSO4 (Seinfeld 1975, p. 218).

You wish to compute the rate of conversion of SO_2 by catalytic oxidation for conditions typical of a natural fog in an urban atmosphere. Assume the concentration of sulfur dioxide is 0.10 ppm, and that of the fog 0.2 g water/m^3 of air. Half the fog droplets contain a catalyst capable of oxidizing SO_2 to H_2SO_4. The catalyst concentration within these droplets is equivalent to 500 ppm of $MnSO_4$. Compute the rate of conversion from SO_2 to sulfuric acid under such conditions.

Solution

Equation (1.6) gives the rate of catalytic oxidation of the SO_2. The gas-phase SO_2 concentration, in μg/m^3, is from Example 1.1 for the same conditions (assuming 300 K and 1 bar).

Basis: 1 m^3 of air at 300 K and 1 bar

From Example 1.1, $[SO_2] = 257$ μg/m^3. Substituting in Eq.(1.6), $R_{SO_2} = 0.0067 \times 257 = 1.722$ μg/(min-mg $MnSO_4$). For the given fog concentration of 0.2 g water/m^3 air, and considering that half of the drops contain the catalyst in a liquid-phase concentration of 500 ppm (because it refers to a liquid solution this concentration is on a mass basis!) the total mass of catalyst in 1 m^3 of air is

$$\left.\frac{0.2\text{ g H}_2\text{O}}{\text{m}^3\text{ air}}\right|\frac{500\text{ g MnSO}_4}{10^6\text{ g H}_2\text{O}}\left|\frac{10^3\text{ mg}}{1\text{ g}}\right|(0.5) = 0.05\text{ mg/m}^3$$

Therefore, the rate of oxidation of the sulfur dioxide is $R_{SO_2} = 1.722 \times 0.05 = 0.086$ μg SO_2 / min-m^3 = 5.17 μg SO_2/hr-m^3. Considering that the initial concentration of SO_2 was 257 μg/m^3, the rate of oxidation is approximately 2% conversion/hour.

Comments

The assumed constant rate of oxidation in this example would require a steady concentration of SO_2. Natural and man-made sources could generate enough SO_2 to maintain constant conditions. Observe that the average atmospheric retention time of SO_2 under these conditions is approximately 50 hours, plenty of time for the prevailing winds to carry it hundreds of kilometers away from the source.

1.4-3 CO_2

A problem of major new concern is the steady increase in the CO_2 content of the atmosphere (see Figure 1.2). Because CO_2 in low concentrations has no short-term toxic or irritating effects, is abundant in the atmosphere, and is necessary to plant life, it has not been considered a pollutant. However, the huge increase in world CO_2 emission from the combustion of fossil fuels, added to worldwide destruction of forests, has resulted in a steady rise in ambient CO_2 concentrations.

Higher concentrations of CO_2 threaten to disrupt the global climate by altering the radiation absorption characteristics of the atmosphere. The earth's atmosphere, which is almost transparent to short-wave solar radiation, contains about 78% nitrogen, 21% oxygen, and a host of trace gases such as water, CO_2 and methane. Water vapor, clouds, and other gases absorb about 20% of the incoming solar radiation. An additional 30% of this radiation scatters or reflects back into space. The atmosphere is opaque to the less energetic, infrared radiation emitted by the earth's surface. About 90% of this heat energy given off by the surface is absorbed by clouds, water vapor, and trace gases such as CO_2, methane, and CFCs that accumulate from human activities.

Once absorbed in the atmosphere, part of this energy radiates back to the surface for further warming. In this way, clouds, water vapor, and other trace gases have the effect of warming the surface. The recycled energy reemitted from the atmosphere to the surface is nearly twice the energy reaching the surface from the Sun (Gates et al. 1990). This phenomenon is known as the *greenhouse effect.*

Atmospheric CO_2, already 25% higher than before the industrial revolution, continues to rise at an increasing rate. Methane and CFCs are also on the rise. These changes in composition will affect the planet's climate.

The global average climate was relatively stable for several thousand years before the industrial period. Since then, global average temperatures have risen about 0.5 ^{o}C, suggesting that enhancement of the greenhouse effect has already started to have a warming effect. There is general agreement that the climate will be further changed as greenhouse gas emissions continue to increase, but there is no consensus on the specifics such as how fast and how much regional temperatures and precipitation will change.

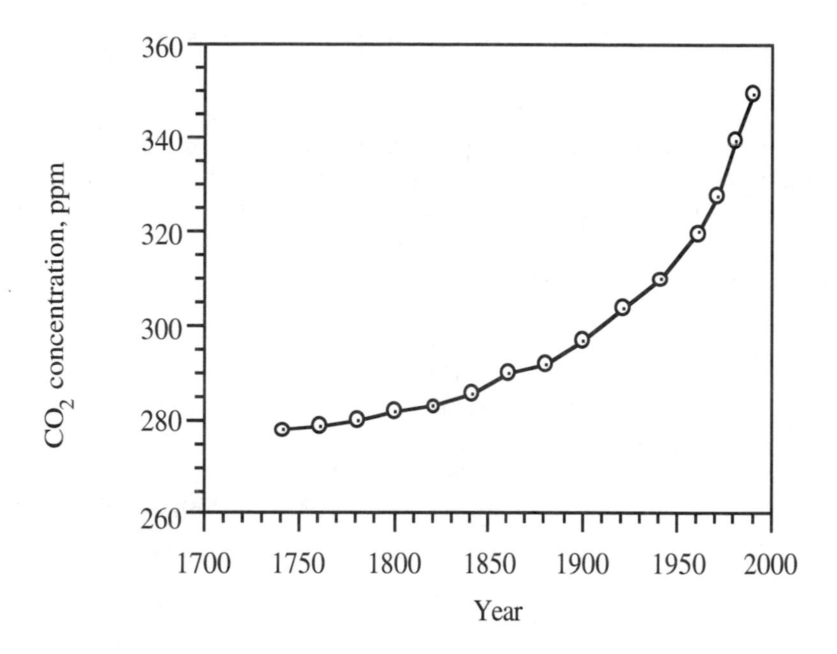

Figure 1.2 Growth in atmospheric CO_2 concentration (from U. S. EPA "Policy Options for Stabilizing Global Change," Draft Report to Congress, Feb. 1989).

Hansen et al. (1981) predicted a doubling of the concentration of CO_2 in the atmosphere by the middle of the next century. A recent model of the global climate (Gates et al. 1990) predicts that such an increase will result in an increase of 4 ^{o}C in the earth's average temperature, with some areas in the Northern Hemisphere experiencing increases of up to 9 ^{o}C. The effects on the climate would be dramatic.

An accelerated rise in the sea level caused by global warming would be an environmental threat of unprecedented proportions. During the next century expansion of surface waters, and the melting of glaciers and ice caps would raise the level of the oceans. The U.S. EPA estimates the rise in sea level will be from 0.5 to 2 m. Hekstra (1989) of the Ministry of Housing, Physical Planning, and Environment in the Netherlands asserts that a rise of 1 m could affect all land of less than 5 m in elevation. He estimates that 5 million km^2 are at risk. This is only about 3% of the earth's surface, but it is a third of all cropland and home to a billion people.

Example 1.3 Thermal Equilibrium of Earth-Atmosphere System (Seinfeld 1975, p. 139)

(a) If the earth-atmosphere system is assumed to be in radiative equilibrium with the sun (it emits as much energy as it receives), it is possible to estimate a value for the temperature T_e of the earth-atmosphere system. If the planetary albedo (the fraction of incoming solar radiation that is reflects or scatters back to space) is denoted by α, and the solar constant by S, show that T_e is given by:

$$T_e = \left[\frac{(1-\alpha)S}{4\sigma}\right]^{0.25}$$

where σ is the Stefan-Boltzmann constant (5.669×10^{-8} W/m^2-K^4). Calculate the value of T_e for $\alpha = 0.34$.

Solution

In thermal equilibrium, the fraction of the total incoming solar energy absorbed by the system earth-atmosphere must be equal to the energy radiated by this system back to space. The solution assumes that the earth-atmosphere system behaves as a blackbody. The solar constant is the solar radiation flux on a unit area normal to the beam at the outer edge of the atmosphere.

$$\text{Absorbed incoming radiation} = S(1-\alpha)\pi\frac{D^2}{4}$$

$$\text{Emitted outgoing radiation} = \sigma T_e^4 \pi D^2$$

where D is the system diameter. Equating these two expressions, and solving for the equilibrium temperature we obtain the desired expression. Substituting $S = 1395$ W/m^2,

$$T_e = 252\ \text{K}\ (-21\,^\circ\text{C})$$

(b) This value is obviously too low. The real temperature is about 300 K. The solution did not take consider the greenhouse effect. Not all radiation emitted by the earth is lost to outer space; a significant fraction, denoted by β, is absorbed by the atmosphere. Estimate the value of β, assuming the equilibrium temperature is 300 K.

Solution

The modified form of the term that describes the emitted outgoing radiation is

$$\text{Emitted outgoing radiation} = \sigma T_e^4 \pi D^2 \left(1 - \beta\right)$$

Equating this expression to the absorbed incoming radiation, and solving for β, gives

$$\beta = 1 - \frac{S\left(1 - \alpha\right)}{4\sigma T_e^4}$$

Substituting numerical values, the value of β is 0.50.

Comments

This simplified analysis of radiation equilibrium highlights the importance of the greenhouse effect in controlling the average global temperature.

1.4.4 CO

CO is extremely hazardous. It is a colorless, odorless, tasteless gas that reacts with the hemoglobin in blood to prevent oxygen transport. Its effects on humans range from slight headaches to nausea to death, depending on the concentration and the time of exposure.

The toxicity is due solely to the interaction of CO with blood hemoglobin. When a mixture of air and CO is breathed, both are transferred through the lungs to the blood and adsorb onto hemoglobin, but the equilibrium coefficient for CO is approximately 210 times that for oxygen. Thus, the equilibrium ratio of carboxyhemoglobin (HbCO) to oxyhemoglobin (HbO_2) is

$$\frac{\text{HbCO}}{\text{HbO}_2} = 210\frac{p_{\text{CO}}}{p_{\text{O}_2}} \qquad (1.8)$$

where p_{CO} and p_{O_2} are the partial pressures of CO and oxygen in the incoming air. Equilibrium is not instantaneous, and the process is reversible. When the inhaled air is free of CO, HbCO slowly breaks down, allowing CO to be expelled from the lungs.

Example 1.4 Effect of CO on the Oxygen Transport Capacity of the Blood (Cooper and Alley 1986, p. 38).

Estimate the percentage of HbCO in the blood of a traffic officer exposed to 40 ppm CO for several hours. Assume the HbCO contents reaches 60% of its equilibrium saturation value.

Solution

The concentration of HbCO at saturation is obtained from Eq. (1.8). At a total pressure of 1 atm and a mol fraction of 21% oxygen, the saturation ratio is

$$\frac{HbCO}{HbO_2} = 210\frac{40 \times 10^{-6}}{0.21} = 0.04$$

At 60% saturation:

$$\frac{HbCO}{HbO_2} = 0.04 \times 0.60 = 0.024$$

The fraction of hemoglobin unavailable for oxygen transport is

$$\text{Percent HbCO} = \frac{HbCO}{HbCO + HbO_2} \times 100 = \frac{\dfrac{HbCO}{HbO_2}}{1 + \dfrac{HbCO}{HbO_2}} \times 100$$

Substituting: Percent HbCO = 2.3%

Comments

The effects of carbon monoxide on human health depend on the fraction of hemoglobin that is tied up. Likewise, the carboxyhemoglobin fraction depends on the CO concentration in the inhaled air and the time of exposure. For example, exposure for 8 h to concentrations of 30 ppm produces 5% HbCO in the blood. Cigarette smoke contains 400 to 450 ppm CO. Most smokers average above 5% HbCO. Table 1.2 relates health effects to percent HbCO (Wark and Warner 1981). The traffic officer of Example 1.4 may experience some psychomotor disturbances, which he probably blames on stress.

Table 1.2 Health Effects of CO

Percent HbCO	Health Effects
Less than 1.0	No apparent effects
1.0–2.0	Some evidence of effect on behavioral performance
2.0–5.0	Central nervous system effects; impairment of time-interval discrimination, visual acuity, and other psychomotor functions
5.0–10.0	Cardiac and pulmonary functional changes
10.0–80.0	Headaches, fatigue, drowsiness, coma, and death

Source: Wark and Warner (1981).

1.4.5 VOCs

Compounds in this set have a wide range of reactivity. Some are particularly susceptible to photochemical reactions in the atmosphere; others are inherently stable. Compounds in both extremes of reactivity can cause serious air pollution problems. Particularly reactive VOCs (such as butene or ethylene) combine with the oxides of nitrogen in sunlight to form photochemical oxidants including ozone (O_3) and peroxyacetyl nitrate (PAN). These oxidants are severe eye, nose, and throat irritants. Ozone attacks synthetic rubber, textiles, paints, and other materials. Oxidants extensively damage plant life. Vegetation can suffer at concentrations as low as 0.05 ppm (Cooper and Alley 1986). Eye irritation begins at 0.10 ppm, and severe coughing occurs at 2.0 ppm.

Chlorinated fluorocarbons are at the other extreme of reactivity. These gases are extensively used as the operating fluid in refrigerators and air conditioners, in the manufacture of plastic foams and insulation, as propellants in aerosol spray cans, and as solvents, particularly in the electronic industry. The stability of CFCs makes them persistent in the atmosphere. Their atmospheric lifetimes (defined as the time required for removal of 63% of the compound) are 52 years for CFC-11 and 101 years for CFC-12 (Wuebles and Connell 1990).

CFCs have ample time to migrate to the stratosphere where they deplete the ozone layer by reacting catalytically with it. The earth's atmosphere is divided into various layers. The region from the ground up to about 15 km is known as the troposphere. This region generally has decreasing temperature with altitude and contains all of our weather. Above the troposphere, extending to about 50 km above the earth is the stratosphere. This region generally has increasing temperature with altitude and contains most of the atmospheric ozone. CFCs that penetrate the stratosphere decompose to liberate chlorine free radicals that catalyze the destruction of ozone. Because the reaction between CFCs and ozone is catalytic, each molecule entering the stratosphere can destroy thousands of molecules of ozone.

Ozone in the troposphere is considered an air pollutant; ozone in the stratosphere is vital to life as we know it on earth. Such an important role is due to its ability to absorb virtually all of the high-energy, short-wave radiation coming from the sun: cosmic radiation, gamma and X-rays, and ultraviolet radiation. All are ionizing and can cause skin cancer and mutations. Without the screening effect of the ozone layer, most of this radiation would reach the surface of the earth.

Observations in the last decade indicate large decreases in upper stratospheric ozone (see Figure 1.3). About half of the change can be attributed to the chlorine released from dissociation of CFCs; the other half is related to solar flux variations (Wuebells and Connell 1990).

Disturbing evidence of a damaged ozone layer continues to surface. In 1985, the satellite Nimbus 7 revealed a gaping hole over Antarctica that spreads from the South Pole each year at the beginning of austral spring and engulfs the entire Antarctic continent by early October (Cogan 1989). More than half of the ozone in that area is destroyed each spring. Every summer its replenishment from lower latitudes reduces the ozone layer over the entire Southern Hemisphere.

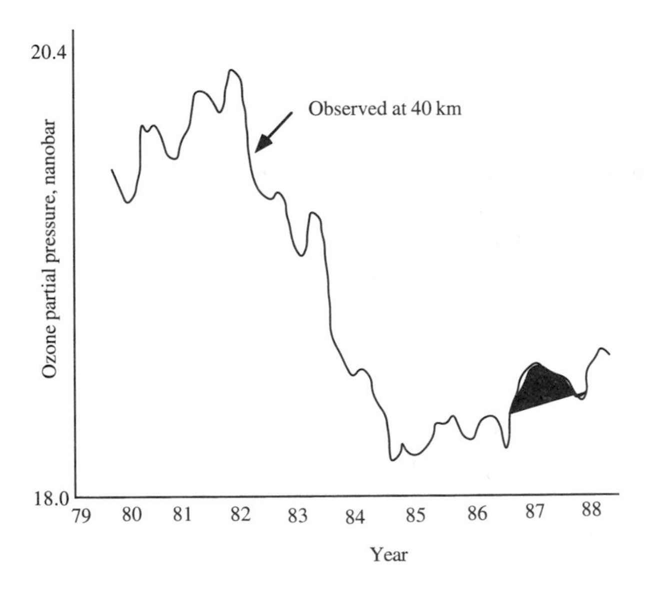

Figure 1.3 Trends in upper stratospheric ozone (Wuebbles and Connell 1990). Reprinted with permission from the Lawrence Livermore National Laboratory operated by the University of California for the U. S. Department of Energy

Disturbing evidence of a damaged ozone layer continues to surface. In 1985, the satellite Nimbus 7 revealed a gaping hole over Antarctica that spreads from the South Pole each year at the beginning of austral spring and engulfs the entire Antarctic continent by early October (Cogan 1989). More than half of the ozone in that area is destroyed each spring. Every summer its replenishment from lower latitudes reduces the ozone layer over the entire Southern Hemisphere.

Although the detailed mechanism for the development of an ozone hole is still not known, three necessary elements have been identified (Cogan 1989). First, the atmosphere over one of the poles needs to wind up in a tight spiral, known as a polar vortex, that is virtually impenetrable by outside air masses. Second, stratospheric air temperatures inside the vortex need to fall below 193 K to freeze out droplets of nitric acid and below 188 K to turn water vapor into ice. The resulting particles, thought to be nitric acid trihydrate wrapped in a thick coat of ice, fall out of the ozone layer before evaporating in the troposphere several kilometers below. That leaves the lower stratosphere practically defenseless against ozone depletion. If the third element is in place—elevated levels of chlorine, and without nitric acid to sequester chlorine in inactive reservoirs of chlorine nitrate and hydrogen chloride, chlorine free radicals create havoc among ozone molecules. This process allows one chlorine atom to destroy tens of thousands of ozone molecules before the polar vortex breaks up.

1.5 FEDERAL AIR POLLUTION LEGISLATION IN THE UNITED STATES

Your objectives in studying this section are to

1. Be familiar with the major pieces of legislation created by the U.S. federal government to deal with air pollution problems.
2. Define National Ambient Air Quality Standards (NAAQS).
3. Be familiar with New Source Performance Standards (NSPS).
4. Be familiar with the New Source Review (NSR) process.

Two dramatic episodes in the late 1940s and early 1950s spurred the U.S. federal government into legislative action. In 1948 in Donora, Pennsylvania, a 4-day smog made 7,000 people sick and killed 20. In 1952, when a 3-day smog covered London, England, there were an estimated 4,000 deaths attributed to that air pollution episode. Those with bronchitis, emphysema, and heart trouble were particularly vulnerable (Cooper and Alley 1986).

The history of federal legislative efforts begins with the *Air Pollution Control Act of 1955*. This act provided funds only for research and technical assistance. The Clean Air Act of 1963 replaced the 1955 act and established federal authority to address interstate air pollution problems.

The *Clean Air Act Amendments of 1970* are powerful landmarks of environmental legislation. One of their major objectives was to attain clean air by 1975. The Act required the recently created EPA to establish *National Ambient Air Quality Standards (NAAQS),* both primary standards (to protect public health) and secondary standards (to protect public welfare). The act also required states to submit *State Implementation Plans (SIPs)* for attaining and maintaining the national primary standards within three years.

The *Clean Air Act Amendments of 1977* significantly modified the 1970 act. The EPA was required to review and update the NAAQS as of January 1, 1980, and at 5-year intervals thereafter. This set of amendments mandated the "prevention of significant deterioration" (PSD) of air quality in regions cleaner than the NAAQS. Before the 1977 amendments, it was theoretically possible to locate air pollution sources in such regions and pollute clean air up to the limits of the ambient standards. However, the new act defined class 1 (pristine areas), class 2 (almost all other areas), and class 3 (industrialized areas). Under the PSD provisions, the ambient concentration of pollutants are allowed to rise very little in class 1 areas, by specified amounts in class 2 areas, and by larger amounts in class 3 areas (Flagan and Seinfeld 1988).

Two types of standards emerged from the federal legislation. The first are the NAAQS. The second are the source performance standards—those that apply to emissions of pollutants from specific sources. NAAQS are always expressed in terms of concentrations, such as micrograms per cubic meter or parts per million. Source performance standards are written in terms of mass emissions per unit time or unit of production, such as grams per minute, or kilograms of pollutant per ton of product.

Table 1.3 presents the current NAAQS. The particulate matter entry in this table requires some explanation. In 1987 the EPA proposed the following relative to the particulate matter standard:

> **1. That total suspended particulate matter (TSP) as an indicator for particulate matter be replaced for the primary standard by a new indicator that includes only particles with an aerodynamic diameter smaller or equal to a nominal 10 μm (PM_{10}).**
>
> **2. That the level of the 24-hour primary standard be 150 μg/m^3 and the deterministic form of the standard be replaced with a statistical form that permits one expected exceedance of the standard level per year.**
>
> **3. That the level of the annual primary standard be 50 μg/m^3, expressed as an expected annual arithmetic average.**

For comparison purposes, the original primary standard based on total suspended particulate matter was 260 μg/m^3 for the 24-hour average, and 75 μg/m^3 for the annual geometric mean. This change of standard was a belated recognition by EPA that particles bigger than 10 μm are virtually harmless to human health. Recent epidemiological data suggests that even the revised standard is dangerously lax because increased mortality has been observed at concentrations as low as a third of the current standard (American Lung Association 1992).

Table 1.3 National Ambient Air Quality Standards (Primary)

Pollutant	Averaging Time	Standard
SO_2	Annual average	860 μg/m^3
	24 h	365 μg/m^3
NO_2	Annual average	100 μg/m^3
CO	8 h	10 mg/m^3
	1h	40 mg/m^3
Ozone	1 h	0.12 ppm
Particulate matter (PM_{10})	Annual geometric mean	50 μg/m^3
	24 h	150 μg/m^3

Source: 40 CFR (Code of Federal Regulations) 50, 1982.

Table 1.4 presents some examples of NSPS. These apply to sources constructed after promulgation of the rules. Existing sources often require special handling. Emission standards specify the maximum rate of emissions permitted at a new source. They constitute the basis for estimating the performance required of pollution control equipment.

Example 1.5 Emissions Allowed from Coal-Fired Power Plant.

Calculate the daily emissions of particulates, SO_2, nitrogen oxides, and CO_2 permitted by the NSPS for a proposed 300 MW coal-fired power plant. The coal contains 77% C and 3% sulfur. It has a heating value of 27,850 kJ/kg. The thermal efficiency of the plant is 35%.

Solution

The thermal load and coal requirements are calculated from the plant rated capacity, the coal heating value, and the cycle thermodynamic efficiency.

Table 1.4 Some New Source Performance Standards

Source	Standard
Steam electric power plants	
Particulate matter	13 g/10^6 kJ
NO_x	
Gaseous fuel	86 g/10^6 kJ
Liquid fuel	130 g/10^6 kJ
Coal	260 g/10^6 kJ
SO_2	
Gas or liquid fuel	86 g/10^6 kJ
Coal	At least 70% removal depending on conditions (see Figure 1.4 for details)
Solid waste incinerators	
Particulate matter	0.18 g/dscm[a] corrected to 12% CO_2 (3-h average)
Sewage sludge incinerators	
Particulate matter	0.65 g/kg sludge input (dry basis)
Iron and steel plants	
Particulate matter	50 mg/dscm
Primary copper smelters	
Particulate matter	50 mg/dscm
SO_2	0.065% by volume
Sulfuric acid plants	
SO_2	2 kg/metric ton of 100% acid
Portland cement plants	
Particulate matter	0.15 kg/metric ton of feed (maximum 2-h average)

[a] Dry standard cubic meter.

Source: 40 CFR (Code of Federal Regulations) 60, 1982.

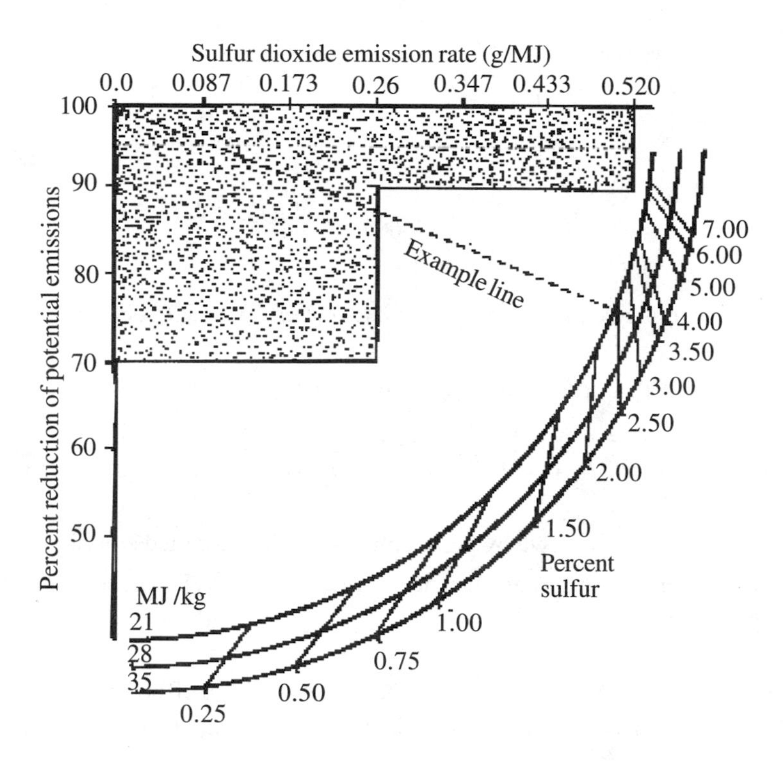

Figure 1.4 Graphical interpretation of the 1980 NSPS for SO_2 emissions from coal-fired power plants (From Molburg, J. *JAPCA*, **30**:172, 1980; reprinted with permission from JAPCA)

To use this figure, draw a straight line from the origin of the graph to the point within the curved lines that best represents the properties of the coal. The point where the line intersects the shaded *admissible region* defines the operating point, which consists of a required removal efficiency and an allowable emission-rate.

$$\left|\frac{300\text{ MW}}{0.35}\right|\frac{1{,}000\text{ kW}}{1\text{ MW}}\left|\frac{1\text{ kJ}}{1\text{ kW-s}}\right|\frac{3{,}600 \times 24\text{ s}}{1\text{ day}} = 7.41 \times 10^{10}\ \frac{\text{kJ}}{\text{day}}$$

$$\frac{7.41 \times 10^{10}}{27{,}850} = 2.66 \times 10^{6}\ \frac{\text{kg of coal}}{\text{day}}$$

Calculate the permitted emissions of particulate matter, NO_x, SO_2, and CO_2.

Particulate: $13 \times 7.41 \times 10^{10} / (10^{6} \times 10^{3}) = 963$ kg/day.

NO_x : $260 \times 7.41 \times 10^{10} / (10^{6} \times 10^{3}) = 19{,}300$ kg/day.

SO_2: From Figure 1.4, as shown by the example line, the permitted emission rate is 260 g/10^6 kJ, corresponding to a removal efficiency of about 87%.Then, the daily emissions of SO_2 are 19,300 kg/day.

CO_2: Because there are no regulations for this gas, there is no effort to reduce its emissions. The rate of emission is from a material balance, assuming complete combustion:

$$\left|\frac{2.66 \times 10^{6}\text{ kg coal}}{\text{day}}\right|\frac{0.77\text{ kg C}}{\text{kg coal}}\left|\frac{44\text{ kg }CO_2}{12\text{ kg C}}\right| = 7.51 \times 10^{6}\ \frac{\text{kg }CO_2}{\text{day}}$$

Comments

Even when a power plant of this size satisfies EPA standards, it is a major source of air pollutants. Although the role of CO_2 as a greenhouse gas is well established, there have been no efforts by EPA to regulate it. That may soon change. If there ever are legal restrictions on CO_2 emissions, there will be a shift toward renewable, environmentally sound sources of energy.

New major stationary sources of air pollution and major modifications to major stationary sources are required by the Clean Air Act to obtain an air pollution permit before commencing construction. The process is called *New Source Review (NSR)* and is required whether the major source or modification is planned for an area where the NAAQS are exceeded (nonattainment areas) or an area where air quality is acceptable. Permits for sources in attainment areas are referred to as PSD permits, whereas in nonattainment areas they are referred to as NAA permits.

The PSD and NAA requirements are pollutant specific. Although a facility may emit many air pollutants, only one or a few may be subject to the PSD or NAA permit regulation, depending on the magnitude of the emissions of each pollutant. A source may have to obtain both PSD and NAA permits if it is located in an area that is designated nonattainment for one or more of the pollutants.

On August 7, 1980, EPA extensively revised both the PSD and NAA regulations. A major stationary source was defined as any source type belonging to a list of 28 categories (see Table 1.5) that emits or has the potential to emit, after pollution control, 91 metric tons per year or more of any pollutant subject to regulation under the Clean Air Act. Any other stationary source, not included in the list, with the potential to emit such pollutants in amounts equal to or greater than 227 metric tons per year is also considered a major source.

Example 1.6 NSR Applicability to a Portland Cement Plant

A new Portland cement plant is proposed for a locality where the PM_{10} annual geometric mean ambient air concentration is 25 $\mu g / m^3$. The kiln is designed for continuous operation at a maximum feed rate of 100 metric tons/hour. Determine whether this plant must be submitted to the NSR procedure, and, if so, what type of preconstruction permit must it obtain?

Solution

From Table 1.4, the NSPS for Portland cement plants is 0.15 kg of particulate matter/metric ton of feed. Therefore, the potential to emit of this source is $100 \times 24 \times 365 \times 0.15 = 131{,}400$ kg/year = 131.4 metric tons/year. Virtually all of these emissions will be PM_{10} because the pollution control equipment needed to satisfy the NSPS will remove almost all of the bigger particles. According to Table 1.5, Portland cement plants have a major source threshold of 91 metric tons per year of any regulated pollutant; therefore, the source is subject to NSR. Because the ambient air PM_{10} concentration is below the corresponding NAAQS, the plant needs a PSD preconstruction permit.

Table 1.5 Source Categories with 91-Metric Ton per Year Major Source Thresholds

1. Fossil fuel-fired steam electric plants of more than 73-MW heat input
2. Coal-cleaning plants (with thermal dryers)
3. Kraft pulp mills
4. Portland cement plants
5. Primary zinc smelters
6. Iron and steel mill plants
7. Primary aluminum ore reduction plants
8. Primary copper smelters
9. Municipal refuse incinerators capable of burning 227 metric tons per day
10. Hydrofluoric acid plants
11. Sulfuric acid plants
12. Nitric acid plants
13. Petroleum refineries
14. Lime plants
15. Phosphate rock-processing plants
16. Coke oven batteries
17. Sulfur recovery plants
18. Carbon black plants (furnace plants)
19. Primary lead smelters
20. Fuel-conversion plants
21. Sintering plants
22. Secondary metal production plants
23. Chemical process plants
24. Fossil fuel boilers (or combinations) totaling more than 73-MW heat input
25. Petroleum storage and transfer units with a total capacity of more than 300,000 barrels
26. Taconite ore-processing plants
27. Glass fiber-processing plants
28. Charcoal production plants

Source: U.S.EPA, "New Source Review Workshop Manual" (Draft), Office of Air Quality and Standards, Research Triangle Park, NC (October 1990).

The NSR requirements for major new sources, or major modifications, locating in attainment or PSD areas include (U.S. EPA 1990a)

1. Apply the *best available control technology (BACT)*. **A BACT analysis is done on a case-by-case basis, and considers energy, environmental, and economic impacts in determining the maximum degree of reduction achievable for the proposed source or modification.** *In no event can the determination of BACT result in an emission limitation which would not meet any applicable NSPS.*
2. Conduct an ambient air quality analysis to demonstrate that the new emissions would result in violations of the NAAQS or the applicable PSD increments.
3. Do not adversely impact a class 1 area.
4. Analyze impacts to soil, vegetation and visibility.
5. Undergo adequate public participation. Specific public notice requirements and a public comment period are required before the PSD review agency takes final action on a PSD application.

On December 1, 1987, EPA issued a memorandum implementing the "top-down" method for determining BACT. The top-down process provides that all available control technologies be ranked in descending order of control effectiveness. The PSD applicant first examines the most stringent—or "top"—alternative. That alternative is established as BACT unless the applicant demonstrates that technical considerations, or energy, environmental, or economic impacts justify a conclusion that the most stringent technology is not "achievable" in that case. Then, the next most stringent alternative is considered, and so on.

Average and incremental cost effectiveness are the two economic criteria that are considered in the BACT analysis. The average cost effectiveness (ACE) of a control option is defined as

$$\text{ACE} = \frac{\text{Control option annualized cost}}{\text{Uncontrolled emission rate} - \text{Control option emission rate}} \tag{1.9}$$

The incremental cost effectiveness (ICE) compares the costs and emissions performance level of a control level to those of the next most stringent option. The final decision regarding elimination of a control alternative based on economic grounds will be made by the review authority considering previous regulatory decisions. Study cost estimates used in BACT are typically accurate to ± 20% to 30%. Therefore, control cost options that are within that range of each other are considered indistinguishable when comparing options.

Example 1.7 Top-down BACT economic impact analysis

A peak load 75-MW gas turbine is proposed for an area that is in compliance for all regulated pollutants. Uncontrolled emissions of NO_x would be 282 metric tons per year. The following table summarizes the potential NO_x control technologies identified as part of the top-down BACT analysis.

Alternative	Emissions[a]	Total Annual Cost ($/yr)
Option 1	22	1,717,000
Option 2	42	593,000
Option 3	70	356,000
NSPS	156	288,000
Uncontrolled	282	0

[a] Metric tons per year.

Calculate the average and incremental cost effectiveness for each control option. Determine if there is evidence to eliminate Option 1 based on economic grounds.

Solution

Cost effectiveness results are tabulated as follows:

Alternative	Emissions Reduction[a]	ACE ($/ton)	Incremental Annual Cost ($/yr)	Incremental Emissions Reduction[a]	ICE ($/ ton)
Option 1	260	6,600	1,124,000	20	56,200
Option 2	240	2,470	237,000	28	8,464
Option 3	212	1,680	68,000	86	791
NSPS	126	2,285			

[a] Metric tons/year.

The average cost effectiveness for Option 1 is significantly higher than for the other alternatives. The difference is even more dramatic when the alternatives are compared in terms of the incremental cost effectiveness. These results strongly suggest that Option 1 should be eliminated based on a negative economic impact.

Comments

It is very important in BACT analyses to apply a cost estimation methodology for total annual costs calculations that is accepted by EPA (see, e.g., U.S. EPA 1990b). All cost estimation calculations in this book follow those guidelines.

The preconstruction NAA review requirements differ from the PSD requirements. First, the emissions control requirement for NAA, the *lowest achievable emission rate* (LAER), is defined differently from the BACT emissions control requirement. (LAER is defined as the most stringent emission limitation achieved in practice by such class or category of source.) Second, the source must obtain *emissions reductions (offsets)* of the nonattainment pollutant from other sources that impact the same area as the proposed source and show that there will be progress toward achievement of the NAAQS.

To help air pollution control professionals make control technology determinations, EPA established the *BACT/LEAR Clearinghouse*, a technology transfer resource that contains emissions limits and required control technology for various source types throughout the country (U.S. EPA 1991a).

The *Clean Air Act Amendments of 1990* (see Appendix C) constitute a completely new approach to air pollution control legislation. It is probably the most comprehensive environmental legislation ever passed to control air pollution. The amendments direct the U.S.to implement strong policies and regulations that "will ensure cleaner air for all Americans" (U.S. EPA 1991b, p.1). To achieve this goal, EPA's implementation strategy will not only employ traditional approaches for controlling air pollution, but also strive to harness the power of the marketplace, encourage local initiatives, and emphasize pollution prevention. These efforts will be structured to achieve environmental benefits in a cost-effective manner, ensuring consistency with national energy and economic policies. The EPA intends to review continually all activities undertaken to implement the amendments, assess their effectiveness, and make modifications as necessary.

1.6 CHEMICAL ENGINEERING PRINCIPLES IN SOLUTION OF AIR POLLUTION PROBLEMS

Your objectives in studying this section are to

1. Apply some of the skills you have acquired as a chemical engineering student to the solution of typical air pollution problems.
2. Define such terms as air-to-fuel ratio, equivalence ratio, and flue gas desulfurization (FGD).

Controlling air pollution consists of a final treatment of all the plant streams that enter the atmosphere. That treatment might be physical or chemical, or both. It could be any of the unit operations of chemical processing, with the design requiring the application of the principles of material balances; energy balances; stoichiometry; momentum, heat, and mass transport; thermodynamics; and reaction kinetics. The examples here illustrate the implementation of these concepts in solving problems in the control of air pollution.

Example 1.8 Material Balances Applied to Combustion.

Coal that contains 77.2% C, 5.2% H, 1.2% N, 2.6% S, 5.9% O, and 7.9% ash is burned with dry air in an electric power plant. The flue gases (on a dry basis) have 15.95 % CO_2, 2.65% O_2, and 2,000 ppm SO_2. Calculate the following:
(a) Air-to-fuel ratio (on a mass basis)
(b) Equivalence ratio, ϕ.

Solution

(a) There are four streams in this problem: coal, air, dry flue gases, and the water formed by combustion (see Figure 1.5). The compositions of all four streams are known. With a basis of 100 mol of dry flue gas, material balances on three atomic species give the amount of water produced, and the quantities of coal and air used.

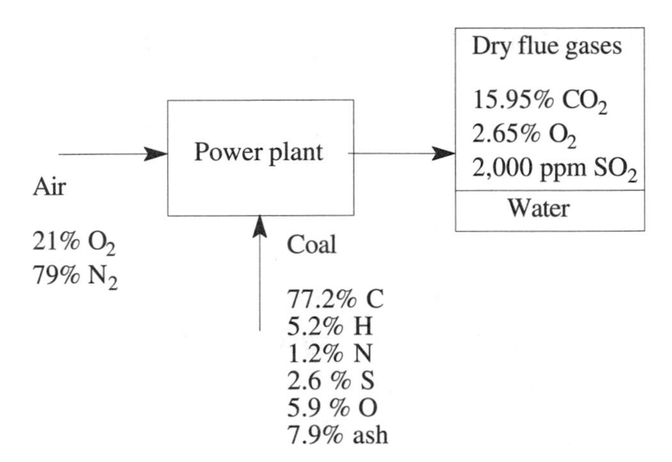

Figure 1.5 Schematic diagram of the power plant of Example 1.8

Basis: 100 kmol of dry flue gases.

Let

F = mass of coal used, in kg
A = mass of air used, in kg
W = kmol of water formed

Carbon is a tie-element between the dry flue gases and the coal

$$\frac{0.772F}{12} = 15.95 \qquad F = 247.9 \text{ kg of coal}$$

Atomic hydrogen is a tie-element between the coal and the water

$$\frac{0.052F}{1.0} = 2W \qquad W = 6.45 \text{ kmoles of } H_2O$$

A material balance on atomic oxygen gives the amount of combustion air supplied

$$\frac{0.21\times 2\times A}{29} + \frac{0.059\times F}{16} = 15.95\times 2 + 2.65\times 2 + 0.2\times 2 + 1\times W$$

(The average molecular weight of air is 29). Substituting the values of F and W already calculated,

$$A = 2{,}951 \text{ kg air.}$$

Therefore,

$$\frac{A}{F} = \frac{2{,}951}{247.9} = 11.9 \frac{\text{kg air}}{\text{kg coal}}$$

b) The equivalence ratio, ϕ, is defined by

$$\phi = \frac{(A/F)_s}{(A/F)}$$

where $(A/F)_s$ is the air to fuel ratio for stoichiometric combustion. For a new basis of 100 kg of coal, the theoretical amount of oxygen required (subtracting that supplied by the coal) is

$$\frac{(77.2)(1)}{12} + \frac{(5.2)(1/4)}{1} + \frac{(2.6)(1)}{32} - \frac{(5.9)(1/2)}{16} = 7.63 \text{ kmol}$$

Therefore, the stoichiometric air required is $7.63 \times 29 / 0.21 = 1{,}053$ kg, and

$$\left(\frac{A}{F}\right)_s = 10.53 \frac{\text{kg air}}{\text{kg coal}}$$

Combining this with the result of part a,

$$\phi = 10.53/11.9 = 0.88$$

Comments

The air to fuel and equivalence ratios determine the quantity and type of air pollutants generated by combustion. Power plants always operate with an excess of air to give equivalence ratios less than unity. This type of combustion is also called *fuel lean.* Excess air minimizes emissions of VOCs and CO, but favors the formation of NO. Internal combustion engines with equivalence ratios slightly greater than 1.0 (*fuel rich*) enhance CO and VOCs production, but tend to suppress the formation of NO.

Example 1.9 Thermodynamics: Chemical Reaction Equilibria

Atmospheric air (79% N_2 and 21% O_2) is heated to 2,000 K. Calculate the equilibrium NO concentration, in ppm.

Solution

The equilibrium constant for the reaction

$$N_2 + O_2 \leftrightarrow 2NO$$

from Seinfeld (1975) at 2,000 K is:

$$K_p = \frac{p_{NO}^2}{p_{N_2} p_{O_2}} = \frac{y_{NO}^2}{y_{N_2} y_{O_2}} = 21.9 \exp\left(-\frac{21,741.3}{T}\right) = 4.0 \times 10^{-4}$$

where y_i refers to the equilibrium mole fraction of component i, and T is in K. Combining this expression for the equilibrium constant with material balances gives the equilibrium composition.

Basis: 1 kmol atmospheric air

The number of moles of each species in the equilibrium mixture expressed in terms of the initial number of moles and the reaction coordinate, ε, are (Smith and Van Ness 1987) :

$$n_{N_2} = 0.79 - \varepsilon = y_{N_2}$$

$$n_{O_2} = 0.21 - \varepsilon = y_{O_2}$$

$$n_{NO} = 2\varepsilon = y_{NO}$$

In this case, the number of moles of each species is equal to its corresponding mole fraction because the total number of moles remains equals to unity. Substituting into the expression for the equilibrium constant gives

$$4 \times 10^{-4} = \frac{4\varepsilon^2}{(0.79 - \varepsilon)(0.21 - \varepsilon)}$$

Solving for the value of the reaction coordinate in equilibrium,

$$\varepsilon = 4 \times 10^{-3} \text{ kmol} \qquad y_{NO} = 8{,}000 \text{ ppm}$$

Comments

Some strategies for controlling NO_x emissions exploit the sensitivity of the equilibrium concentration of NO in combustion mixtures to temperature and oxygen availability.

Example 1.10 Mass Transfer Operations and Design Optimization (Modified Version of Problem 10.3, Peters and Timmerhaus 1980)

A small coal-fired electric power plant will use FGD to satisfy the corresponding new source performance standard for SO_2. The proposed system consists of a packed tower in which the flue gases flow countercurrent to an aqueous sodium sulfite solution. The SO_2 is removed from the gases through absorption and reaction by the liquid phase. The flue gas flows at a rate of 71.8 kg/s. Under the specified design conditions, the number of transfer units required is a function of the entering mass velocity of the gas (based on the cross sectional area of the empty tower) is given by

$$N_{tOG} = 1.05 G_s^{0.18}$$

where G_s is in kg/m2-s. The height of a transfer unit, H_{tOG}, is constant at 4.57 m. The cost for the installed tower is \$1,500/m3 of inside volume, and annual fixed charges amount to 20% of the initial cost. Variable operating charges for gas blower and liquid pumping power, and other items are represented by

$$\frac{\$}{\text{hr}} = 0.0975\, G_s^2 + \frac{0.11}{G_s} + \frac{0.0244}{G_s^{0.8}} + 5.0$$

The unit is to operate 8,000 hr/yr. Determine the height and diameter of the tower that result in the minimum total annual cost.

Solution

The approach is to express each term in the equation for total annual cost (TAC) in terms of the cross sectional area of the empty tower, A, and then find the value of A that minimizes TAC. Figure 1.6 is a schematic diagram of the proposed FGD system.

Basis: 1 yr (8,000 hr)

The packed height, Z, is given by:

$$Z = N_{tOG} H_{tOG} = (4.57)(1.05 G_s^{0.18}) = 4.8\ G_s^{0.18}$$

Because $G_s = 71.8/A$, and the packed volume, $V = AZ$:

$$Z = 4.8 \left[\frac{71.8}{A}\right]^{0.18} = \frac{10.36}{A^{0.18}} \qquad V = 10.36 A^{0.82}$$

The total installed cost (TIC) for the system is TIC = \$1,500 × $10.36A^{0.82}$ = \$15,540$A^{0.82}$. The fixed charges are 0.2 × 15,540$A^{0.82}$ = \$3,108$A^{0.82}$/yr. Variable operating costs are expressed in terms of A, and on a yearly basis they are given by

$$\text{Annual variable costs, \$/yr} = 4.02 \times 10^6/A^2 + 12A + 6.4A^{0.8} + 40{,}000$$

The total annual cost is the sum of the fixed charges and the variable costs:

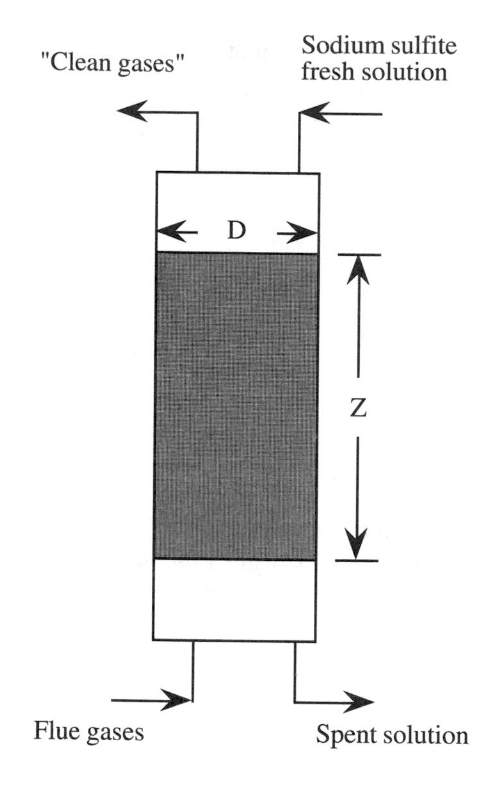

Figure 1.6 FGD system of Example 1.10

$$\text{TAC} = 3{,}108A^{0.82} + 4.02 \times 10^6/A^2 + 12A + 6.4A^{0.8} + 40{,}000$$

To determine the optimum design, differentiate with respect to A and set the derivative to zero. The optimum design and annual cost is

$A = 17.4\ \text{m}^2$ $D = 4.7$ m $Z = 6.2$ m TAC = \$86,000/yr

1.7 CONCLUSION

Air pollution problems are no longer limited to local concerns. Regional and global threats to our atmospheric environment include acid rain, the greenhouse effect, and the depletion of the ozone layer. Combustion-dependent activities such as electric power generation and the use of the automobile generate most air pollutants. In the United States, federal legislation has helped to alleviate the problem, but there is still a long way to go. Engineers must effectively and economically remove pollutants before they reach the atmosphere. Chemical engineers, skilled in many disciplines relevant to air pollution control equipment design, have a distinct advantage.

REFERENCES

American Lung Association, *J. Air Waste Manage. Assoc.,* **41**:708 (1992).

40 CFR 60. Code of Federal Regulations, Title 40, Part 60, "Standards of Performance for New Stationary Sources," Office of Federal Register, Washington, D.C., July 1 (1982).

Cheng, R. T., Corn, M., and Frohliger, J. O. *Atmos. Environ.,* **5**:987 (1971).

Cogan, D. G. *Environ. Sci. Technol.,* **23**:20 (1989).

Cooper, C. D., and Alley, F.C. *Air Pollution Control; A Design Approach,* Prindle, Weber & Schmidt, Boston, MA (1986).

Davison, R. L., Natusch, D. F. S., Wallace, J. R., and Evans, C. A. *Environ. Sci. Technol.,* **8**:1107 (1974).

Flagan, R. C., and Seinfeld, J. *Fundamentals of Air Pollution Engineering,* Prentice Hall, Englewood Cliffs, NJ (1988).

Friedlander, S. K. *Smoke, Dust and Haze,* Wiley, New York (1977).

Fulkerson, W., Judkins, R. R., and Sanghvi, M. K. *Scientific American,* p. 129 (September 1990).

Gaffney, J. F., Streit, G. E., Spall, W. D., and Hall, J. H. *Environ. Sci. Technol.,* **6**:519 (1987).

Gates W. L., MacCracken, M. C., and Potter G. L. *E&TR,* May-June, p. 56 (1990).

Hansen, J., Johnson, D., Lacis, A., Lebedeff, F., Lee, P., Rind, D., and Russell, G. *Science,* **213**:4511, August 28 (1981).

Hekstra, G. P. *The Ecologist,* January/February (1989).

Molburg, J. *JAPCA,* **30**:172 (1980).

National Research Council. "Airborne Particles." Subcommittee on Airborne Particles, Committee on Medical and Biological Effects of Environmental Pollutants. University Park Press, Baltimore, MD (1979).

Peters, M. S., and Timmerhaus, K. D. *Plant Design and Economics for Chemical Engineers,* 3rd ed., McGraw-Hill, New York (1980).

Seinfeld, J. *Air Pollution Physical and Chemical Fundamentals,* McGraw-Hill, New York (1975).

Smith, J. M., and Van Ness, H.C. *Introduction to Chemical Engineering Thermodynamics,* 4th ed., McGraw-Hill, New York (1987).

U. S. Environmental Protection Agency. "Air Quality Criteria for Particulate Matter and Sulfur Oxides," Report No. EPA-600/8-82-029 (1982).

U. S. Environmental Protection Agency. "National Air Pollutant Emission Estimates, 1970-1981, Report No. EPA-450/4-82-913 (1982b).

U. S. Environmental Protection Agency. "New Source Review Workshop Manual" (Draft), Office of Air Quality Planning and Standards, Research Triangle Park, NC (1990a).

U. S. Environmental Protection Agency. "OAQPS Control Cost Manual," 4th ed., EPA 450/3-90-006 (1990b).

U. S. Environmental Protection Agency. "RACT/BACT/LEAR Clearinghouse: A Compilation of Control Technology Determinations," (First Supplement to 1990 Edition), EPA 450/3-91-015 (1991a).

U. S. Environmental Protection Agency. "Implementation Strategy for the CAAA of 1990," Office of Air and Radiation, Research Triangle Park, NC (1991b).

Wark, K., and Warner C. F. *Air Pollution its Origin and Control,* Harper and Row, New York (1981).

Wuebles, D. J., and Connell, P. S. *E&TR*, May-June, p. 49 (1990).

PROBLEMS

The problems at the end of each chapter have been grouped into four classes (designated by a superscript after the problem number)

Class a: Illustrates direct numerical application of the formulas in the text
Class b: Requires elementary analysis of physical situations, based on the subject material in the chapter
Class c: Requires somewhat more mature analysis
Class d: Requires computer solution

1.1^b. Relationship between ppm and micrograms per cubic meter for gaseous pollutants

Demonstrate Eq. (1.4) for gaseous pollutants.

1.2^a. Conversion of micrograms per cubic meter to ppm

The secondary NAAQS for SO_2 is 1,300 $\mu g/m^3$ (for a 3-h averaging time). Calculate the equivalent concentration in ppm at 25 ^oC and 1 atm.

Answer: 0.50 ppm

1.3^a. Conversion of ppm to micrograms per cubic meter

Ozone concentration sometimes reaches a value of 295 $\mu g/m^3$ over a 1-h period in urban areas with photochemical smog problems. Determine by what percentage this level exceeds the primary NAAQS for ozone. Conditions are 298 K and 1 atm.

Answer: 25 %

1.4[a]. Conversion of ppm to milligrams per cubic meter

The concentration of carbon monoxide in cigarette smoke reaches levels of 400 ppm or higher. For this particular value, determine the concentration in milligrams per cubic meter at 25 °C and 1 atm.

Answer: 458 mg/m^3

1.5[a]. The Greenhouse Effect

Consider Example 1.3b. If the global average temperature of the system earth-atmosphere is to increase by 4 °C, determine the percent increase required in the parameter β. Assume the solar constant and planetary albedo remain constant.

Answer: 4.9%

1.6[a]. Carboxyhemoglobin in blood

Consider that cigarette smoke contains an average of 450 ppm of CO. If the average oxygen content in the air inside the lungs is 20%, calculate the percent of the hemoglobin in the form of carboxyhemoglobin when saturation is reached.

Answer: 32%

1.7[c]. Carboxyhemoglobin in blood

Assume that there are 5,000 mL of blood in the body and that blood normally contains 20 mL of O_2 (in the form of HbO_2) per 100 mL of blood. A person is breathing at a volume rate of 3.6 L/min, and the air contains 600 ppm of CO. Determine the time required, in minutes, for the blood to reach 40% saturation with CO if the initial percent carboxyhemoglobin is 5%. Assume that the average oxygen concentration of the air in the lungs is 20%, and that all the CO inducted into the lungs is absorbed.

Answer: 92.6 min

1.8d. HbCO levels as a function of exposure

The following table relates carboxyhemoglobin levels in blood to CO concentration, in ppm, and time of exposure, in hours . Fit these data to an equation of the form

$$y = ax^b z^c$$

where:

y = % HbCO

x = CO concentration, in ppm

z = exposure time in hours

a, b, c are parameters to be estimated

CO concentration, ppm	Exposure time, hr	Percent HbCO
160	0.10	1.25
30	1.00	1.25
10	4.20	1.25
290	0.10	2.00
52	1.00	2.00
10	9.50	2.00
850	0.10	5.00
150	1.00	5.00
29	10.00	5.00

Source: NAPCA. "Air Quality Criteria for Carbon Monoxide," AP-62, Washington, D.C. HEW, (1970).

Answer: $a = 0.0722$

$b = 0.8405$

$c = 0.6173$

1.9b. Carboxyhemoglobin in blood

From the results of Problem 1.8, calculate the time required for the blood of a person exposed to a concentration of CO of 100 ppm to reach 50% of saturation. Assume 20% oxygen content in the air.

Answer: 1.81 h

1.10a. SO_2 NSPS for coal-fired power plants

A large electric power plant burns coal with a sulfur content of 2.25%, and a heating value of 30,171 kJ/kg. If it is to satisfy the NSPS for SO_2, determine the allowable emission rate, in g/10^6 kJ, and the required scrubber efficiency.

Answer: 260 g/million kJ, 83% efficiency

1.11b. SO_2 daily emissions from a coal-fired power plant

a) If the coal in Problem 1.10 is burned in a 40% efficient 400 MW power plant, estimate the allowable rate of SO_2 emissions, in kilograms per day.

Answer: 22,000 kg/d

b) If the FGD system converts the SO_2 removed to $CaSO_3$, calculate the amount of this waste generated, in kilograms per day.

Answer: 203,000 kg/d

1.12b. Equilibrium NO concentration in atmospheric air

Compute and plot the equilibrium concentration of NO in atmospheric air at various temperatures from 1000 K to 4000 K. Use a semi-log plot to accommodate the great variations in NO concentrations in this temperature range.

1.13b. Combustion material balances

Methanol (CH_3OH) is burned in dry air at an equivalence ratio of 0.75. Calculate

(a) Air-to-fuel ratio

Answer: 8.62 kg air/kg fuel

(b) Composition of the combustion products on a wet basis

Answer: 9.04% CO_2

1.14b. Combustion material balances

A fuel oil with composition 86% C and 14% H is burned in dry air. Analysis of the combustion products, on a dry basis, is 1.5% O_2 and 600 ppm CO. Calculate the equivalence ratio.

Answer: 0.93

1.15b. Particulate deposition in the lungs

The average quantity of air inhaled by an adult is 500 mL per breath. If 15 breaths per minute are taken and the air contains particulate matter at a concentration of 200 $\mu g/m^3$, what is the total hourly mass of particulate reaching the alveoli if the average diameter of the particles is 0.5 μm? It has been found that approximately 80% of particles this size are able to penetrate all the way into the alveoli.

Answer: 72 $\mu g/h$

1.16c. VOCs removal using refrigerated condensers

Refrigerated condensers are used to remove VOCs from waste gas streams by cooling the streams to temperatures as low as 190 K (see Figure 1.7). A refrigerated system must be installed to control gasoline vapor emissions generated by the loading of tank trucks at a bulk terminal. The total volume of waste gases generated is 15,140 m^3/day at 293 K and 1 atm, and with a VOC loading of 0.36 kg of gasoline vapor per cubic meter of air. The refrigeration unit must operate continuously 8,400 h/yr removing 90% of the gasoline vapor by cooling the stream to 190 K. The initial installed cost of this type of condenser is given by (Vatavuk, W. M., *Estimating Costs of Air Pollution Control*, Lewis, Chelsea, Michigan, 1990)

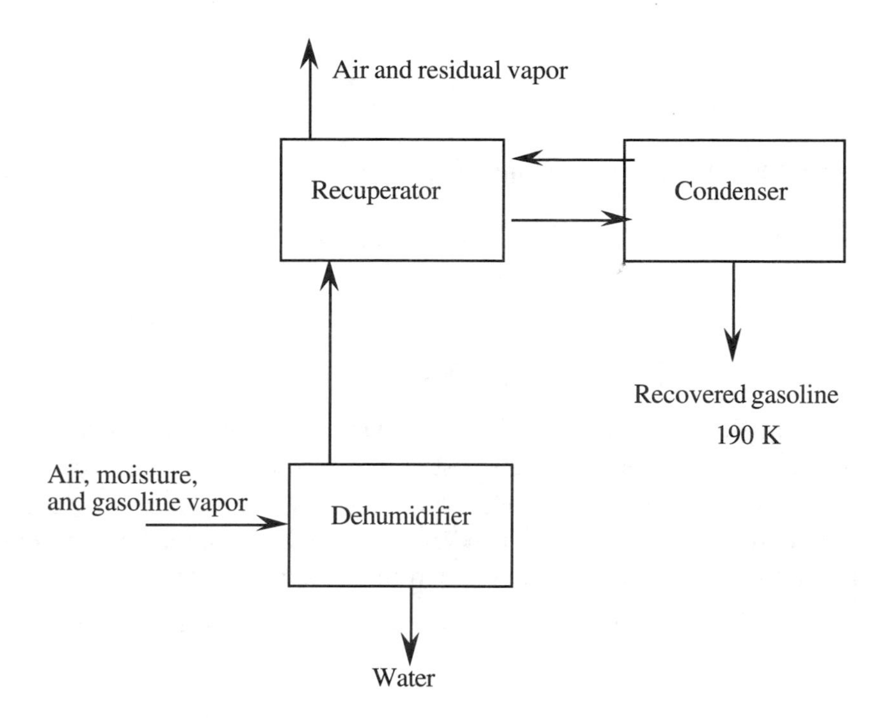

Figure 1.7 Schematic diagram of a refrigerated VOC condenser

$$TIC = 676.19\ Q^{0.791} + 16{,}642$$

where TIC is the total initial, installed cost of the equipment (in dollars of 1988), and Q is the flow rate of waste gases, in cubic meters per day measured at initial conditions. This correlation is valid for flow rates from 12,000 to 18,000 m^3/d. Fixed annual charges are 13% of the TIC, and annual variable charges are $176,340. There is a credit for the gasoline recovered, at $0.25/L. The density of the liquid gasoline is 0.84 g/mL. Estimate the total annual cost, TAC, for this system.

Answer: TAC = –$154,500/yr (profit!)

1.17b. VOCs removal through condensation

Air at 300 K and 1 atm is saturated with CCl_4 vapor. To what temperature must the mixture be cooled in order to condense 95% of the vapor, at constant pressure? Antoine's equation for CCl_4 is:

$$\ln p^* = 15.8742 - \frac{2808.19}{T - 46}$$

where p^* is the vapor pressure, in mm Hg, and T is in K.

Answer: 248.3 K

1.18d. Effect of the cost of electricity on optimal design

Refer to Example 1.10. In the equation for variable operating charges there is a term proportional to G_s^2 that represents the cost of the electricity required to operate the fan and the pump. The numerical value of the coefficient in that term (0.0975) corresponds to a unit cost of electricity of $0.06 /kW-h. Calculate the optimal dimensions of the FGD column for other electricity unit costs, from $0.03 to $0.12 per kW.

Answer: Diameter = 5.32 m @ $0.12/kW-h

1.19c. Accidental VOCs releases (Problem No. 71, ***Safety, Health, and Loss Prevention in Chemical Processes: Problems for Undergraduate Engineering Curricula,*** **© 1990 by AIChE, reproduced by permission of The Center for the Chemical Process Safety of AIChE)**

A tank truck has overturned and a large pool of benzene has formed on the ground.The terrain is fairly flat, and the benzene has spread into a pool that is approximately 20 m in diameter. The wind blows across the pool at a velocity of 7 m/s.

Benzene is considered to be a carcinogen, and worker exposure is to be limited to no more than 1 ppm. We wish to estimate the rate of benzene evaporation. The temperature of the air is 30 °C, and that of the liquid pool is about 18 °C. Use a flat plate turbulent boundary layer model for heat transfer, and then estimate the mass transfer coefficient from the Chilton-Colburn analogy.

Answer: 0.71 kg/s

1.20b. Average cost effectiveness of FGD system

The flue gases of Example 1.10 (average molecular weight = 29) contain 500 ppm of SO_2 at the FGD system inlet. The unit is designed for 70% SO_2 removal. Estimate the average cost effectiveness of the FGD system.

Answer: $55/metric ton

1.21a. European carbon taxes

By early 1991, the U.S. was still delaying a global warming policy, stating that the greenhouse effect had not been scientifically proven. In fact, projections at that time called for a 15% rise in U.S. CO_2 emissions by 2000 (Gilges, K. *Chemical Engineering,* January 1991, p. 52A). Conversely, several European nations were already instituting or proposing target CO_2 emissions reductions using economic instruments to encourage change. A tax on carbon CO_2 (carbon tax) was the preferred tool. Countries that had already implemented carbon taxes included Sweden, Denmark, Finland, and the Netherlands. The recently unified Germany was proposing a tax of $6.60 per metric ton of CO_2, very similar to the Finnish tax of $6.10/metric ton.

If a carbon tax of $6.6/metric ton were imposed on the CO_2 emissions of the power plant in Example 1.5, estimate the annual fee, assuming that the plant operates 350 days per year.

Answer: $17.3 million

1.22b. Average cost effectiveness of a Venturi scrubber

A variety of *wet-scrubbing* devices are used for particulate matter removal from waste gases. *Venturi scrubbers,* such as the one shown in Figure 1.8, remove particles in the size range of 0.5 to 5.0 μm with efficiencies above 95%. Water is injected into the throat of a venturi through which the waste gases flow at velocities of 45 to 150 m/s. The water is atomized by the high gas velocity. The liquid droplets collide with the particles in the gas stream, and the water and particles fall down for later removal. Power costs are relatively high for this device because of the high inlet gas velocity.

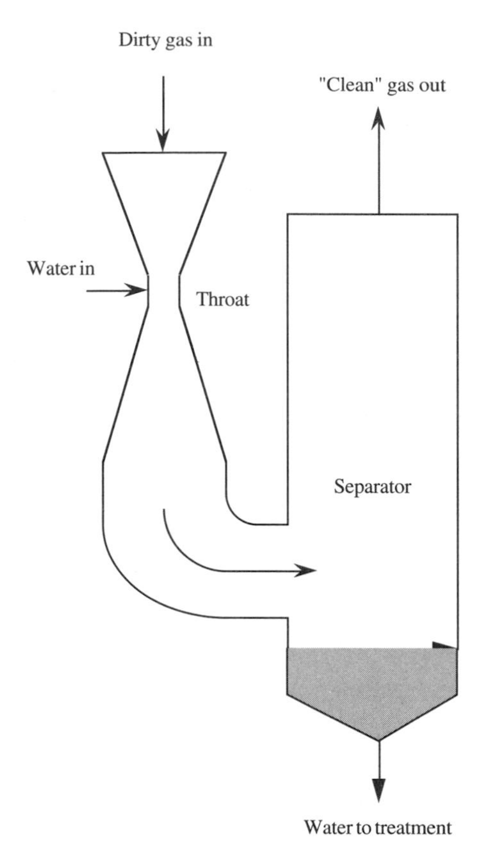

Figure 1.8 Schematic diagram of a Venturi scrubber

A steel mill plant contemplates the use of a Venturi scrubber to remove iron oxides particles from their waste gases. The waste gases flow at the rate of 110 m^3/s at 1,000 K and 102 kPa, and contain 0.0026 kg/m^3 of particulate matter. The water content of the gases is 8%. A Venturi scrubber was designed to satisfy the particle NSPS for steel plants. It is estimated that the total annual cost for the system is $1,224,000. Calculate the average cost effectiveness for the system if it operates 365 days per year.

Answer: $136/metric ton

1.23b. Incremental cost effectiveness

A base load 75-MW gas turbine is proposed for an area that is in compliance for all regulated pollutants. Uncontrolled emissions of NO_x would be 1,410 metric tons per year. The following table summarizes the potential NO_x control technologies identified as part of the top-down BACT analysis.

Alternative	Emissions[a]	Total annual cost, $/year
Option 1	75	3,380,000
Option 2	210	1,730,000
Option 3	350	883,000
NSPS	780	805,000
Uncontrolled	1,410	0

[a] Metric tons per year.

Calculate the average and incremental cost effectiveness for each control option. Determine if there is evidence to eliminate Option 1 based on economic grounds.

1.24[a]. Environmental costs of electricity

A recent report reviews studies that quantify the externality costs of environmental damage caused by electric power services (Ottinger et al., *Environmental Costs of Electricity,* Oceana, New York, 1990). The damage cost of CO_2 emissions—the potential for global warming— was estimated at 1.5¢/kg of CO_2.Estimate the damage cost of CO_2 emissions for the coal-fired power plant of Example 1.5. Give your result in ¢/kW-h generated. Compare the damage costs to the carbon taxes of Problem 1.21.

Answer: 1.56 ¢/kW-h

2

Cost Estimation Methodology

2.1 INTRODUCTION

Most air pollution control design problems are open-ended, meaning that there are several viable solutions. A major decision point in the design process is the selection of the most economically feasible solution alternative. This, usually, means choosing the alternative with the highest profitability potential. However, when dealing with air pollution control processes there is seldom any chance for making a profit, because the benefits to be obtained from such processes are intangible, while the costs involved are indeed very tangible. In such cases, cost would dictate the technique selected. Therefore, it is very important for the engineer involved in the design of air pollution control equipment to develop a reliable cost estimation methodology that will allow him to choose wisely between viable alternatives. Such is the objective of this chapter. On our way toward that objective, we will review some basic concepts of engineering economics and explore their applications to air pollution control equipment design.

2.2 CAPITAL INVESTMENT ESTIMATION

Your objectives in studying this section are to

1. Define total capital investment, depreciable investment, and nondepreciable investment.
2. Use average cost factors to estimate total capital investment for air pollution control devices based on equipment cost.

Before an air pollution control device can be put into operation, a significant amount of money must be supplied to purchase and install the necessary equipment. Frequently, land and service facilities must be obtained. In addition, it is necessary to have money available for the payment of expenses involved in the device operation. The sum of all these costs, and others, make up the total capital investment (TCI). The TCI is comprised, in turn, of depreciable and non-depreciable investments (Vatavuk 1990). The former refers to those costs that can be depreciated during the useful life of the control device. They include equipment and facilities-related costs plus any other costs that can be legitimately claimed as such on income tax returns. (Depreciation, a very important concept in cost estimation, will be discussed in detail later in the chapter). The latter refers to such costs as land and working capital. They are nondepreciable because they are recoverable when the control device ceases operation.

Table 2.1 lists the items comprising the total capital investment. Note that the total depreciable investment has two components: the battery limits and offsite facilities costs. The latter consist of those facilities that, though not an integral part of the control system, are needed to support its operation. (An electrical substation erected to supply power for an energy-intensive control system would be an example.) Add-on control systems typically require no offsite facilities, but others, such as FGD systems, typically do.

The battery limits cost is broken down into direct and indirect costs. Direct costs cover the cost of equipment (including freight and applicable sales taxes), such "hard" installation costs as piping and electrical work, site preparation, and buildings. The indirect ("soft") installation costs comprise engineering costs, construction and field expenses, contractor fees, startup and performance tests (to get the control system running and to verify that it meets the vendor's guarantees), and contingencies (Vatavuk 1990).

Cost analyses for complete industrial plants have shown that preliminary estimates of direct and indirect installation costs can be developed from the purchased cost of major equipment items. The costs of equipment installation, piping, instrumentation, and so forth are estimated as percentages of the purchased equipment cost. A similar approach is widely used for estimating the installed cost of air pollution control equipment. Table 2.2 presents, as an example, average cost factors used for thermal and catalytic incineration devices (widely used for control of VOCs).

Example 2.1 TCI for Thermal VOC Incinerator

The cost of buying a thermal incinerator for the control of VOCs in 26.3 m^3/s (at 300 K and 1 atm) of air was $142,000 in April of 1986. This cost includes all auxiliary equipment, instrumentation and controls. Using average cost factors for this type of device, estimate the TCI required. Presume no additional land is needed. A working capital of $30,000 is specified, mostly for a 30-d initial supply of fuel. No special site preparation or buildings are needed.

Solution

To estimate the TCI, we must calculate the depreciable and nondepreciable investments. Because no land is required, the nondepreciable investment involved is equal to the working

Table 2.1 Components of TCI

TCI = Depreciable Investment + Nondepreciable Investment

Nondepreciable investment consists of land and working capital

Depreciable investment consists of

- Offsite facilities
- Battery limits
  - Total indirect cost
    - Indirect installation costs
      - Engineering and supervision
      - Construction and field expenses
      - Construction fee
      - Startup
      - Performance tests
      - Contingencies
  - Total direct cost
    - Site preparation
    - Buildings
    - Purchased equipment cost
      - Control device
      - Auxiliary equipment
      - Instrumentation
      - Sales taxes
      - Freight
    - Direct installation cost
      - Foundations and supports
      - Handling and erection
      - Electrical
      - Piping
      - Insulation
      - Painting

Source: Adapted from Vatavuk, W. W. *Estimating Costs of Air Pollution Control*, p. 19, Lewis, Chelsea, MI. Copyright 1990. Reprinted with permission.

Table 2.2 Average Cost Factors for VOC Incinerators

Cost Item	Cost Factor
Direct costs	
1) Purchased equipment cost	
Incinerator and auxiliary equipment	1.00 *A*
Instrumentation and controls	0.10 *A*
Taxes and freight	0.08 *A*
Total purchased equipment cost	B = 1.18 *A*
2) Direct installation costs	
Foundations and supports	0.08 *B*
Erection and handling	0.14 *B*
Electrical	0.04 *B*
Piping	0.02 *B*
Insulation and painting	0.02 *B*
Site preparation (SP) and building (Bldg)	As required
Total installation direct costs	0.30 *B*+ SP + Bldg
Indirect costs	
Engineering and supervision	0.10 *B*
Construction, field, and fee	0.15 *B*
Start up and performance tests	0.03 *B*
Contingencies	0.03 *B*
Total direct plus indirect costs	1.61 *B* + SP + Bldg

Source: Adapted from Katari, K. S., Vatavuk, W. M., and Wehe, A. H. *JAPCA*, **37**:198 (1987). Reprinted with permission from *JAPCA*.

capital. The depreciable investment consists of the total direct plus indirect costs of purchasing and installing the equipment. We use cost factors from Table 2.2 to estimate this total. To calculate the total purchased equipment cost (B) we must consider the fact that the equipment cost given includes instrumentation and controls. Therefore,

$$1.1\,A = \$142{,}000 \qquad A = \$129{,}100$$

$$B = 1.18\,A = \$152{,}300$$

Because no site preparation or buildings are required:

$$\text{TCI} = 1.61\,B + \$30{,}000 = \$275{,}250 \text{ (April 1986)}$$

Comments

This example shows that, even for a relatively simple device such as a thermal incinerator, the total capital investment required can be much higher than the equipment cost, a fact often overlooked by neophytes in the area of cost estimation. Another point that should be emphasized is that any cost estimate is meaningless unless its date is clearly specified. Updating costs will be the subject of next section.

2.3 COST INDICES

Your objectives in studying this section are to

1. Define cost index.
2. Identify the cost indices most often used in air pollution control equipment cost estimation.
3. Use cost indices to update air pollution control equipment costs.

Most of the equipment cost data available are from several months to several years out of date and must be adjusted for the effect of inflation. Extrapolation of equipment costs to a later date is accomplished by using one of several published cost indices. A cost index is defined as the ratio of the cost of an item or equipment group at a specific time to the cost of the item at a base time in the past. In equipment cost estimation, an equipment or process cost index is used to update equipment cost as follows:

$$\text{Cost}_{\text{present}} = \text{Cost}_{\text{past}}\left(\frac{\text{present index}}{\text{past index}}\right) \tag{2.1}$$

No published index is ideal for updating the prices of air pollution control equipment. The variety of equipment comprising control systems is so great, no single index can accurately track the effect of inflation on all of them. However, three published indices can be used in some instances. These are the Chemical Engineering Plant Cost Index (CEP), the Marshall and Swift Equipment Cost Index (M&S), and the Producer Price Index (PPI). A brief description of each follows.

2.3.1 Chemical Engineering Plant Cost Index

This index is published in *Chemical Engineering*, and is updated monthly. Construction costs for chemical plants form its basis. The index consists of a composite ("CE Index") and 66 components, each weighted according to its contribution to the composite. The composite is often used to update total process plant costs, and the components are applied to updating the costs of individual items, both equipment and labor. The four major components (and their corresponding weights) are equipment, machinery, and supports (61%); construction labor (22%); buildings (7%); and engineering and supervision (10%).

The equipment component of the CE Index is often used to update air pollution control equipment costs. This component does include a mix of equipment that at least partly corresponds to control systems paraphernalia. It is updated monthly, so that it can reflect short-time price changes. Table 2.3 summarizes the composite CE Index and the equipment component from January 1985 to June 1990. Even though the index is updated monthly, for the sake of brevity, we present only quarterly and annual average values.

2.3.2 Marshall and Swift Equipment Cost Index

Originally known as the "Marshall and Stevens" Index, it is also published in *Chemical Engineering* where it is updated quarterly. The M & S Index compiles separate cost indices for 47 commercial, industrial, and housing activities. Reported is a composite—the average of the 47 individual indices—and two components: "process industries, average," and "related industries."

The M & S composite index is often used to update air pollution control equipment costs, even though it is really industry cost based. The variety of equipment costs reflected by the 47 industries making up the M & S composite could, in a way, reflect the diversity of pollution control equipment used, not only in the chemical process industry, but in related industries as well. Table 2.4 lists the M & S composite and process industries component from 1985 to June 1990.

Table 2.3 Chemical Engineering Plant Cost Index

Year	Month	CE Index	Equipment
1985	March	324.8	346.9
	June	324.8	347.0
	September	326.1	347.2
	December	326.1	348.1
	Annual average	325.3	347.2
1986	March	317.4	336.9
	June	316.2	333.4
	September	319.3	336.6
	December	319.2	335.7
	Annual average	318.4	336.3
1987	March	318.7	337.8
	June	321.9	340.4
	September	325.2	345.2
	December	332.5	357.2
	Annual average	323.8	343.9
1988	March	336.5	364.0
	June	341.6	371.6
	September	346.2	377.6
	December	349.2	383.2
	Annual average	342.5	374.1
1989	March	354.2	390.7
	June	355.6	392.4
	September	357.0	392.1
	December	357.3	390.0
	Annual average	355.4	391.3
1990	March	354.7	388.9
	June	356.9	391.8

2.3.3 Producer Price Index

The PPI is calculated and published monthly by the Bureau of Labor Statistics of the U.S. Department of Labor. It measures average changes in prices received by domestic commodity producers at all "stages of processing." Nearly 7,000 individual indices are calculated, based on price reports supplied by producers and the Federal government.

Table 2.4 Marshall and Swift Equipment Cost Index

Year	Quarter	M & S Index	Process Industries
1985	First	786.6	811.7
	Second	788.7	812.9
	Third	791.1	814.2
	Fourth	792.0	814.9
	Annual average	789.6	813.4
1986	First	793.5	815.1
	Second	797.8	817.4
	Third	798.3	816.5
	Fourth	801.0	818.7
	Annual average	797.7	816.9
1987	First	803.7	821.5
	Second	808.8	825.1
	Third	814.8	831.3
	Fourth	827.0	843.6
	Annual average	813.6	830.4
1988	First	835.3	851.4
	Second	846.7	864.8
	Third	856.5	875.0
	Fourth	869.5	889.1
	Annual average	852.0	870.1
1989	First	884.7	902.7
	Second	894.7	913.8
	Third	897.0	916.7
	Fourth	903.9	923.6
	Annual average	895.1	914.2
1990	First	905.8	925.3

Source: *Chemical Engineering.*

The individual indices are aggregated in three different ways: (1) stage of processing; (2) commodity classification system; and (3) Standard Industrial Classification. Of these, the commodity classification system is the best suited for updating some specific air pollution control equipment components. We will illustrate later the use of the "metal products" and "finished fabrics" PPIs in updating equipment components.

Table 2.5 summarizes the metal products and finished fabrics PPIs from 1985 to May 1990. Even though the indices are updated monthly, for the sake of brevity, we tabulate them on a quarterly and annual average basis.

2.3.4 Cost Indices Compared

We have discussed in some detail three cost indices used in air pollution control engineering: the equipment component of the CE Index; the M & S Composite Index; and the metal products and finished fabrics PPIs. Figure 2.1 shows the trends exhibited by these indices from 1984 to 1989. The equipment component of the CE Index and the M&S Composite Index show very similar trends during this period, except for the unusual years of 1986 and 1987. During those years, the process equipment market was in a slump because of a slowdown in plant construction and competition from overseas fabricators. The CE Index reflected these market conditions, reaching values lower than those corresponding to 1984. Conversely, the M & S Composite grew steadily during the same period, although the increases for the years 1985, 1986, and 1987 were modest. The metal products PPI followed the CE Index very closely, including the dip between 1985–87, up to 1987. During the years 1988–89, it grew considerably faster than the CE Index.

Based on those considerations, and the suggestions by Vatavuk (1990), we have chosen the equipment component of the CE Index to update the costs of major pieces of air pollution control equipment. Pieces of equipment that consist mainly of fabricated metal parts (such as cyclones) are cost updated using the metal products PPI; the costs of fabric bags used in filters for particulate matter removal are updated with the finished fabrics PPI.

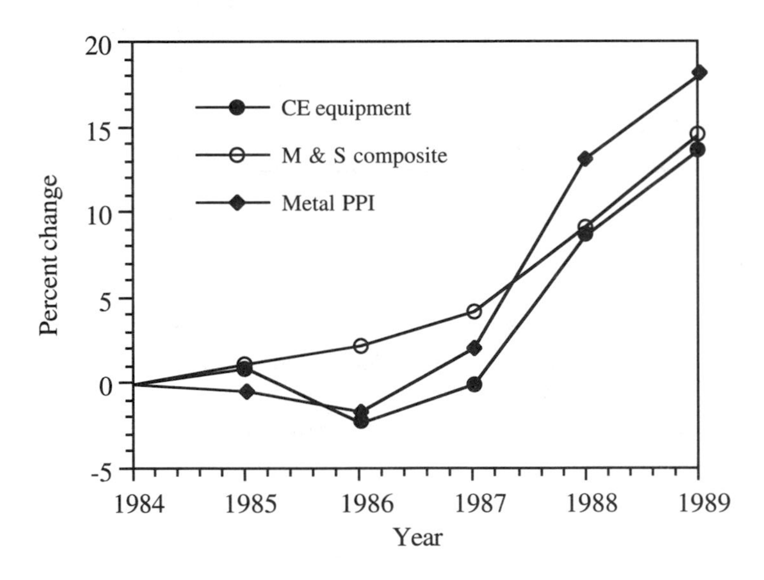

Figure 2.1 Trends in cost indices from 1984–89

Table 2.5 Selected Producer Price Indices

Year	Month	Metal Products PPI	Finished Fabrics PPI
1985	March	105.0	102.0
	June	104.4	100.9
	September	104.2	101.3
	December	103.9	101.1
	Annual average	104.4	101.4
1986	March	103.2	101.3
	June	103.0	101.5
	September	103.3	101.4
	December	103.3	101.3
	Annual average	103.2	101.4
1987	March	104.0	103.1
	June	105.8	104.0
	September	108.8	104.7
	December	112.9	106.2
	Annual average	107.1	104.2
1988	March	115.4	109.0
	June	118.0	109.6
	September	120.3	109.8
	December	124.0	110.8
	Annual average	118.7	109.4
1989	March	126.2	112.5
	June	123.7	113.8
	September	123.5	114.4
	December	121.6	115.1
	Annual average	123.9	114.0
1990	April	123.0	116.1
	May	123.0	116.0

Source: *Producer Price Indices Data for May 1990* (and various other issues for 1988 and 1989), Bureau of Labor Statistics, U.S. Department of Labor (July 1990).

Example 2.2 Equipment Cost Estimation Using CE Equipment Index

Consider the thermal incinerator of Example 2.1. Estimate the cost of buying the same equipment in June 1990.

Solution

We will use Eq. (2.1) to solve this problem. In this equation $cost_{past}$ refers to the cost of the equipment in April 1986, $cost_{present}$ refers to the cost updated to June 1990, present index is the value of the CE Equipment Index in June 1990, and past index is the value for April 1986. We obtain the corresponding values from Table 2.3 (values for the month of April are not tabulated; therefore, we will use the value for the month of March).

Cost present = \$142,000 (391.8/336.9) = \$165,140 (June 1990)

Comments

In this example, we used the CE Equipment Index to update the cost of a major piece of air pollution control equipment. It is left as an exercise for the curious reader to show that using the M & S Composite Index the result would have been very similar.

Example 2.3 Equipment Cost Estimation Using CE Equipment Index and the Producer Price Indices

Fabric filters are gaining widespread acceptance for the removal of particulate matter from gaseous streams. The cost of buying one of these devices is usually divided into four components: (1) the cost of the basic baghouse unit, (2) additional cost owing to insulation, (3) stainless steel construction expenses for corrosive applications, and (4) the cost of the fabric bags. The total cost of buying a fabric filter system to process 23.6 m3/s of air at 436 K, contaminated with corrosive particulate matter, was \$221,000 during the third quarter of 1986. This cost could be divided as follows:

Basic baghouse unit	\$132,600
Insulation	11,400
Stainless steel construction	58,900
Fabric bags	18,100

Update the equipment cost to June 1990.

Solution

The solution uses Eq. (2.1), with the appropriate cost indices. The basic baghouse unit and insulation costs can be lumped and updated using the CE Equipment Index. Stainless steel construction expenses are updated using the metal products PPI, and the fabric bags cost is updated using the finished fabrics PPI. The corresponding values of the indices are obtained from Tables 2.3 and 2.5. Third quarter of 1986 is understood to mean September of that year.

Basic baghouse plus insulation = (132,600 + 11,400)(391.8/336.6) = \$167,600
Stainless steel expenses = 58,900 (123.0/103.3) = \$70,100
Fabric bags cost = 18,100 (116.0/101.4) = \$20,700
Therefore, the updated total cost of buying the system is \$258,400 (June 1990).

Comments

This example illustrates the simultaneous use of various cost indices to update different components of a fairly complex air pollution control device. Unfortunately, there are no indices published that are specific to the air pollution control equipment industry, therefore we have to rely on such mixed-index estimates. As a general rule, Vatavuk (1990) suggests limiting this approximation to cost estimation over periods not exceeding 5 years.

2.4 TOTAL ANNUAL COST

Your objectives in studying this section are to

1. Define total annual cost, depreciation, capital recovery cost, and recovery credits.
2. Identify and estimate the different components of the total annual cost for an air pollution control application.

Determination of the capital investment required to implement a given pollution control technology is only one part of a complete cost estimate. Another equally important part is the estimation of annual costs for operating the process. TAC refers to the sum of all those expenses incurred every year during the life of the control equipment.

TAC is comprised of two major elements: the direct annual cost (DAC) and the indirect annual cost (IAC). These costs are offset, in some happy instances, by recovery credits (RC). These credits are due to material and energy recovered by the control system, such as the gasoline vapors recovered in Problem 1.16. Recovery credits are considered revenue or income.

The following equation relates the four concepts defined previously:

$$\text{TAC} = \text{DAC} + \text{IAC} - \text{RC} \quad (2.2)$$

where the units of each term are dollars per year.

Direct annual costs are those that tend to be proportional or partially proportional to the level of usage given to the control equipment. Direct costs are, in turn, divided into two categories: variable, and semivariable. Variable costs are those that are proportional to the level of usage, such as utilities (electricity, fuel, steam, water, etc), raw materials, waste treatment, and final disposal. Semivariable costs depend on level of usage but only partially. Labor costs (operating, supervision, and maintenance) payroll overhead, maintenance materials, and recurring replacement parts are examples of semivariable costs. Even at zero level of usage, semivariable costs are from 20% to 40% of their value at full usage.

Conversely, indirect annual costs are totally independent of level of usage. Even if the control system were shut down, the totality of these expenses would be incurred. They include plant overhead, property taxes, insurance, general and administrative expenses, and capital recovery costs. Table 2.6 summarizes the various components of TAC. Each of these components will be discussed in some detail

2.4.1 Utilities

The quantities and type of utilities required by pollution control systems are calculated based on the operating and design features of each particular application. We will devote a significant portion of subsequent chapters to this task. With some systems, the cost for utilities can overshadow all other components of the TAC. Thermal VOC incinerators are a good example. For systems with no waste heat recovery, fuel can account for 75% or more of TAC.

2.4.2 Raw Materials

The kinds and quantities of raw materials depend on the capacity and type of the pollution control technology employed. Some, such as most particulate control equipment , require little or no raw materials. Others, such as most FGD systems, consume large amounts. The amounts required for each specific application will be calculated from process considerations through material and energy balances.

Table 2.6 Components of TAC

DAC			
	Variable		
		Raw materials	
		Utilities	
			Electricity
			Steam
			Fuel
			Water
			Compressed air
			Others
		Waste treatment	
		Final disposal	
	Semivariable		
		Labor	
			Operating
			Maintenance
			Supervision
		Maintenance materials	
		Replacement parts	
		Payroll overhead	
IAC			
	Capital recovery		
	Plant overhead		
	Property taxes		
	Insurance		
	Administrative charges		
Recovery credits			
	Materials		
	Energy		

2.4.3 Labor

Practically all pollution control devices require at least some operating and maintenance labor. This can range from a few minutes per shift to the requirement of several full-time employees. In subsequent chapters we will provide some estimates of these labor requirements for the various pollution control devices to be studied in detail.

In addition to operating and maintenance labor, Vatavuk (1990) suggests adding from 10% to 20% of operating labor to account for the cost of supervisory labor (i.e., foremen, shift supervisors). Typically the supervisory, operation, and labor rates ($/h) are different. Values of labor rates can be found in such publications as the Monthly Labor Review (U.S. Department of Labor, Bureau of Labor Statistics).

2.4.4 Maintenance

Maintenance costs are divided into maintenance labor (already covered) and maintenance materials. Maintenance materials include small tools, lubricants, wire, tape, and other items consumed during the year. Their cost is usually estimated as 100% of the maintenance labor cost.

2.4.5 Replacement Parts

Replacement parts is a cost component not covered under the annual maintenance charge for two reasons: (1) this cost is incurred less frequently than annually, and (2) it is typically much larger in magnitude than the annual maintenance cost (Vatavuk 1990). Some examples of pollution control devices requiring replacement parts are given in Table 2.7.

Table 2.7 Examples of Replacement Parts

Pollution Control Device	Replacement Parts
Catalytic incinerator	Catalyst
Carbon adsorbers	Activated carbon
Fabric filters	Bags
Thermal incinerators	Refractory lining

The cost of replacing these parts should incude such items as applicable sales taxes, freight, and replacement labor, besides the cost of the parts themselves. As a first approximation, replacement labor can be estimated as 100% of the replacement parts cost (Vatavuk 1990).

Because the replacement parts cost is incurred periodically, but with periods longer than a year, it must be treated as a capital expenditure with a useful life equivalent to the replacement period. The following equation is used to calculate an annualized replacement cost:

$$C_{rep} = P_{rep}\, CRF_{rep} \tag{2.3}$$

where:

C_{rep} is the annualized replacement cost, in \$/year

P_{rep} is the total cost of replacement parts and labor (including taxes and freight)

CRF_{rep} is a capital recovery factor, to be defined in Section 2.4.9.

2.4-6 Waste Treatment and Disposal

Most often, the air pollutants captured by the control system cannot be sold or reused and, therefore, must be disposed of. Ultimately, disposal involves hauling the material to some kind of depository, or incinerating it, depending on the characteristics of the waste. The costs associated with these practices can be quite high and are very site-specific.

2.4.7 Overhead

Overhead is usually divided into two categories: payroll overhead and plant overhead. The former includes expenses associated with operating, supervisory, and maintenance labor (such as Social Security, insurance premiums, pension fund contribution, vacations, and other fringe benefits). It is generally computed as a fraction of the total labor cost.

Plant overhead includes those expenses not necessarily tied to the operation and maintenance of the control technology but that must be allocated in some way to it. They include facility security, plant offices, parking area, lighting, cafeterias, locker rooms, and so on. It is usually estimated as a fraction of total labor cost plus maintenance materials.

For most cost estimation purposes it is sufficiently accurate to combine both payroll and plant overhead into a single factor, estimated as a percentage of total labor. Peters and Timmerhaus (1980) suggest a range of 50% to 70% of total labor plus maintenance materials.

2.4.8 Property Taxes, Insurance, and Administrative Charges

These indirect costs are fixed charges. Property taxes and insurance premiums are self-explanatory; administrative charges include sales costs when applicable, research and development costs, accounting, and other home office expenses.

These three components of the TAC are usually estimated as a fraction of TCI. For air pollution control applications , 4% of TCI has been suggested (Vatavuk 1990), broken down as follows: property taxes, 1%; insurance, 1%; and administrative costs, 2%.

2.4.9 Capital recovery and depreciation

Capital recovery and depreciation are two different approaches to the same objective: to arrive at an annual expenditure that will correctly account for the initial depreciable capital investment including an acceptable rate of return on it. Actually, depreciation is, as we shall see, a special case of capital recovery used for income tax purposes.

Grant et al. (1982) explain the classical method for calculating capital recovery. For each year of the project's life a certain amount of money, or payment, is set aside for the purpose of capital recovery. Each yearly payment earns interest at the rate the firm deems "minimally attractive" for its capital investments. This rate changes from one firm to another, depending, basically, on the financial stability of the concern. The sum of these payments over the life of the project, plus the interest they earn, plus the final salvage value, if any, must equal the initial depreciable investment (less the replacement parts, which have been already accounted for as an annualized cost) plus the interest this capital investment would have earned had it been invested elsewhere at the same rate of return. Sounds confusing? It is. Expressed in mathematical formulas it is easier to understand than in conceptual terms.

Let CRC be the annual payment set aside for capital recovery at the end of each of the n years which represent the useful life of the control device. Let i represent the "minimum attractive" rate of return specified by the firm for capital investments. The first payment of CRC is made at the end of the first year and will bear interest for $n - 1$ years. Therefore, at the end of the useful life of the device this first payment will have grown to an amount of $CRC(1 + i)^{n-1}$. The second payment is made at the end of the second year and will bear interest for $n - 2$ periods giving an accumulated amount of $CRC(1 + i)^{n-2}$. Similarly, each annual payment will give an additional accumulated amount until the last payment is made at the end of the useful life of the device. The cumulative amount at the end, including the salvage value S, is given by

$$\text{Cumulative amount} = \text{CRC}\left[(1+i)^{n-1} + \cdots + (1+i) + 1\right] + S \tag{2.4}$$

Conversely, if the total depreciable investment (TDI), excluding the cost of replacement parts (P_{rep}), had been invested elsewhere at the same rate of return, and for the useful life of the device, the total accumulated at the end would be $(TDI - P_{rep})(1 + i)^n$. According to the concept of capital recovery, the two accumulated totals must be equal.

$$\left(\text{TDI} - \text{P}_{\text{rep}}\right)(1+i)^n = \text{CRC}\left[(1+i)^{n-1} + \cdots + (1+i) + 1\right] + S \tag{2.5}$$

Multiplying each side of Eq. (2.5) by $(1 + i)$, and substracting Eq. (2.5) from the result gives

$$\text{CRC} = \left(\text{TDI} - \text{P}_{\text{rep}}\right)\frac{i(1+i)^n}{\left[(1+i)^n - 1\right]} - S\,\frac{i}{\left[(1+i)^n - 1\right]} \tag{2.6}$$

Equation (2.6) can be expressed in terms of a capital recovery factor (CRF) and a sinking fund factor (SFF) defined as

$$\text{CRF} = \frac{i(1+i)^n}{\left[(1+i)^n - 1\right]}, \qquad \text{SFF} = \frac{i}{\left[(1+i)^n - 1\right]} \tag{2.7}$$

$$\text{CRC} = \left(\text{TDI} - P_{rep}\right)\text{CRF} - (S)(\text{SFF}) \tag{2.8}$$

Example 2.4 Capital Recovery Calculations.

The total depreciable investment for a fabric filter system for particulate removal is $412,000. The fabric bags must be replaced every two years, and the total cost of replacement (including sales tax, freight, and labor) is $17,000. The useful life of the system is 10 years, and the salvage value of the equipment is $25,000. According to company policy, the minimum "attractive rate of return" is 12%.

(a) Calculate the annualized cost of replacing the bags.
(b) Calculate the capital recovery annual cost.

Solution

a) The solution uses Eqs. (2.3) and (2.7) to calculate the annualized cost of replacing the bags. The useful life of the bags, n, is 2 years; the minimum attractive rate of return, i, is 0.12; the replacement cost is $17,000.

$$\text{CRF} = \frac{0.12(1.12)^2}{\left[(1.12)^2 - 1\right]} = 0.592$$

$$C_{rep} = (\$17{,}000)(0.592) = \$10{,}600/\text{yr}$$

(b) Use Eqs. (2.7) and (2.8) to solve this part. For this problem, $n = 10$ and $i = 0.12$; therefore, CRF = 0.177, SFF = 0.057

$$\text{CRC} = (412{,}000 - 17{,}000)(0.177) - (25{,}000)(0.057) = \$68{,}600/\text{yr}$$

Comments

It should be emphasized than in Eq. (2.8) the cost of replacement parts is substracted from the total depreciable investment to avoid double-counting. Remember that when calculating an annualized replacement parts cost, the useful life in the equation for calculating the corresponding capital recovery factor refers to the time interval between replacements (e.g., 2 years for this example).

Although engineering economists favor the capital recovery factor method, accountants prefer to use depreciation techniques to account for capital expenditures recovery. Depreciation is a form of capital recovery applicable to a property with 2 or more years' lifespan in which an appropriate portion of the asset's value is periodically charged to current operations, and which is acceptable for income tax purposes accounting.

The Internal Revenue Service (IRS) establishes the rules of the game for depreciation calculations. Those rules are changed frequently, and somewhat arbitrarily. Therefore, for the rest of this book we will use the capital recovery factor approach and leave depreciation considerations to the accountants. Regardless of which method is used, "the total of depreciation plus interest [charges] is intended to serve the same purpose as [the] annual cost of capital recovery with a return" (Grant et al. 1982).

Current EPA guidelines recommend that income tax considerations be excluded from cost analyses. Income taxes generally represent transfer payment from one segment of society to another and as such are not properly part of economic costs (U.S. EPA 1990).

Example 2.5 TAC for Fabric Filter System

Consider the fabric filter system of Example 2.4. The following additional annual costs have been determined for this operation:

Operating labor	\$26,000
Maintenance labor	14,000
Electricity	48,000
Compressed air	8,000
Waste disposal	150,000

Estimate the TAC for this operation. Use the results of Example 2.4 for capital recovery cost and annualized replacement parts charges.

Solution

Solve this problem in tabular form, following the general outline of Table 2.6.

Direct annual costs	
Raw materials	
Utilities	
Electricity	$48,000
Compressed air	8,000
Waste disposal	150,000
Labor	
Operating	26,000
Supervisory (15% of operating)	3,900
Maintenance	14,000
Maintenance materials (100% maintenance labor)	14,000
Replacement parts (see Example 2.4)	10,600
Total DAC	274,500
Indirect annual costs	
Capital recovery cost (see Example 2.4)	68,600
Overhead [0.6(26000 + 3900 + 14000 + 14000)]	34,700
Property taxes (0.01 TCI = 0.01 × 412,000)	4,120
Insurance (0.01 TCI)	4,120
Administrative charges (0.02 TCI)	8,240
Total IAC	119,780
Recovery Credits	—
TAC	$394,280

Comments

In this example, payroll and plant overhead were lumped together under indirect annual costs. This is sufficiently accurate for most cost estimation purposes.

A close look at each component of the total annual cost tabulation reveals that, for this particular problem, the cost of waste disposal is, by far, the most important disbursement. The particulate matter collected in this instance is a nuisance, and the company is charged $22/metric ton for final disposal. Conversely, if there happened to be a use for the particulate matter, such as filling material for concrete blocks, the company could end up receiving recovery credits for the material instead of paying for final disposal. The effect on TAC could be dramatic!

2.5 CHOICE BETWEEN VIABLE ALTERNATIVES

Your objectives in studying this section are to

1. Define cash flow, present worth (PW), profitability index (PI), and equivalent uniform annual cash flow (EUAC).
2. Choose the best among several viable design alternatives using either the PW, the PI, or the EUAC measures of merit.

Several equally effective control techniques are often available to reduce the emissions from an air pollution source . There are several measures of merit that cost analysts use to compare competing control technologies. This section covers three of the most often used: PW, PI, and EUAC. Before doing so, a very basic concept of engineering economics must be understood: cash flow.

During the useful life of an air pollution control device, expenditures are made, and, sometimes, revenues are received. The capital expenditures usually occur at or before the startup of the system, and in every year thereafter the various annual costs are spent. The amounts and timing of these costs and revenues comprise the project cash flows. If income tax is not considered, the net cash flow (NCF) for a given year is the difference between revenues and total annual costs (excluding the capital recovery charge).

$$\text{NCF} = \text{Revenues} - (\text{TAC} - \text{CRC}) \tag{2.9}$$

Figure 2.2 is a typical cash flow diagram for an air pollution control device with a useful life of 8 years. It shows nine cash flows: the TCI made at year "0" and one NCF for each of the 8 years of the project. Notice that the NCF is the same from year 1 to year 7. In year 8, however, the NCF increases (i.e., becomes less negative). The explanation for this is that at the end of the project the working capital, land, and equipment salvage are added in as revenues, thus increasing the corresponding NCF. The following example will help to clarify these concepts.

Example 2.6 Cash Flow Calculations

Calculate the annual cash flows for the fabric filter system of Example 2.5.

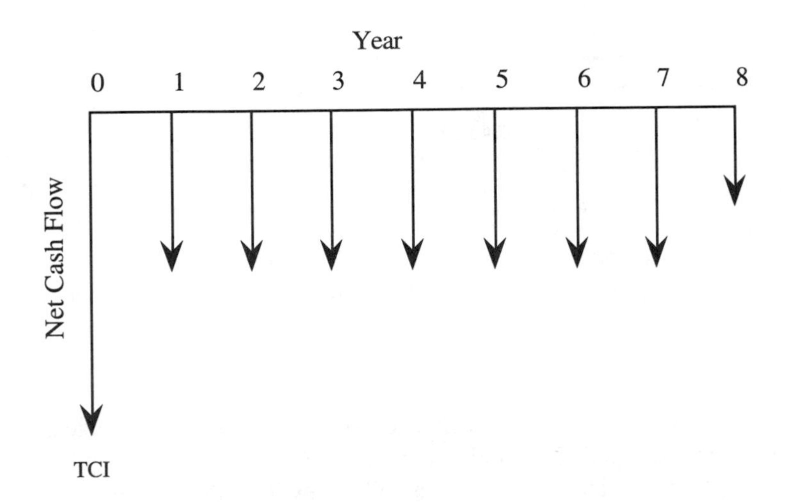

Figure 2.2
Typical cash flow diagram

Solution

The solution uses Eq. (2.9) to calculate annual NCFs. The following information is obtained from Example 2.5:

TCI = \$412,000 TAC = \$394,280/yr

CRC = \$68,600/yr S = \$25,000

The cash flow at time "0" is the TCI, or –\$412,000 (the negative sign indicates it is a disbursement). For years 1 to 9, the NCF is

NCF = 0 – (\$394,280 – 68,600) = –\$325,680/yr

For year 10, the NCF is calculated with the same equation, but the salvage value of the equipment (S) is incorporated in the calculation as a revenue:

NCF = \$25,000 – (\$394,280 – 68,600) = –\$300,680/yr

The following table summarizes the cash flows:

Year	Cash Flow
0	–$412,000
1 to 9	–325,680/yr
10	–300,680

After this short introduction to the concept of cash flow, we are in a position to understand the ideas behind the measures of merit to be considered when comparing control techniques.

2.5.1 PW

Choosing between viable alternatives usually involves determining what is economical in the long run, that is, over a considerable period. In such problems it is necessary to recognize the time value of money. Because of the existence of interest, a dollar now isworth more than the prospect of a dollar next year or at some other later date. In this context, interest may be thought of as the return obtainable by the productive investment of capital.

Because the annual NCFs for a given project occur at different points in time, the only way to add them while considering the time value of money is to "discount" all of them to the same reference time. If the reference time is taken as "0," this discounting process gives us the PW of each cash flow. By "discounting" we mean multiplying each NCF by the discount factor $1/(1 + i)^k$, where i is the minimum attractive rate of return, and k is the year in which the corresponding cash flow happens.

The sum of these discounted cash flows, when added to the total capital investment, yields the net present worth (NPW) of the project. When comparing control technologies, the one with the highest positive (or lowest negative) NPW is the one to select.

Example 2.7 NPW of Air Pollution Control Device

Calculate the net present worth of the fabric filter system of Example 2.6.

Solution

The solution discounts each of the annual net cash flows of Example 2.6 to calculate their present worths. The value of the minimum attractive rate of return for the firm has been specified in previous examples as 12%. Table 2.8 presents the results of the calculations and shows that the net present worth of the system is –$2,244,350.

Table 2.8 PW Calculations for Example 2.7

Year, k	Cash Flow	Discount Factor	PW
0	-$412,000	1.0	-$412,000
1	-325,680	0.893	-290,790
2	-325,680	0.797	-259,630
3	-325,680	0.712	-231,890
4	-325,680	0.636	-207,130
5	-325,680	0.567	-184,800
6	-325,680	0.507	-165,000
7	-325,680	0.452	-147,320
8	-325,680	0.404	-131,540
9	-325,680	0.361	-117,440
10	-300,680	0.322	- 96,810
Net present worth			-2,244,350

Comments

The results of the previous calculations indicate that the net financial effect of this pollution control system is equivalent to an initial loss of $2,244,350. It is easy to understand, then, the reluctance of most profit-oriented concerns to invest their precious capital on such propositions.

Example 2.8 Choice between Alternatives Based on PW

In the control of gaseous pollutants emissions from a certain source, two alternative devices are under consideration. We shall call them Plans A and B. Economic evaluation of the two plans yields the following results:

	Plan A	Plan B
TCI	$150,000	$200,000
Useful life	5 yr	5 yr
Annual NCF	–$100,000	–$80,000

Based on the PW measure of merit, determine which plan to choose if the minimum attractive rate of return is 10%.

Solution

The solution discounts all cash flows to time "0" and calculates the NPW of both plans. Choose the plan with the highest (least negative) NPW. The results are tabulated as follows:

Year	Discounted Cash Flows Plan A	Plan B
0	–$150,000	–$200,000
1	–90,910	–72.727
2	–82,645	–66,116
3	–75,131	–60,105
4	–68,301	–54,641
5	–62,092	–49,674
NPW	–529,079	–503,263

Therefore, Plan B should be chosen because it results in a higher (less negative) NPW.

Comments

Even though Plan B requires a higher TCI, the resulting reduction in annual costs is enough to recover the incremental investment with a rate of return higher than the minimum attractive rate of 10%, as will be shown in Example 2.10

In those instances in which installation and operation of an air pollution control device results in profits, or when the selection of a more capital-intensive alternative results in annual savings because of reduced operation and maintenance expenses, there is a method for choosing between the different alternatives that involves evaluating the PI. The PI is the rate of return that will make the NPW of the project exactly equal to zero. In other words, when each of the annual cash flows is discounted to time "0" (using PI as the rate of return) and added to the TCI, the net result is zero. Clearly, this is a special case of PW analysis in which the NPW is specified, and the rate of return must be calculated, by trial and error usually. When comparing alternatives that result in profits (positive annual cash flows), the one with the highest PI should be chosen.

If all the alternatives have negative annual cash flows, usually the case in air pollution control equipment design, they are compared pairwise on a mutually exclusive basis in terms of incremental capital investment and incremental annual cash flow. The alternative with the lowest TCI requirement is the basis for the initial comparison. An incremental investment is justified only if the resulting PI is higher than the specified minimum attractive rate of return. The following examples illustrate these concepts.

Example 2.9 Profitability Index: Special Case of PW Analysis

Consider the refrigerated condenser of Problem 1.16. The total capital investment is $1,386,100, and the capital recovery cost is $180,200/yr. The useful life of the system is 15 years, and there is no salvage value. There is a net yearly profit of $154,500. Calculate the PI for this investment.

Solution

The first step in solving this problem is to calculate the NCF for each of the 15 years covered by the project. In calculating the annual profits, all costs were accounted for including the capital recovery cost. To calculate the NCF, the capital recovery cost must be added to the profits. Because there is no salvage value, the NCF for each of the 15 years is $154,500 + $180,200, or $334,700/yr. The next step is, then, to discount the 15 net cash flows to time "0" using PI as the rate of return. Adding the present worths of these cash flows to the TCI and equating the net present worth to zero, we obtain an equation to solve for PI. If we define a new variable $x = 1/(1 + \text{PI})$, the discount factor for year k is simply x^k. The NPW then is

$$\text{NPW} = 0 = -1{,}386{,}100 + 334{,}700\left(x + x^2 + x^3 + \cdots + x^{15}\right)$$

Simplifying and rearranging,

$$F(x) = x^{15} + x^{14} + \cdots + x - 4.141 = 0$$

This is a 15th-degree polynomial. It has 15 roots, however, the one we are looking for should be located between 0 and 1 (remember the definition of x). To find the solution, we introduce a root-finding algorithm that is a slight modification of the method suggested by Press et al. (1989). It combines the convergence characteristics of the Newton-Raphson method with the robustness of bisection to find the value of a root that has been previously bracketed (the bracketing algorithm has, already, been incorporated into the method). The function under consideration, and an expression for its first derivative, are supplied by the user in a subroutine. Two distinct initial estimates of the root are also required. The desired accuracy of the estimated root is supplied by the user.We ran the program for this example, using as initial estimates 0.8 and 0.5, and specifying an accuracy of 0.00001. The result obtained was $x = 0.8125$, corresponding to a PI = 23.1%.

Comments

The resulting profitability index for this project is very attractive indeed. Recovering the gasoline vapors is a good way of making money while simultaneously preserving the atmospheric environment.

The root-finding program introduced in this example is quite general, and guaranteed to converge to a bracketed root. The only part of the program that needs to be modified to use it for other problems is the SUBROUTINE FUNCD.

```
      PROGRAM RTFIND
C
C  USING A COMBINATION OF NEWTON-RAPHSON AND BISECTION, FIND
C  THE ROOT OF A FUNCTION. X1 AND X2 ARE INITIAL ESTIMATES OF
C  THE ROOT, WHICH IS REFINED TO AN ACCURACY XACC SPECIFIED
C  BY THE USER. FUNCD IS A USER- SUPPLIED SUBROUTINE WHICH
C  RETURNS BOTH THE FUNCTION VALUE (F) AND THE FIRST DERIVATIVE (DF)
C
      PARAMETER (FACTOR=1.6 , NTRY=50)
      EXTERNAL FUNCD
      PRINT *, 'ENTER FIRST ESTIMATE OF ROOT  '
      READ *, X1
      PRINT *, 'ENTER SECOND ESTIMATE OF ROOT  '
      READ *, X2
      PRINT *, 'ENTER DESIRED ROOT ACCURACY  '
      READ *, XACC
C
C  ROOT BRACKETING ALGORITHM
C
      CALL FUNCD (X1, F1, DF1)
      CALL FUNCD (X2, F2, DF2)
      DO 11 J=1, NTRY
        IF (F1*F2 .LT. 0.) GO TO 12
        IF (ABS(F1) .LT. ABS(F2)) THEN
            X1=X1+FACTOR*(X1-X2)
            CALL FUNCD (X1, F1, DF1)
        ELSE
            X2=X2+FACTOR*(X2-X1)
            CALL FUNCD(X2, F2, DF2)
        END IF
 11   CONTINUE
C
C  END OF BRACKETING
```

```
C
 12  ROOT=RTSAFE (FUNCD,X1,X2,XACC)
     PRINT *
     PRINT *, 'THE VALUE OF THE ROOT IS  ', ROOT
     END
C
C
     SUBROUTINE FUNCD(X,F,DF)
     F=X**15+X**14+X**13+X**12+X**11+X**10+X**9+X**8+X**7+X**6
    *  +X**5+X**4+X**3+X**2+X-4.141
     DF=15*X**14+14*X**13+13*X**12+12*X**11+11*X**10+10*X**9
    *  +9*X**8+8*X**7+7*X**6+6*X**5+5*X**4+4*X**3+3*X**2+2*X+1
     RETURN
     END
      FUNCTION RTSAFE(FUNCD,X1,X2,XACC,CA,CB)
```



```
     EXTERNAL FUNCD
     PARAMETER (MAXIT=100)
     CALL FUNCD(X1,FL,DF,CA,CB)
     CALL FUNCD(X2,FH,DF,CA,CB)
     IF(FL*FH.GE.0.) PAUSE 'root must be bracketed'
     IF(FL.LT.0.)THEN
       XL=X1
       XH=X2
     ELSE
       XH=X1
       XL=X2
       SWAP=FL
       FL=FH
       FH=SWAP
     ENDIF
     RTSAFE=.5*(X1+X2)
     DXOLD=ABS(X2-X1)
     DX=DXOLD
     CALL FUNCD(RTSAFE,F,DF,CA,CB)
     DO 11 J=1,MAXIT
       IF(((RTSAFE-XH)*DF-F)*((RTSAFE-XL)*DF-F).GE.0.
    *     .OR. ABS(2.*F).GT.ABS(DXOLD*DF) ) THEN
         DXOLD=DX
         DX=0.5*(XH-XL)
         RTSAFE=XL+DX
         IF(XL.EQ.RTSAFE)RETURN
       ELSE
```

```
        DXOLD=DX
        DX=F/DF
        TEMP=RTSAFE
        RTSAFE=RTSAFE-DX
        IF(TEMP.EQ.RTSAFE)RETURN
      ENDIF
      IF(ABS(DX).LT.XACC) RETURN
      CALL FUNCD(RTSAFE,F,DF,CA,CB)
      IF(F.LT.0.) THEN
        XL=RTSAFE
        FL=F
      ELSE
        XH=RTSAFE
        FH=F
      ENDIF
 11   CONTINUE
     PAUSE 'RTSAFE exceeding maximum iterations'
     RETURN
     END
```

Example 2.10 Choice between Alternatives Based on PI

Consider the alternative plans proposed in Example 2.8. If Plan B is chosen over Plan A, determine the resulting PI. If the minimum attractive rate of return is 10%, determine which plan should be chosen.

Solution

Of the two alternatives under consideration, Plan A requires the smaller initial investment. Plan B requires an incremental TCI of \$50,000, but results in a net reduction of annual costs (i.e., an incremental annual cash flow) of \$20,000/year. Choosing Plan B over Plan A, then, is equivalent to investing \$50,000 in a project that will yield annual profits of \$20,000 for the next five years. A tabulation of the cash flows follows.

		Cash Flows	
Year	**Plan A**	**Plan B**	**B – A**
0	–\$150,000	–\$200,000	–\$50,000
1	–100,000	–80,000	20,000
2	–100,000	–80,000	20,000
3	–100,000	–80,000	20,000
4	–100,000	–80,000	20,000
5	–100,000	–80,000	20,000

Discounting the column labeled B – A to time "0" and setting its NPW to zero, yields the following equation for PI:

$$-50{,}000 + 20{,}000\left(x + x^{2} + x^{3} + x^{4} + x^{5}\right) = 0$$

where $x = (1 + \text{PI})^{-1}$. Using the computer program introduced in Example 2.9, the solution is $x = 0.7773$ corresponding to PI = 26.65%.

Comments

The PI calculated is much higher than the specified minimum attractive rate of return. Therefore, it is justified to invest the additional $50,000 required for Plan B, and it should be chosen over Plan A.

Example 2.11 Alternative Analysis by PI

A chemical plant must install air pollution control equipment to reduce its emissions of VOCs. The following viable alternatives are under consideration:

Plan	A	B	C	D
TCI	$100,000	$160,000	$200,000	$260,000
Annual NCF	–85,000	–73,000	–66,500	–53,000

If the minimum attractive rate of return is 12%, determine which alternative should be chosen. The useful life of all the alternatives is 10 yr, with no salvage value.

Solution

Plan A, with the lowest TCI, is chosen as the basis for the initial comparison. Compare Plan B, with the lowest incremental TCI, with Plan A.

Year	NCFs (B – A)
0	–$60,000
1–10	12,000/yr

Annual savings of $12,000 for the next ten years results by making an additional investment of $60,000. Thus, the PI on the incremental investment is calculated as 15%. Because this is higher than the minimum attractive rate of return, Plan B is chosen over Plan A. The next step is to compare Plan C with Plan B.

Year	NCFs (C – B)
0	–$40,000
1–10	6,500/yr

The incremental TCI of $40,000 results in annual savings of $6,500 for a PI of 10%. Because this value is lower than the minimum attractive rate of return, Plan B is chosen over Plan C. The next step is to compare Plan D withPlan B.

Year	NCFs (D – B)
0	–$100,000
1–10	$20,00/yr

The PI for the incremental TCI of $100,000 is calculated as 15%. Therefore, Plan D is chosen over Plan B and is the alternative that should be recommended.

When comparing alternatives through their NPW or PI, the cash flow series must relate to the provision of a needed service for the same number of years. If two alternatives have different lives, a study period must be chosen that is the least common multiple of the lives of the various assets involved. For example, if one alternative had a 10-yr life and the other a 25-yr life, it would be necessary to consider a 50-yr period. If lives were, say, 13 and 20 yr, a 260-yr period would be needed before reaching the point where the alternatives gave equal years of service. Generally, the assumption is made that the costs for replacement of all assets at the end of their useful lives will be the same as the initial costs that have been forecast (Grant et al. 1982).

Example 2.12 NPW of Alternatives with Different Lives

Consider two alternative investments, which we shall call Plans D and E. Estimates for the plans are

Alternative	Plan D	Plan E
TCI	$50,000	$120,000
Life	20 yr	40 yr
Salvage value	$10,000	$20,000
Annual NCF	–$9,000	–$6,000

Determine which alternative should be chosen based on NPW if the rate of return must be, at least, 11%.

Solution

Because the alternatives have different lives, a study period must be chosen that is the least common multiple of their lives. For this particular case, the study period should be 40 yr, corresponding to two life cycles for Plan D and one for Plan E. Assume that at the end of the first life cycle Plan D can be renovated by investing $50,000, and the annual NCF for the next life cycle will still be -$9,000. Figure 2.3 illustrates the cash flows for both alternatives. The next step is to calculate the net present worth for each of the alternatives. Consider first Plan D. The net disbursement required for renewal in 20 yr will be the TCI required minus the salvage value of the assets ($50,000 – $10,000). The corresponding discount factor (20 yr, 11%) is 0.1240. The PW of this cash flow is –$4,960. For the 40-yr period, there is a uniform annual disbursement (called an annuity) of $9,000. It can be shown (see Problem 2.8) that the present worth of an annuity can be calculated multiplying the annual cash flow by the series present worth factor (SPWF) given by:

$$\text{SPWF} = \left[\frac{(1+i)^n - 1}{i(1+i)^n} \right]$$

For this case, SPWF = 8.951, and the PW of the annuity is –$80,560. The PW of the salvage value of $10,000 received after 40 yr is $10,000 (0.0154) = $150. The NPW of Plan D is, then, (–50,000 – 4960 – 80,560 + 150) = –$135,370.

Consider Plan E. The present worth of the uniform annual cash flows is –$6,000 (8.951) = –$53,710. The PW of the salvage value is $20,000 (0.0154) = $310. Therefore, the net present worth of Plan E is (–120,000 – 53,710 + 310) = –$173,400. Plan D should be chosen since it has the highest NPW.

Comments

Everything that has been said about comparing alternatives with different lives through the PW method also applies to the PI method.

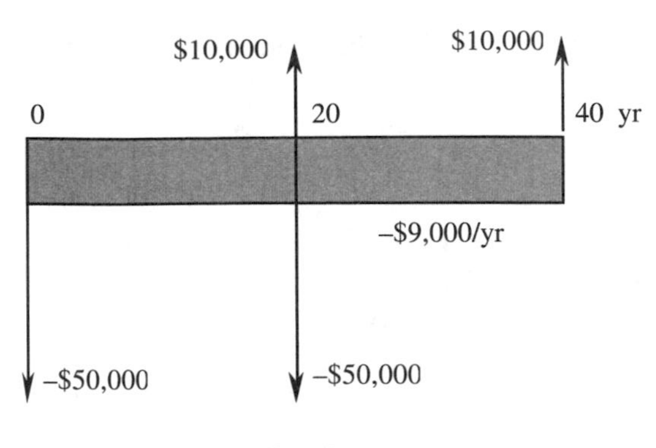

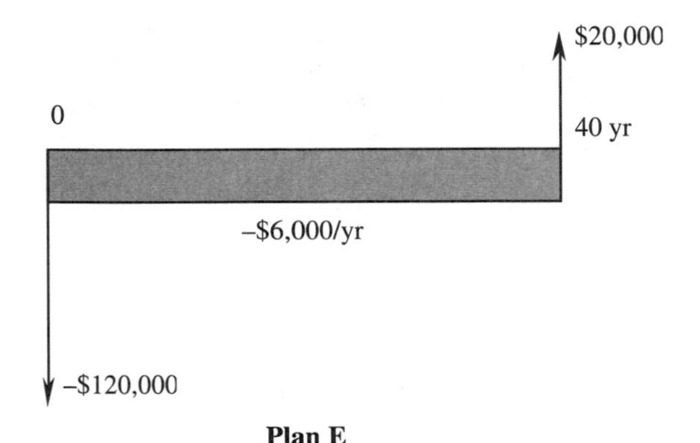

Figure 2.3 Cash flow diagram for Example 2.12

2.5.2 EUAC

At different points in the life of a project there are cash flows that can be either positive or negative. The EUAC method converts every cash flow into an equivalent uniform annual cash flow distributed throughout the entire life of the project. The sum of all those equivalent cash flows gives the net EUAC. When comparing alternatives through this method, the one with the highest EUAC (or the least negative) is chosen. The minimum attractive rate of return is used to calculate the equivalent uniform annual cash flows taking into consideration the time value of money.

Example 2.13 Comparison of Alternatives through EUAC

Consider the alternatives presented in Example 2.8. Determine which alternative to choose based on EUAC.

Solution

Application of the EUAC method requires converting all cash flows to equivalent uniform annual flows. The TCIs of Example 2.9, which happen instantly at time "0", are converted to annual uniform disbursements through the capital recovery factor defined in Eq. (2.7). For this example $n = 5$ years, $i = 0.10$, and CRF = 0.2638.

	Plan A ($/yr)	Plan B ($/yr)
CRC = CRF × TCI	39,570	52,760
Annual NCF	–100,000	–80,000
EUAC	–139,570	–132,760

Plan B should be chosen, because it results in the higher EUAC.

Comments

Notice that the EUAC of a project is the negative of the TAC, when the latter is computed including the capital recovery cost. Even if the different alternatives being considered have different lives, the comparison is done in terms of TAC. The effect of the different lifespans will show up in the respective capital recovery factors. For example, if the useful life of Plan A is 10 years, instead of 5, the corresponding CRF is 0.1628, and its EUAC is –$124,400/yr, making it the best alternative.

Example 2.14 Comparison of Alternatives through EUAC

Consider the data presented in Example 2.11. Choose the best alternative based on EUAC. If all the alternative technologies reduce pollutant emissions by 1,000 metric tons/yr, calculate the average cost effectiveness for each.

Solution

To calculate the EUAC for each alternative, the corresponding capital recovery costs must be subtracted from the annual NCFs. The capital recovery factor corresponding to $n = 10$ years and $i = 0.12$ is CRF = 0.177. The following table summarizes the EUAC and ACE calculations:

Alternative	A	B	C	D
TCI	$100,000	$160,000	$200,000	$260,000
Annual NCF	–85,000	–73,000	–66,500	–53,000
CRC ($/yr)	17,700	28,300	35,400	46,000
EUAC	–102,700	–101,300	–101,900	–99,000
ACE, $/Metric ton	102.7	101.3	101.9	99.0

Alternative D should be chosen, because it has the highest EUAC, corresponding to the lowest value of ACE.

2.5.3 Comparison of Measures of Merit

The three measures of merit discussed for comparison of alternatives (PW, PI, and EUAC) give the same results when applied correctly to an investment situation. If the objective of the analysis is determining not only what alternative to choose but also the expected rate of return in making the right decision, then the profitability index must be calculated.

If it is enough to determine which alternative to choose, the choice between PW and EUAC depends on the relative magnitudes of the initial investment and the annual cash flows. If the initial investment is much higher than the annual disbursements, a PW analysis is more realistic. For annual cash flows of the same order of magnitude of the initial investment, EUAC analysis is adequate.

2.6 CONCLUSION

Most air pollution control design problems have several viable solutions. The decision of which alternative to choose is usually based on economic considerations. This chapter presented a method to estimate the total capital investment of a pollution control device based on the purchased equipment cost. Cost indices were used to update equipment costs. The different components of the TAC were identified, and guidelines presented for their estimation. Three measures of merit for comparing viable alternatives were discussed and illustrated through examples.

The cost estimation methodology presented in this chapter is consistent with the guidelines established by the U.S. EPA for regulatory purposes and will be applied for design purposes throughout the rest of this book. The following chapters will be devoted to sizing and costing specific air pollution control devices.

REFERENCES

Grant, E. L., Ireson, W. G., and Leavenworth, R. S. *Principles of Engineering Economy,* 7th ed., John Wiley and Sons, New York (1982).

Katari, V. S., Vatavuk, W. M., and Wehe, A. H. *JAPCA*, **37**:198 (1987).

Peters, M. S., and Timmerhaus, K. D. *Plant Design and Economics for Chemical Engineers*, 3rd ed., McGraw-Hill, New York (1980).

Press, W. H., Flannery, B. P., Teukolsky, S. A., and Vetterling, W. T. *Numerical Recipes: The Art of Scientific Computing*, Cambridge University Press, New York (1989).

Producer Price Indexes Data for May 1990 (and various other issues for 1988 and 1989), Bureau of Labor Statistics, U S. Department of Labor (July 1990).

U.S. EPA. *New Source Review Workshop Manual,* Office of Air Quality Planning and Standards, Research Triangle Park, NC (1990).

Vatavuk, W. M. *Estimating Costs of Air Pollution Control,* Lewis, Chelsea, MI (1990).

PROBLEMS

The problems at the end of each chapter have been grouped into four classes (designated by a superscript after the problem number)

Class a: Illustrates direct numerical application of the formulas in the text
Class b: Requires elementary analysis of physical situations, based on the subject material in the chapter
Class c: Requires somewhat more mature analysis
Class d: Requires computer solution

2.1[a]. Installed cost of a VOC thermal incinerator

The cost of purchasing a VOC thermal incinerator for a given application was $130,000 in March 1985. Estimate the installed cost of the same piece of equipment in June 1990. Assume that no special building or site preparation is required. Use the equipment component of the CE Index to update costs.

Answer: $279,000

2.2[a]. Components of TCI

Because of the nature of the material incinerated in Problem 2.1, the ash produced constitutes a hazardous waste. An offsite facility is required to process it; the capital investment needed is $75,000 (June 1990). Land must be purchased at a cost of $16,000. A working capital of $30,000 is required for the incinerator.

(a) Calculate the depreciable investment for the combined facilities.

Answer: $354,000

(b) Calculate the nondepreciable investment.

Answer: $46,000

(c) Calculate the TCI required.

Answer: $400,000

2.3[a]. Average cost factors for VOC incinerators

The cost of a VOC incinerator, including auxiliary equipment, taxes and freight, is $350,000.

(a) Calculate the total purchased equipment cost.

Answer: $382,400

(b) A site preparation fee of $50,000 is needed, and a building must be erected at a cost of $35,000. Estimate the total installation direct costs.

Answer: $199,700

(c) Estimate the total installed cost.

Answer: $700,660

2.4[a]. Cost indices

(a)Removal of relatively big particles (greater than 10 μm) is often accomplished using multiple-cyclone systems. These consist, basically, of concentric metal pipes located inside a metal container. Particles are removed by centrifugal force. The cost of one such system was $55,000 in June 1986. Estimate its cost in May 1990 using an appropriate cost index.

Answer: $65,680

(b) The cost of felted Teflon bags for use in pulse-jet fabric filters was $97.40/m^2 in September 1986. Estimate their cost in May 1990 using an appropriate cost index.

Answer: $111.42/m^2

2.5[b]. Installed cost of carbon adsorber

Activated carbon adsorbers are used for VOC control applications. It has been found that the purchased equipment cost for these units is proportional to $Q^{0.70}$ where Q is the volumetric flow rate of gas processed. Total installation costs are estimated as 75%

of the purchased equipment cost. The purchased equipment cost of a carbon adsorber to process 2.36 m^3/s was $50,000 in 1979 (Composite CE Index = 238.7). Based on these data, estimate the total installed cost of a similar unit to process 3.54 m^3/s in June 1990.

Answer: $173,850

2.6[a]. TAC of pollution control device

The total depreciable investment for an air pollution control device which requires no replacement parts is $500,000. The useful life of the system is 15 years, and it has no salvage value. The minimum attractive rate of return for the firm is 15%. The following annual costs have been estimated:

Raw materials	$25,000
Operating labor	30,000
Maintenance labor	15,000
Electricity	70,000
Steam	30,000
Waste treatment	100,000
Recovery credits	185,000

Estimate the total annual cost for this process.

Answer: $248,710/yr

2.7[c]. TAC of carbon adsorber

A carbon adsorber system is designed to remove 90% of the n-pentane in an air stream. Gas flow rate is 2.6 m^3/s at 101.3 kPa and 308 K; the initial n-pentane concentration is 1,872 ppm. The amount of carbon required is 2,650 kg, and it must be replaced every 2 yr. The total pressure drop for the gas flowing through the system is 4.23 kPa. An electric pump is used at the entrance of the adsorber with a mechanical efficiency of 65%. Steam is used to regenerate the carbon beds at a rate of 300 kg/h. Cooling water is used at a rate of 18,000 kg/h to condense the gaseous effluent during regeneration. The system operates continuously 24 h/d, 350 d/yr. Operating labor requirement is 2 h per shift; maintenance requires 1 h per shift.The installed cost of the adsorber (excluding the carbon) is $225,000. The cost of replacing the carbon (including taxes, freight, and labor) is $4.40/kg carbon. Operating labor cost is $15/h; maintenance is $20/h; electricity costs

$0.06/kW-h; cooling water cost is $0.10/m^3; and steam costs $10/1000 kg. Recovered n-pentane is worth $0.25/kg. The useful life of the system is 16 yr with no salvage value. The minimum attractive rate of return specified by the firm is 15%. Calculate the total annual cost.

Answer: $134,160/yr

2.8b. Present worth of an annuity

Show, using the concept of a geometric progression (*Standard Mathematical Tables,* 18th ed. p. 102, CRC, Cleveland, Ohio, 1970), that the PW of an annuity is given by

$$\text{PW} = (\text{ACF})(\text{SPWF})$$

where ACF is the constant annual cash flow, and SPWF is the series present worth factor given by:

$$\text{SPWF} = \left[\frac{(1+i)^n - 1}{i(1+i)^n}\right]$$

2.9a. PW calculations

A piece of equipment will cost $6,000 new and have an expected life of 6 yr, with no salvage value at the end of its life. Total annual disbursements (not including capital recovery costs) are $1,500. What is the present worth of this piece of equipment if the minimum attractive rate of return is 12%?

Answer: $12,167

2.10b. PW of an annuity

A present investment of $50,000 is expected to reduce annual expenditures by a uniform amount for the next 15 yr. If the rate of return on this investment is 11.1%, calculate the yearly savings.

Answer:$7,000/yr

2.11d . Profitability index

Five years ago, a company invested $100,000 on a piece of equipment that reduced its yearly expenses by the following amounts:

Year	Annual Savings
1	$9,500
2	10,000
3	10,500
4	11,000
5	11,500

The company is upgrading its facilities and has just sold the equipment for $107,500. Estimate the profitability index on the original investment.

Answer: 11.58%

2.12a. EUAC calculations

The net present worth of an air pollution control device is –$100,000. The useful life of the device is 10 years, and the minimum attractive rate of return is 9%. Calculate the equivalent uniform annual cash flow for this investment.

Answer: –$15,580/yr

2.13b. Future worth (FW) of an annuity

The FW is the total amount of money accumulated at the end of an annuity. Show that the future worth of an annuity of uniform payments ACF is given by

$$\text{FW} = (\text{ACF})(\text{SCAF})$$

where SCAF is the *series compound amount factor* given by:

$$\text{SCAF} = \left[\frac{(1+i)^n - 1}{i}\right]$$

2.14c. EUAC of uniform arithmetic gradient (Grant et al. 1982)

Engineering economics problems frequently involve disbursements or receipts that increase or decrease each year by varying amounts. For example, the maintenance expense for a piece of mechanical equipment may tend to increase each year. If the increase or decrease is the same every year, the yearly increase or decrease is known as a *uniform arithmetic gradient*. In this case, the payment the second year is greater than the first year by an amount G, the third year is G greater than the second year, and so on. Thus, the incremental payments by year are as follows:

End of Year	Incremental Payment
1	0
2	G
3	$2G$
4	$3G$
.....	.....
$(n-1)$	$(n-2)G$
n	$(n-1)G$

(a) Draw a cash flow diagram for this situation.

(b) From the cash flow diagram, it is evident that a uniform gradient is equivalent to an annuity of G started at the end of the second year, plus an annuity of G started at the end of the third year, and so on. Each of these annuities terminates at the end of the nth year. Using the results of Problem 2.13, show that the future worth of the gradient is given by

$$\text{FW} = \frac{G}{i}\left[(1+i)^{n-1} + (1+i)^{n-2} + \cdots + (1+i) + 1\right] - \frac{nG}{i}$$

c) Show that the *equivalent uniform annual cash flow of a uniform arithmetic gradient* (EUAC_g) is given by:

$$\text{EUAC}_\text{g} = \frac{G}{i}\left[1 - \frac{ni}{(1+i)^n - 1}\right]$$

2.15b. EUAC calculations involving a uniform gradient

Air pollution control device A has a total capital investment of $50,000, an estimated service period of 12 yr, and a salvage value of $14,000 at the end of the 12 yr. Estimated annual disbursements for operation and maintenance are $6,000 for the first year, $6,300 for the second year, and will increase $300 each year thereafter. If the minimum attractive rate of return is 12%, calculate the EUAC for this device (see Problem 2.14).

Answer: –$14,750/year

2.16b Choice between alternatives based on EUAC

Consider the air pollution control device of Problem 2.15. There is an alternative device B that will do exactly the same but at a different cost. It requires an initial investment of $35,000 and has no salvage value at the end of the 12-yr service period. Estimated disbursements for operation and maintenance are $8,000 for the first year, $8,500 for the second year, and will increase $500 each year thereafter. Determine which of the devices should be chosen, based on their EUAC.

Answer: Choose A

2.17b Choice between alternatives based on PW

Repeat Problem 2.16, but choose between alternatives A and B based on their respective NPWs.

Answer: NPW of A is –$91,370

2.18b NPW of alternatives with different lives

Choose between alternatives E and F, based on NPW. The minimum attractive rate of return is 15%. The following table summarizes the economic characteristics of the alternatives:

Alternative	Unit E	Unit F
Total capital investment	$220,000	$360,000
Life in years	6	10
Salvage value	$10,000	$30,000
Annual operation and maintenance expenses	$84,000	$67,500

Answer: Choose F

2.19b. EUAC of alternatives with different lives

Calculate the EUAC for the alternatives of Problem 2.18. Choose between them based on their EUAC.

Answer: –$140,940/yr for E

2.20d PI for alternatives with different lives

Consider the alternatives of Problem 2.18. Determine the PI when alternative F is chosen over alternative E.

Answer: 17.6%

2.21d Profitability Index for VOC refrigerated condenser

Figure 2.4 shows a detailed flow diagram of a refrigerated condenser used to recover 99.5 percent of the acetone in a gaseous waste stream. The total capital investment for the system is $2,500,000 and its useful life is 10 years. The recovered acetone has a value of $0.50/kg. Estimate the PI for this investment. The following cost data applies:

System operates 8,150 h/yr
Annual operating labor expenses, $40,800
Annual maintenance labor expenses, $25,500
Cost of electricity, $0.06/kW-h
Cost of cooling water, $0.13/m^3;

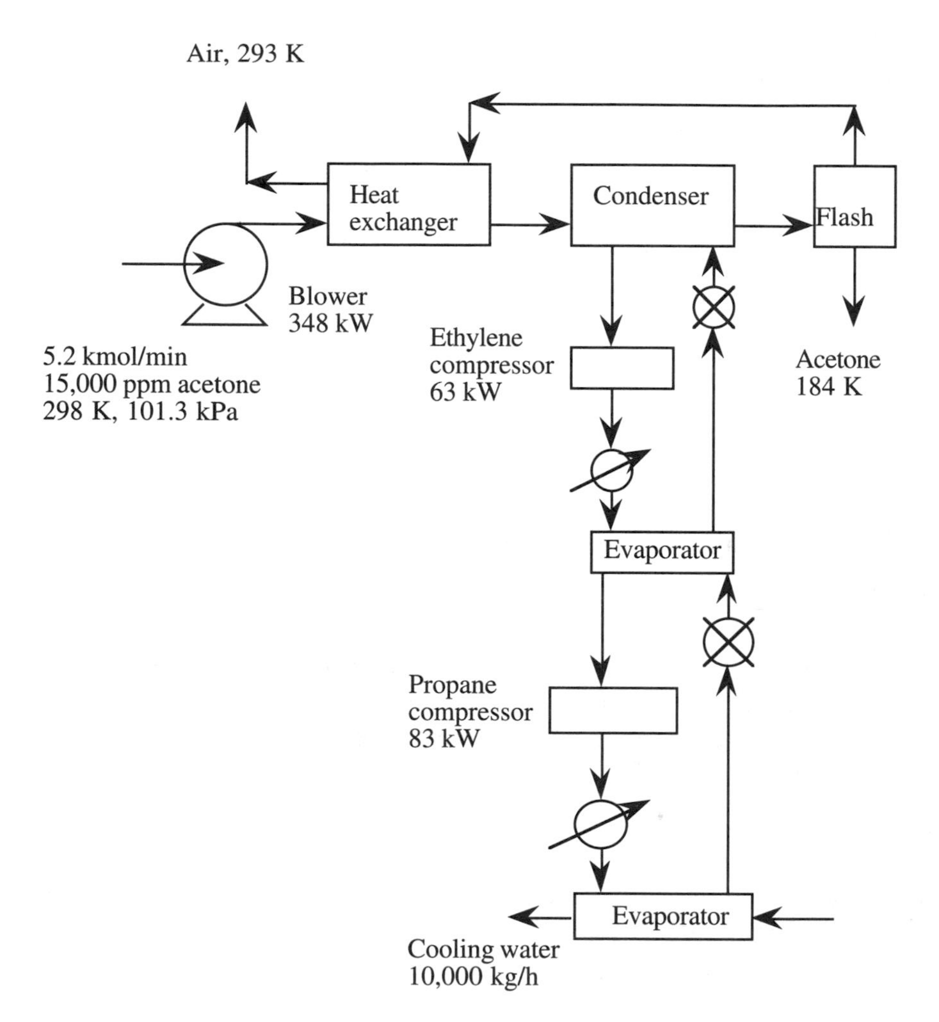

Figure 2.4 Acetone refrigerated condenser (Douglas, J. M. *Conceptual Design of Chemical Processes*, p. 495, McGraw-Hill, New York, © 1988; reprinted with permission from McGraw-Hill, Inc.)

3

Incineration for Control of VOC Emissions

3.1 INTRODUCTION

Incineration, a controlled oxidation process, is a technique for destruction of VOC emissions from industrial waste gases. In the process, the VOCs react at high temperatures with oxygen to form CO_2 and water while liberating heat. Three parameters—temperature, residence time, and turbulence (the "three Ts")—have an interrelated effect on the final combustion performance. To achieve good destruction rates, emission effluents must be held for sufficient residence times at combustion temperatures ranging from 1,000 to 1,200 K. Turbulent flow conditions must be maintained in the incinerator for good mixing of the VOCs and oxygen.

This chapter reviews incineration techniques for control of VOCs emissions It presents an overview of the process including theoretical fundamentals and design considerations. Both thermal and catalytic incinerators are considered. Another incinerator type that is considered is the regenerative thermal oxidizer (RTO), a variant of the thermal design. The importance of heat recovery is emphasized throughout. Capital and annual operating cost estimates for the three types of incinerators are included.

3.2 WASTE GAS CATEGORIZATION

Your objectives in studying this section are to

1. Define lower and upper explosive limits.
2. Determine whether incineration is suitable for a given waste gas.
3. Determine whether auxiliary fuel or air is required.

The suitability of incineration, and the auxiliary fuel and air requirements for application of the technique to a particular waste gas are determined by the gas composition.

Most industrial gases incinerated are dilute mixtures of VOCs, air, and inert gas. Their VOC content is very low, and usually their oxygen content exceeds that required for complete combustion of both the VOC and the auxiliary fuel. The suitability of incineration as a control technique for waste gases depends on the "flammability" of the VOCs involved. Flammability is characterized by two limits: the lower explosive level (LEL) and the upper explosive level (UEL). These limits represent, respectively, the smallest and largest amounts of VOCs, which, when mixed with air, will burn without a continuous application of external heat. To avoid potential explosions, the VOC content of industrial waste gases released to the atmosphere is normally outside the flammability limits. Table 3.1 presents the flammability limits for some selected VOCs. Major reference books, such as *Lange's Handbook of Chemistry*, present additional flammability data.

A waste gas with a VOC content more than 25% of the LEL is usually diluted to below 25% LEL before incineration by adding outside air (thus increasing the waste gas volume flow rate to be treated!). However, a waste gas with a VOC content between 25% and 50% LEL can be incinerated without adding dilution air, provided the VOC levels in the system are continuously monitored via LEL monitors to satisfy fire protection regulations.

A low-oxygen content waste gas could disrupt the burner flame stability Therefore, when a waste gas with less than 13% to 16% oxygen content is incinerated, auxiliary air is required. Only in rare cases the waste gas is a rich VOC stream that can support combustion without auxiliary fuel. Such a rich VOC waste gas is treated as a fuel and burned. Figure 3.1 is a flow chart for categorizing a waste gas to determine its suitability for incineration and to establish its auxiliary fuel and oxygen requirements.

3.3 THEORETICAL CONSIDERATIONS

Your objectives in studying this section are to:

1. Review the stoichiometry of combustion reactions.
2. Predict the kinetics of VOCs incineration.

Table 3.1 Flammability Limits for Selected Organics in Air (298 K and 101.3 kPa)

VOC	LEL (Percent by Volume)	UEL (Percent by Volume)
Acetaldehyde	4.0	6.0
Acetone	2.6	12.8
Acetonitrile	4.4	16.0
Benzene	1.3	7.1
n-Butane	1.9	8.5
n-Butanol	1.4	11.2
Butyl acetate	1.7	7.6
Carbon disulfide	1.3	50.0
Chlorobenzene	1.3	7.1
Cyclohexane	1.3	8.0
Diethyl amine	1.8	10.1
Ethane	3.0	12.5
Ethanol	3.3	19.0
Ethyl acetate	2.2	11.0
Ethylene oxide	3.6	100
Heptane	1.0	6.7
Hexane	1.1	7.5
Isobutane	1.8	10.0
Isopentane	1.4	7.6
Isopropanol	2.0	12.0
Methane	5.4	15.0
Methanol	6.7	36.0
Methyl acetate	3.1	16.0
Methyl cyclohexane	1.2	6.7
Octane	1.0	6.5
Pentane	1.5	7.8
Propane	2.2	9.5
Toluene	1.2	7.1
Trichloroethylene	12.5	90.0
Xylene, m- and p-	1.1	7.0
Xylene, o-	1.0	6.0

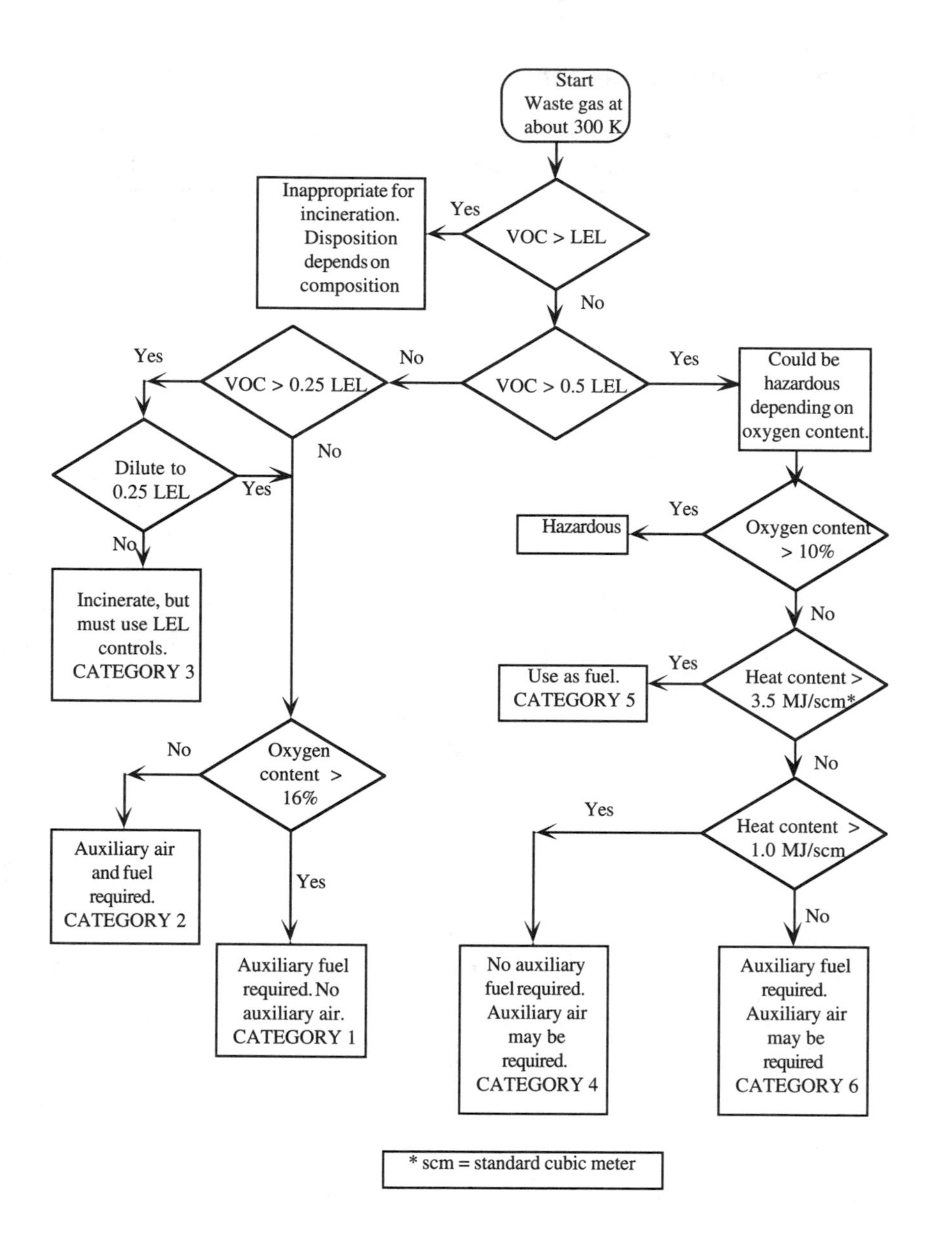

Figure 3.1 Flow chart for categorization of waste gases (From Katari et al. *JAPCA*, **37**:91, 1987; reprinted with permission from *JAPCA*).

3.3.1 Combustion chemistry

For simplicity, consider the case of a premixed dilute stream of a pure hydrocarbon in air. The stoichiometry of complete combustion is:

$$C_xH_y + \left(x + \frac{y}{4}\right) O_2 \rightarrow x\,CO_2 + \frac{y}{2} H_2O \tag{3.1}$$

Note that the formation of nitrogen oxides is not accounted for in Eq. (3.1). Furthermore, if sulfur were present in the VOC, sulfur oxides would be formed.

Example 3.1 Incineration Stoichiometry

A waste gas from an industrial process contains 3,000 ppm toluene, 18% O_2, and 81.7% N_2. It will be incinerated with the addition of methane as auxiliary fuel. The methane will be supplied at a rate of 3 mol/100 mol of waste gas.
(a) Calculate the percent excess oxygen available.
(b) Assuming complete combustion, calculate the composition of the flue gases.

Solution

Two oxidation reactions occur

$$C_7H_8 + 9O_2 \rightarrow 7CO_2 + 4H_2O \tag{3.2}$$

$$CH_4 + 2O_2 \rightarrow CO_2 + 2H_2O \tag{3.3}$$

Basis: 100 mol of waste gas

(a) The following table presents the molar quantities of each chemical species entering the incinerator:

Component	Moles in
C_7H_8	0.3
CH_4	3.0
O_2	18.0
N_2	81.7

Theoretical oxygen required: 0.3(9) + 3.0(2) = 8.7 mol
Oxygen available: 18.0 moles

$$\text{Percent excess oxygen} = \frac{18 - 8.7}{8.7} \times 100 = 107\% \tag{3.4}$$

b) To calculate the concentration of the flue gases, write material balances in tabular form including all the chemical species present in the flue gases.

Species	Moles In	Consumed	Formed	Out	Molar Percent
O_2	18.0	8.7	0.0	9.3	9.00
N_2	81.7	0.0	0.0	81.7	79.10
CO_2	0.0	0.0	5.1	5.1	4.94
H_2O	0.0	0.0	7.2	7.2	6.96

Comments

There is plenty of oxygen in the waste gas for complete combustion of the toluene and the methane added as auxiliary fuel. According to the flowchart of Fig. 3.1, this waste stream belongs in Category 1, meaning that it needs auxiliary fuel, but no auxiliary air for incineration. Recall from Table 3.1 that, for toluene, LEL = 1.2%, or 12,000 ppm.

3.3.2 Kinetics of Oxidation Reactions

The kinetics of oxidation reactions is as important as the stoichiometry. The actual detailed mechanisms of combustion are complex and do not occur in a single step as might be inferred from Eq. (3.1). To simplify kinetic models for air pollution design work, several authors have taken the approach of developing "global models" (Cooper and Alley 1986). A global model ignores many of the details of mechanistic models, and ties the kinetics to the main stable products and reactants. Because CO is a very stable intermediate, the simplest global model for the oxidation of hydrocarbons is a two-step model as follows:

$$C_xH_y + \left(\frac{x}{2} + \frac{y}{4}\right) O_2 \rightarrow x\,CO + \frac{y}{2} H_2O \tag{3.5}$$

$$xCO + \left(\frac{x}{2}\right)O_2 \rightarrow xCO_2 \tag{3.6}$$

Several authors have shown the importance of atoms and radicals such as O, H, CH_3, and HO_2. Water vapor enhances oxidation reactions. Nevertheless, Eqs. (3.5) and (3.6) successfully establish a global model of the overall kinetics of VOC incineration. On the basis of on these reactions, a kinetic model that is of the first order in each reactant results in the rate equations:

$$r_{HC} = -k_1[HC][O_2] \tag{3.7}$$

$$r_{CO} = -xk_1[HC][O_2] - k_2[CO][O_2] \tag{3.8}$$

where

r_i is the rate of formation of component i, mol/m^3-s
[ ] is concentration, in mol/m^3
HC is a generic symbol for any hydrocarbon
k is a rate constant, s^{-1} or m^3/mol-s (as appropriate)

In the presence of excess oxygen, the rate equations reduce to:

$$r_{HC} = -k_1[HC] \tag{3.9}$$

$$r_{CO} = -xk_1[HC] - k_2[CO] \tag{3.10}$$

A third equation can be written for the generation of CO_2.

$$r_{CO_2} = k_2[CO] \tag{3.11}$$

Equations (3.9), (3.10), and (3.11) represent a special case of a general set of consecutive first-order irreversible reactions represented by

$$A \underset{k_1}{\rightarrow} R \underset{k_2}{\rightarrow} S \tag{3.12}$$

When the initial concentrations are $C_A(0) = C_{A0}$, $C_{R0} = C_{S0} = 0$, the solution of the set of equations is:

$$\frac{C_A}{C_{A0}} = e^{-\tau} \tag{3.13}$$

$$\frac{C_R}{C_{A0}} = \frac{1}{1-\rho}\left[e^{-\rho\tau} - e^{-\tau}\right] \tag{3.14}$$

$$\frac{C_S}{C_{A0}} = 1 - e^{-\tau} - \frac{1}{1-\rho}\left[e^{-\rho\tau} - e^{-\tau}\right] \tag{3.15}$$

where

$$\tau = k_1 t \qquad \rho = \frac{k_2}{k_1} \tag{3.16}$$

Figure 3.2 is representative of the behavior of this reaction system. The time response of the intermediate species is highly dependent on the ratio of the reaction rate constants.

Such behavior is representative of the VOC-CO-CO2 system: CO can be present in small or large concentrations in an incinerator, depending on the temperature of operation and the residence time. Figure 3.3 illustrates the formation of CO in a VOC incinerator as a function of residence time for a given temperature. Of the many intermediate species—or products of incomplete combustion (PICs)—formed during incineration of a VOC, CO is the most stable. If the design temperature and residence time are such that the CO concentration is within acceptable limits, then it is usually assumed that the emissions of other PICs from the incineration process are also acceptable.

3.3.3 The three T's of incineration.

The importance of the three T's of incineration—temperature, time, and turbulence—has been recognized for many years. Danielson (1973) suggested that, for good destruction, incinerators should be designed for temperatures of 800 to 1,100 K, for residence times of 0.3 to 0.5 s, and flow velocities (to promote turbulent mixing) of 6 to 12 m/s.

Because the kinetic constants in Eqs. (3.9) and (3.10) increase exponentially with temperature, VOC destruction rates are very sensitive to temperature. However, sufficient time must

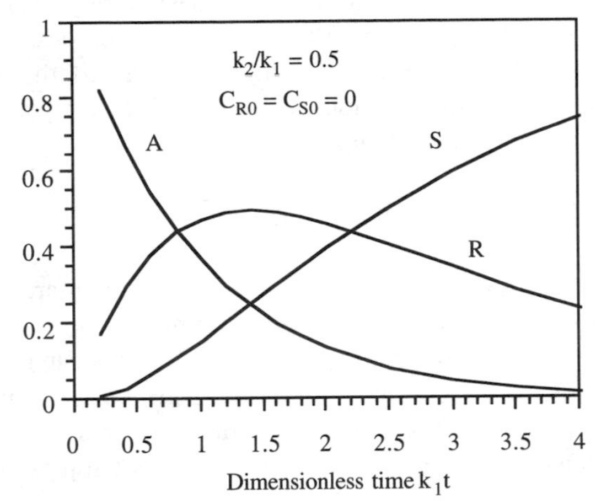

Figure 3.2
Concentration versus residence time in a plug flow reactor for a 1:1:1 series reaction.

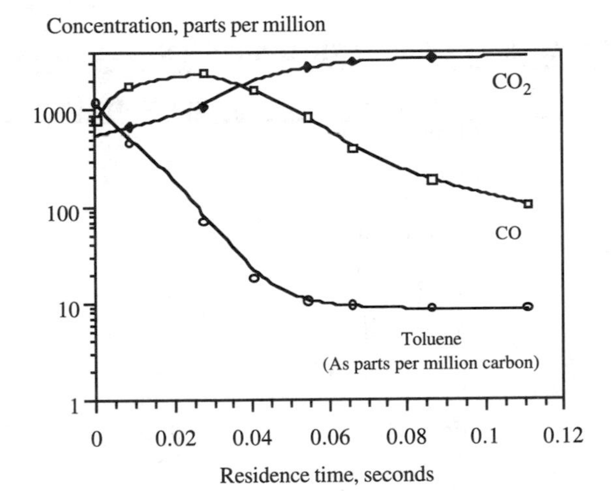

Figure 3.3
Concentration versus residence time in a plug flow reactor for toluene incineration at 1103 K. (Figure 11 from "Fume Incineration Kinetics and its Applications," by Klaus H. Hemsath and Peter E. Susey, *AIChE Symposium Series*, No. 137, Vol. 70, pp. 439-449, 1974). Reproduced by permission of the American Institute of Chemical Engineers © 1974 AIChE.

be provided at the desired temperature to allow the reactions to reach the desired degree of completion. Turbulence ensures sufficient mixing of oxygen and VOCs during the process. At the temperatures of most incinerators, as long as a reasonable flow velocity is maintained, the mixing process is not the rate-controlling step of the destruction reactions. However, as the temperature is raised, at some point the overall rate could be limited by mixing.

3.3.4 Predicting the kinetics of VOCs incineration.

Although kinetics are important to the design of an incinerator, kinetic data are scarce, and are difficult and costly to obtain by pilot studies. Past methods for determining the design and operating temperatures of incinerators were rough approximations. Ross (1977) suggested that the design temperature should be several hundred degrees F above the VOC auto ignition temperature. The auto ignition temperature is the temperature at which combustibles mixtures of the VOC in air ignite without an external spark or flame. Some auto ignition temperatures are presented in Table 3.2.

Table 3.2 Auto ignition temperatures of selected substances in air

Substance	Temperature (K)	Substance	Temperature (K)
Acetone	738	Ethylene oxide	702
Ammonia	924	Heptane	488
Benzene	833	Hexane	498
n-Butane	678	Hydrogen	673
1-Butene	659	Isopentane	693
Carbon disulfide	363	Methane	813
Carbon monoxide	882	Methanol	658
Chlorobenzene	913	Methyl acetate	775
Cyclohexane	518	Propane	723
o-Dichlorobenzene	921	Propene	733
Ethane	788	Toluene	753
Ethanol	638	Trichloroethylene	693
Ethyl acetate	700	Vinyl chloride	734
Ethyl benzene	705	Xylene, m- and p-	803
Ethylene	763	Xylene, o-	738

More recently, methods for the quantitative prediction of kinetic data and design temperatures have been proposed. Cooper, Alley, and Overcamp (1982) combined collision theory with empirical data and proposed a method for predicting an effective first order rate constant, k, for hydrocarbon incineration over the range from 940 to 1140 K. Once k is found, the design temperature can be obtained. According to Arrhenius law, the rate constant is

$$k = A \exp\left(-\frac{E}{RT}\right) \tag{3.17}$$

where

E = activation energy, J/mol
A = pre-exponential factor, s^{-1}
R = ideal gas law constant, 8.314 J/mol-K
T = absolute temperature, K

The pre-exponential factor can be calculated by

$$A = \frac{Z' S y_{O_2} P}{R''} \tag{3.18}$$

where

Z' = collision rate factor
S = steric factor
y_{O_2} = mole fraction of oxygen in the incinerator
P = absolute pressure, atm
R'' = ideal gas constant, 0.08205 L-atm/mol-K

The steric factor in Eq. (3.18) accounts for some collisions that are not effective in producing reactions because of molecular geometry. It can be calculated from

$$S = \frac{16}{\text{MW}} \tag{3.19}$$

where MW is the molecular weight of the hydrocarbon.

The activation energy and collision rate factor were correlated by Cooper et al. (1982) with molecular weight through

$$E = 193{,}020 - 40.45\,(\text{MW}) \tag{3.20}$$

$$Z = \left(0.5 + \frac{MW}{32}\right) 10^{11} \qquad \text{Alkanes} \tag{3.21}$$

$$Z = (0.25 + 0.03\ \text{M.W.})\ 10^{11} \qquad \text{Alkenes} \tag{3.22}$$

$$Z = (-0.60 + 0.0375\ \text{M.W.})\ 10^{11} \qquad \text{Aromatics} \tag{3.23}$$

Once A and E have been estimated, k can be calculated at any temperature. In an isothermal plug flow reactor, the HC destruction efficiency, the rate constant, and the residence time are interdependent, and are related as

$$\eta = 1 - \frac{[\text{HC}]_{out}}{[\text{HC}]_{in}} = 1 - \exp(-k\tau_r) \tag{3.24}$$

where

η = HC destruction efficiency
τ_r = residence time

Example 3.2 Temperature Required for VOC Thermal Incineration

A waste gas contains 1,000 ppm of toluene. Estimate the temperature required in an isothermal plug flow incinerator with a residence time of 0.5 s to give 99.9% destruction of toluene (MW = 92). Assume an oxygen mole fraction of 0.074 and a pressure of 101.3 kPa.

Solution

The required value of k is from Eq. (3.24).

$$k = -\frac{\ln(1 - 0.999)}{0.5} = 13.8\ \text{s}^{-1} \tag{3.25}$$

Use Eqs. (3.19), (3.20), and (3.23) to calculate S, E, and Z'.

$S = 0.174$, $E = 189{,}300$ J/mol, $Z' = 2.85 \times 10^{11}$

Calculate A from Eq. (3.18).

$$A = \frac{2.85(10)^{11}(0.174)(0.074)(1.0)}{0.08205} = 4.47(10)^{10}\ \text{s}^{-1} \tag{3.26}$$

Rearrange Eq. (3.17) and solve for T.

$$T = -\frac{E}{R\ln(k/A)} = -\frac{189{,}300}{(8.314)\ln\dfrac{13.8}{4.47(10)^{10}}} = 1{,}039\ \text{K} \tag{3.27}$$

3.4 DESIGN CONSIDERATIONS FOR THERMAL INCINERATORS

Your objectives in studying this section are to

1. Understand the importance of the concept of heat recovery.
2. Calculate the amount of auxiliary fuel required for a given incineration application.
3. Calculate the dimensions of a thermal incinerator.
4. Understand the operation of a regenerative thermal oxidized (RTO).

The design of a VOC incinerator involves specifying a temperature of operation and a residence time, and then sizing the device to achieve the desired residence time with the proper flow velocity. Material and energy balances are performed on the device to calculate the flow rate of auxiliary fuel required to maintain the desired temperature of operation and the flow rate of the exhaust gases.

3.4-1 Heat recovery

Figure 3.4 is a simplified schematic diagram of a thermal incinerator system with heat recovery. Obviously, because fuel is expensive, it is desirable to recover heat from an incinerator. This heat is usually recovered in recuperative heat exchangers, in which the hot flue gases exchange heat with the cold waste gas.

The waste gas temperature at the incinerator inlet (T_2) is equal to the temperature at the heat exchanger outlet. As this temperature increases, the auxiliary fuel requirement decreases. However, at the same time, the size and cost of the recuperative heat exchanger increases, driving up the total capital investment of the system. Thus, there is a trade-off between capital and operating costs, the extent of which depends on the value of T_2 selected. To avoid preignition of the waste gas before it reaches the combustion chamber, T_2 should not exceed the auto ignition temperature of the VOC (800 to 900 K).

The waste gas temperature entering the incinerator is determined by the waste gas inlet temperature (T_1) and the effectiveness of the heat exchanger. The heat exchanger effectiveness is

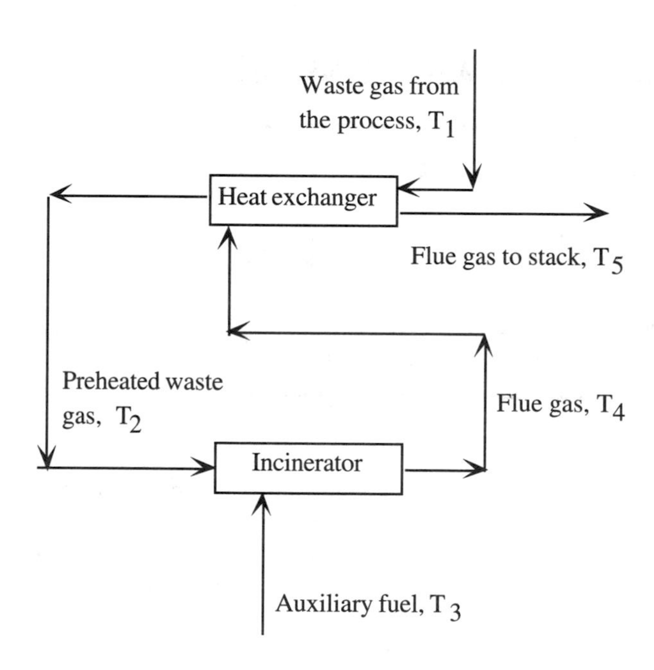

Figure 3.4
Schematic diagram of a thermal incinerator with heat recovery

defined as (Holman 1990):

$$\theta = \frac{\text{Actual heat transferred}}{\text{Maximum possible heat transfer}} \tag{3.28}$$

For the system illustrated in Figure 3.4

$$\theta = \frac{T_2 - T_1}{T_4 - T_1} \tag{3.29}$$

The following values of heat transfer effectiveness are commonly reported for typical modular heat exchangers: 35% to 40% for one-pass units, 45% to 50% for two-pass units, and 65% to 70% for three-pass units.

Example 3.3 Maximum Heat-Transfer Effectiveness

The waste gas of Example 3.2 enters the incinerator system at 300 K. It will be preheated by heat exchange with the flue gases coming out of the incinerator. Calculate the maximum effectiveness of the recuperative heat exchanger if the auto ignition temperature of the VOC should not be exceeded before the waste gas enters the combustion chamber.

Solution

The solution uses Eq. (3.29) to calculate the required effectiveness. The waste gas inlet temperature, T_1, is given as 300 K. The maximum allowable temperature at the exit of the preheater, T_2, is the auto ignition temperature of toluene, given in Table 3.2 as 753 K. The temperature of the flue gases, T_4, was calculated in Example 3.2 as 1,039 K. Substituting in Eq. (3.29), the answer is 61.3%.

Example 3.4 Energy balance around pre-heater

Make a rough estimate of the temperature of the gases going to the stack, T_5, in Example 3.3.

Solution

Make an energy balance around the pre-heater in Figure 3.4. Assume no heat losses. Neglect all energy forms, except enthalpy.

$$H_2 - H_1 = H_4 - H_5 \tag{3.30}$$

where H_i is the total enthalpy per unit time of stream i. To relate the "enthalpy balance" to the temperatures of the streams:

$$m_1 C_{p1}(T_2 - T_1) = m_4 C_{p4}(T_4 - T_5) \tag{3.31}$$

where m_1 and m_4 are the mass flow rate of waste gas and flue gases, respectively, and C_{p1} and C_{p4} are the mean heat capacities of waste gas and flue gases, respectively. To obtain a rough estimate of T_5, make the following simplifying assumptions:

1. The mass flow rate of auxiliary fuel is much smaller than the mass flow rate of waste gas, therefore, $m_4 = m_1$.
2. Both the waste gas and the flue gases are mostly nitrogen. Assuming the heat capacities remain relatively constant in the temperature range involved, $C_{p1} = C_{p4}$.

With these simplifications, Eq. (31) becomes:

$$T_5 = T_4 - (T_2 - T_1) = 1{,}039 - (753 - 300) = 586 \text{ K} \tag{3.32}$$

Comments

Katari et al. (1987) suggest that the temperature of the flue gases going to the stack (T5) should be above 500 K to avoid moisture condensation and possible equipment corrosion problems.

3.4.2 Material and Energy Balances; Sizing the Device.

Once waste gas characteristics and temperatures at various stages of the incineration process are identified, material and energy balances around the incinerator establish the flue gas flow rate leaving the system and the amount of auxiliary fuel required. The energy balance around the incinerator (refer to Fig. 3.4) reduces to

$$-\Delta H = H_2 + H_3 - H_4 = q_L \qquad (3.33)$$

where q_L is the rate of heat loss from the incinerator. The enthalpies in Eq. (3.33) must be related to the respective temperatures and molar flows. Table 3.3 gives equations to calculate molar enthalpies, h, of some combustion gases as functions of temperature. The following examples illustrate the calculations involved.

Table 3.3 Enthalpies of Combustion Gases (kJ/kmol)[a]

Compound	a	b	$c \times 10^3$
H_2O	-9,261	29.378	5.88
CO_2	-12,235	37.462	8.04
N_2	-8,242	26.583	2.84
O_2	-8,856	28.460	3.05

[a]Pressure = 1 atm; $h = a + bT + cT^2$; units of h are kJ/kmol ($h = 0$ @ 298 K); units of T are K (range: 298–1,500 K).

Source: Developed from data presented in Himmelblau, D. M. *Basic Principles and Calculations in Chemical Engineering*, 5th ed. Prentice Hall, Englewood Cliffs, NJ (1989).

Example 3.5 Material and Energy Balances around Thermal Incinerator

The waste gas of Example 3.2 (1,000 ppm toluene, 15% O_2, and 84.9% N_2) flows at the rate of 600 m^3/min, at 300 K and 1 atm. It will be incinerated at 1,039 K, with a retention time of 0.5 s to achieve 99.9% destruction of the toluene. Pure methane at 298 K and 1 atm will be used as auxiliary fuel. A recuperative one-pass heat exchanger (35% effectiveness) is recommended. Calculate the flow rate of methane required.

Solution

The first step calculates the temperature at the entrance of the combustion chamber, T_2. Use Eq. (3.29) with $\theta = 0.35$, $T_1 = 300$ K, and $T_4 = 1{,}039$ K. The answer is $T_2 = 559$ K.

Basis : 1 min

Calculate the moles of waste gas (n_W) entering the system. Assume that the ideal gas law applies:

$$n_W = \frac{PV}{RT} = \frac{(101.3)(600)}{(8.314)(300)} = 24.37 \text{ kmol} \tag{3.34}$$

Define x = kmol of methane supplied. Equations (3.2) and (3.3) of Example 3.1 are the balanced oxidation reactions. Express the material balances in tabular form.

Component	kMoles in	Consumed	Formed	Out
C_7H_8	0.0244	0.0244	0	0
CH_4	x	x	0	0
O_2	3.66	$0.2196 + 2x$	0	$3.44 - 2x$
N_2	20.69	0	0	20.69
CO_2	0	0	$0.171 + x$	$0.171 + x$
H_2O	0	0	$0.0976 + 2x$	$0.0976 + 2x$

Equation (3.33) is the energy balance around the combustion chamber. Choose the thermodynamic path shown in Figure 3.5 to calculate the total enthalpy change, ΔH. Because enthalpy is a state function:

$$\Delta H = \Delta H_w + \Delta H^o_{rxn} + \Delta H_p \tag{3.35}$$

where

ΔH_w = enthalpy change of the waste gas when it goes from 559 K to 298 K at a constant pressure of 1 atm

ΔH^0_{rxn} = standard enthalpy change of reaction

ΔH_p = enthalpy change of the products when they go from 298 K to 1,039 K at a constant pressure of 1 atm

To calculate the first term on the right-hand side of Eq. (3.35),

$$\Delta H_w = \sum_{\text{waste}} \left(n \int_{559\ \text{K}}^{298\ \text{K}} C_p \, dT \right) \tag{3.36}$$

where the summation is over all the components of the waste gases. For oxygen and nitrogen, use the information in Table 3.3. The integrals for each of these two gases is equivalent to the negative of the molar enthalpies evaluated at 559 K. For toluene, integrate the heat capacity equation (given in Appendix A). The result is

$$\Delta H_w = 0.0244\ (-37{,}966) + 3.66\ (-8{,}006) + 20.69\ (-7{,}656) = -188{,}640\ \text{kJ}$$

To calculate the standard enthalpy change of reaction,

$$\Delta H^o_{rxn} = -\left(m_{\text{Tol}} \text{LHV}_{\text{Tol}} + m_{\text{CH}_4} \text{LHV}_{\text{CH}_4}\right) \tag{3.37}$$

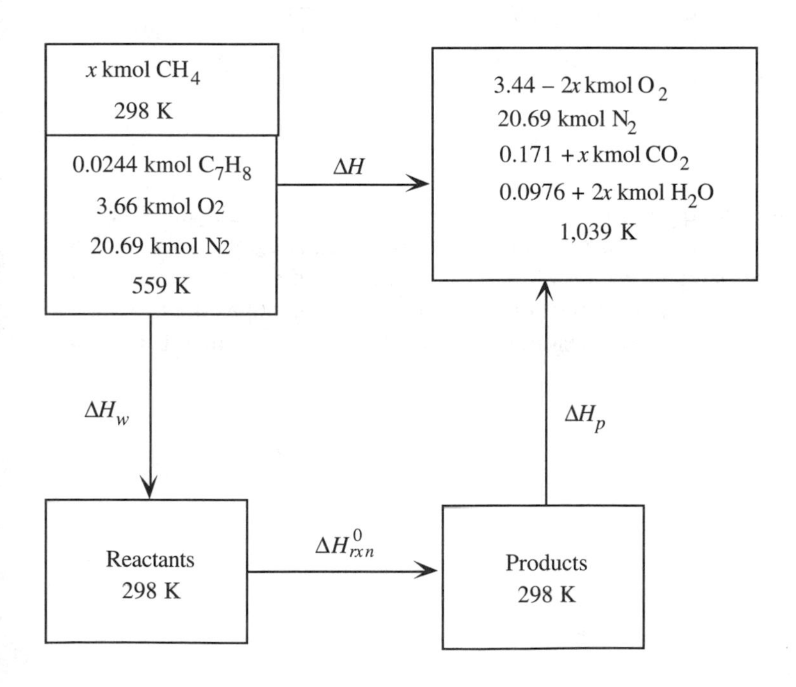

Figure 3.5 Thermodynamic path for enthalpy change calculations

where

m_{Tol} and m_{CH_4} are the masses of toluene and methane burned;

LHV_i is the lower heating value of component i, defined as the negative of the standard heat of combustion on a unit-mass basis at 298 K with $CO_2(g)$ and $H_2O(g)$ as products.

Substituting heating values from Appendix B into Eq. (3.37)

$$\begin{aligned}\Delta H^o_{rxn} &= -[(0.0244)(92)(40{,}950) + x(16)(50{,}150)] \\ &= -(91{,}925 + 802{,}400\,x)\text{ kJ}\end{aligned} \tag{3.38}$$

To calculate the enthalpy change of the products in going from 298 K to 1,039 K, use information from Table 3.3 in

$$\Delta H_p = \sum_{\text{Products}} n(h \text{ @ } 1{,}039\text{ K}) \tag{3.39}$$

$$\begin{aligned}\Delta H_p = (3.44 - 2x)(24{,}006) + (20.69)(22{,}724) + (0.171 + x)(35{,}367) \\ + (0.0976 + 2x)(27{,}610) = 561{,}483 + 42{,}575x\end{aligned}$$

Substituting into Eq. (3.35),

$$\Delta H = -759{,}825x + 280{,}918$$

To calculate the heat losses in Eq. (3.33), details of the dimensions and insulation of the combustion chamber must be known. Clearly, at this stage of the calculations, those details are not known. Most authors suggest that heat losses can be estimated as 10% of the absolute value of the standard enthalpy change of reaction. A forthcoming example explores how reasonable this estimate is, based on heat transfer considerations. Assume it is reasonable, for this example.

$$q_L = 0.10\left|\Delta H^o_{rxn}\right| = 9{,}193 + 80{,}240\,x \tag{3.40}$$

Substitute into Eq. (3.33) and solve for x. The answer is $x = 0.427$ kmol of CH_4. The volume flow rate of methane is from the ideal gas law. $V = nRT/P =$ (0.427)(8.314)(298)/(101.3) = 10.44 m^3/min @ 298 K and 1 atm.

Example 3.6 Sizing of Thermal Incinerator

Calculate the dimensions of the thermal incinerator of Example 3.5. Design for a velocity of the flue gases through the combustion chamber of 10 m/s. The retention time was specified in the previous example as 0.5 s. Determine whether the flow through the combustion chamber is turbulent.

Solution

In the previous example, the moles of each component of the flue gases were expressed in terms of x, the moles of auxiliary fuel supplied per minute. Substituting the value of $x =$ 0.427 kmol/min, the molar flow and composition of the flue gases is:

Component	kMoles/min	Molar Percent
O_2	2.586	10.4
N_2	20.690	82.3
CO_2	0.598	2.5
H_2O	0.952	3.8
Total	24.826	100.00

To calculate the volumetric flow rate of the flue gases, Q, use the ideal gas law $Q = nRT/P =$ (24.826)(8.314)(1039)/(101.3) = 2,117 m^3/min.

Assuming the combustion chamber is cylindrical, its inside diameter, D, is

$$D = \sqrt{\frac{4Q}{\pi u}} = \sqrt{\frac{4 \times 2{,}117}{\pi \times 10 \times 60}} = 2.12 \text{ m} \tag{3.41}$$

To calculate the length, L, of the combustion chamber:

$$L = \tau_r u = 0.5 \times 10 = 5.0 \text{ m} \tag{3.42}$$

To determine the flow regime, calculate the Reynolds number, Re_D, for flow inside a circular pipe

$$\mathrm{Re}_D = \frac{\rho D u}{\mu} \tag{3.43}$$

where ρ and μ are the density and the viscosity, respectively, of the flue gases. The density is estimated with the ideal gas law. The average molecular weight of the flue gases, M_{av}, is 28.4, very close to that of nitrogen.

$$\rho = PM_{av}/RT = (101.3)(28.4)/(8.314)(1039) = 0.333 \text{ kg/m}^3$$

As an estimate of the viscosity of the mixture, use that of pure nitrogen at 1,039 K, $\mu = 4.2 \times 10^{-5}$ kg/m-s (Holman 1990).

$$\mathrm{Re}_D = (0.333)(10)(2.12)/(4.4 \times 10^{-5}) = 1.6 \times 10^5$$

Therefore, the flow is fully turbulent, which guarantees good mixing of the fuel and the oxygen.

Example 3.7 Heat Losses from Thermal Incinerator

Estimate the heat losses from the thermal incinerator of Examples 3.5 and 3.6 through heat transfer calculations. Assume the inside of the combustion chamber is lined with a layer of refractory brick ($k = 1.4$ W/m-K), 10 cm thick. The outside shell is carbon steel ($k = 30$ W/m-K), 1 cm thick. Assume heat is lost to surrounding air at 298 K by natural convection and radiation.

Solution

All the correlations and thermal properties are from Holman (1990). Total heat losses, q_L, are divided into heat losses through the curved cylindrical wall, q_1, and those through the ends of the cylinder, q_2. The resistance to heat transfer of the metal wall is considered negligible in both cases.

Consider first heat losses through the curved wall.

$$q_1 = \frac{T_4 - T_a}{\frac{1}{h_i A_i} + \frac{\ln(r_o/r_i)}{2\pi k_w L} + \frac{1}{(h_o + h_r) A_o}} \tag{3.44}$$

where

T_a is the temperature of the outside air
h_i and h_o are the inside and outside film heat-transfer coefficient, respectively
h_r is the radiation heat-transfer coefficient
A_i and A_o are the inside and outside heat-transfer area, respectively
r_i and r_o are the inside and outside radius of the refractory layer, respectively
k_w is the thermal conductivity of the refractory brick.

To calculate hi, corresponding to cooling of a fluid in fully developed turbulent flow inside a circular pipe, use the Dittus-Boelter equation:

$$Nu_D = 0.023 Re_D^{0.8} Pr^{0.3} = h_i \frac{D}{k} \tag{3.45}$$

All the fluid properties in Eq. (3.45) are evaluated at the gas bulk temperature, 1,039 K in this case. The properties of nitrogen, which constitutes more than 80% of the mixture, are used as an approximation.
$Pr = 0.73$, $k = 0.066$ W/m-K, $Re_D = 1.6 \times 10^5$ (see Example 3.6).
Substituting in Eq. (3.45): $Nu_D = 304.8$, $h_i = k Nu_D / D = (0.066)(304.8)/(2.12)$,
$h_i = 9.5$ W/m^2-K, $(h_i A_i)^{-1} = [\pi(9.5)(2.12)(5.0)]^{-1} = 0.00316$ K/W
To calculate the heat-transfer resistance of the wall,

$$\frac{\ln(r_o/r_i)}{2\pi k_w L} = \frac{\ln(1.16/1.06)}{2\pi \times 1.4 \times 5.0} = 0.00196 \text{ K/W} \tag{3.46}$$

Calculate h_o, corresponding to natural convection from a horizontal cylinder to air at 298 K. The temperature of the metal wall, T_w, is needed for these calculations and for the evaluation of the radiation heat-transfer coefficient, however, it is still not known. An iterative solution procedure is required: assume T_w, calculate h_o and h_r, calculate the total resistance and the heat flow, calculate a new value of T_w, and repeat the procedure until it converges to the actual wall temperature and heat flow.

Assume $T_w = 373$ K. Calculate the film average temperature, T_f.
$T_f = (T_w + T_a)/2 = 336$ K, $\beta = 1/T_f = 0.00298$ K^{-1}
For air at 336 K: $k = 0.028$ W/m-K $\nu = 1.9 \times 10^{-5}$ m^2/s Pr = 0.7

$$\mathrm{Gr}_D\mathrm{Pr} = \frac{g\beta(T_w - T_a)D_o^3\mathrm{Pr}}{\nu^2} \tag{3.47}$$

where D_o is the outside diameter of the cylinder, 2.34 m for this case.
Substituting numerical values, $\mathrm{Gr}_D\mathrm{Pr} = 5.44 \times 10^{10}$. For this value,

$$\mathrm{Nu}_f = \frac{h_o D_o}{k} = 0.13(\mathrm{Gr}_D\mathrm{Pr})^{1/3} = 492 \tag{3.48}$$

$$h_o = 5.9 \text{ W/m}^2\text{-K.}$$

Radiation heat losses are expressed in terms of an equivalent heat transfer coefficient, h_r, defined as

$$h_r = \frac{\varepsilon\sigma(T_w^4 - T_a^4)}{T_w - T_a} \tag{3.49}$$

where ε is the emissivity of the wall, and σ is the Stefan-Boltzmann constant (5.67×10^{-8} W/m^2-K^4). Assuming $\varepsilon = 1.0$, $h_r = 8.67$ W/m^2-K.

$$1/[(h_o + h_r)A_o] = 1/[(8.67 + 5.9)\pi(2.34)(5.22)] = 0.00179 \text{ K/W}$$

Therefore, the heat losses are

$$q_1 = (1{,}039 - 298)/(0.00179 + 0.00196 + 0.00298) = 110{,}104 \text{ W}$$

Check the value of T_w assumed.

$$T_w - T_a = \frac{q_1}{A_o(h_r + h_o)} = 110{,}104 \times 0.00179 = 197 \text{ K} \tag{3.50}$$

The new value of T_w is (197 + 298) = 495 K. New values for the natural convection and radiation heat-transfer coefficients are calculated. Heat losses are corrected, and the procedure repeated until convergence is achieved. The final results are: q_1 = 118,000 W; T_w = 438 K.

To calculate the heat losses through the ends of the cylinder:

$$q_2 = \frac{2A_2(T_4 - T_a)}{\dfrac{1}{h_i} + \dfrac{\Delta x}{k_w} + \dfrac{1}{h_o + h_r}} \tag{3.51}$$

Assume that the film heat-transfer coefficients are the same as those calculated for the curved wall. The area used in Eq. (3.51) is that of a circle with a diameter of 2.34 m. The result is q_2 = 27,807 W.

Therefore, the total heat losses are q_L = 118,000 + 27,807 = 145,807 W or 8,800 kJ/min. Recall from Example 3.5 that the heat liberated during combustion is $\Delta H^0_{rxn} = -91{,}925 - 802{,}400x = -434{,}550$ kJ/min. Total heat losses are, then, approximately 2% of the total heat liberated. This is much lower than the 10% value assumed in Example 2.5.

3.4.3 Regenerative Thermal Oxidizers (RTO).

A variant of the thermal incinerator design is the regenerative thermal oxidizer (Vatavuk 1990) shown in Figure 3.6. In these incinerators, the entering waste gas passes vertically through a chamber filled with a ceramic or stoneware packing (bed #1) where it is warmed by heat stored in the packing from a previous cycle. Some of the VOCs burn in this preheat chamber, while the

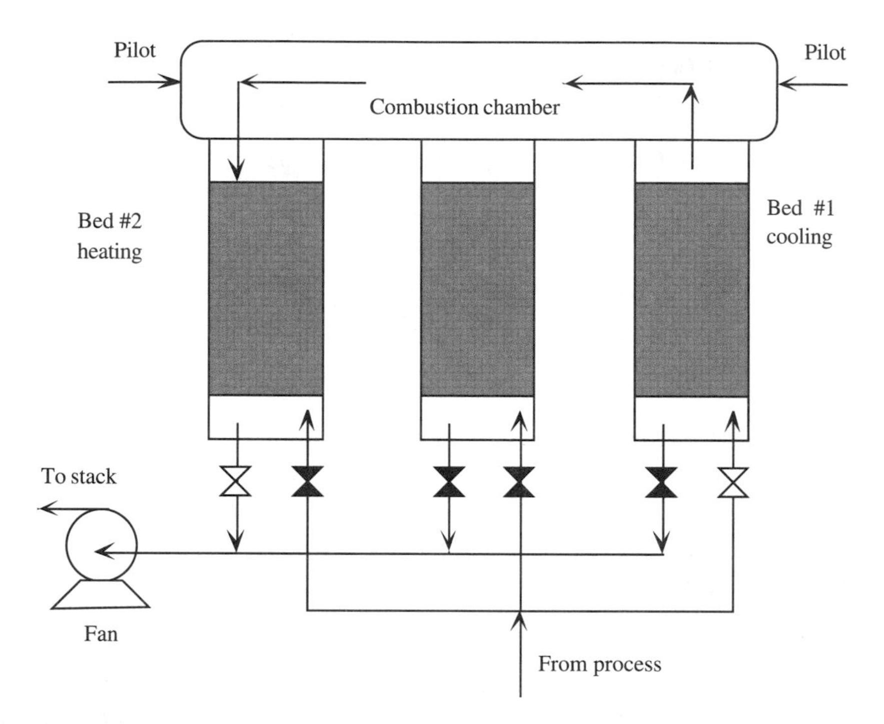

Figure 3.6 Schematic diagram of a regenerative thermal oxidizer

rest is burned in a central combustion chamber at temperatures ranging from 1,000 to 1,500 K. Following combustion, the flue gas exits through another packed bed (bed #2) to which it contributes most of its enthalpy. The beds are then switched, and the incoming waste gas now enters through bed #2 for preheating. Each cooling-heating cycle lasts a few seconds, so that each bed—and the entire RTO itself—approaches steady-state operation.

The great advantage of RTOs over conventional thermal incinerators is that a much higher heat recuperation effectiveness is possible, up to 95% compared with the maximum of around 70% for thermal incinerators. As a result, auxiliary fuel requirements are drastically reduced. Material and energy balances, identical to those illustrated for conventional thermal incinerators, give the amount of auxiliary fuel required.

3.5 CATALYTIC INCINERATORS

Your objectives in studying this section are to

1. Understand the advantages and disadvantages of catalytic incinerators compared with their thermal counterparts.
2. Design catalytic incinerators based on mass transfer considerations.
3. Define space velocity.

Catalytic incineration is essentially a flameless combustion process, wherein a catalyst bed initiates the combustion reaction at much lower temperatures. Historically, the greatest reason for a catalytic incinerator has been to reduce auxiliary fuel consumption. Another advantage of the lower operating temperature is a reduction in the amount of NOx formed during combustion.

However, because the application of either a thermal or catalytic incinerator system without heat recovery is now rare, the fuel savings associated with catalytic incinerators are less significant. Also, this fuel economy is partially offset by increased operating costs for maintenance (i.e., periodic cleaning and replacement of catalyst). Catalysts undergo a gradual loss of activity through thermal aging, fouling, and erosion of their surfaces. Certain poisonous contaminants, such as phosphorus, arsenic, antimony, lead, and zinc, also cause catalyst deactivation (Katari, et al. 1987). Catalyst incineration is not recommended for waste gases containing significant amounts of particulate matter that cannot be vaporized.

The catalysts for catalytic incineration of gaseous VOCs are usually precious or base metals or their salts. They may be supported on inert carriers, such as alumina or porcelain, or unsupported. Precious metal oxide catalysts are less brittle and more expensive than base metal types, and are used in lesser amounts per unit of waste gas volume. Of the precious metal oxide catalysts, platinum-palladium oxides are preferred. Others include rhodium, nickel, and gold. Manganese dioxide is the most common base metal oxide catalyst.

In catalytic incinerators, the waste gas temperature is typically raised in a preheat chamber to 525 to 600 K. The thoroughly mixed gaseous effluent from the preheat chamber passes through specially designed units containing catalyst elements, on the surface of which oxidation occurs at an accelerated rate at temperatures of 650 to 750 K. The heat of reaction from the oxidation of the VOCs in the catalyst bed causes the gas temperature to increase as it passes across the catalyst bed. The amount of VOCs in the waste gas determines the temperature increase. The maximum temperature to which the catalyst bed can be exposed continuously is limited to about 900 K.

Depending on the catalyst bed temperature swings (i.e., the frequency at which it is subjected to extreme temperature excursions), the operating and maintenance practices, and the particulate matter and specific catalyst poisons encountered, the catalyst has an effective life of 2 to 10 yr.

The overall rate of catalytic oxidation depends on both the rate of diffusion of the VOC to the surface of the catalyst and the reaction kinetics. At temperatures above 525 K, mass transfer usually controls the rate of the process. Thus, the design of catalytic units reduces to specification of the proper length of bed to permit sufficient residence time, based on mass transfer rates, to achieve the desired degree of VOC destruction.

3.5.1 Design of Catalytic Incinerators Based on Mass Transfer Considerations

When mass transfer of reactants is the limiting step in catalytic combustion, the design procedure is simplified to a number-of-transfer-units method (Retallick 1981). The reaction of two fluids over a solid catalyst takes place in two steps: transport of reactants to the catalyst surface and reaction at the surface. At steady state the rates of reaction and mass transfer are the same. Yet the rate may be limited by either of the two steps.

Consider a channel in a catalytic reactor, as shown on Figure 3.7. When mass transfer limits the rate, the surface concentration of the reactants is nearly zero. For that case, a material balance of reactant A in a differential reactor volume is

$$\frac{d[\mathrm{A}]}{dl} = -\frac{ak_c[\mathrm{A}]}{u} \tag{3.52}$$

where:

l is the longitudinal distance along the channel
a is the surface area per unit volume of reactor
k_c is the mass-transfer coefficient
u is the bulk velocity of the fluid.

Integrating from 0 to l,

$$[\mathrm{A}]_l = [\mathrm{A}]_0 \exp\left(-\frac{ak_c l}{u}\right) \tag{3.53}$$

For a given reactor of length L, the concentration at the reactor outlet, $[\mathrm{A}]_L$, is

$$[\mathrm{A}]_L = [\mathrm{A}]_0 \exp\left(-\frac{ak_c L}{u}\right) \tag{3.54}$$

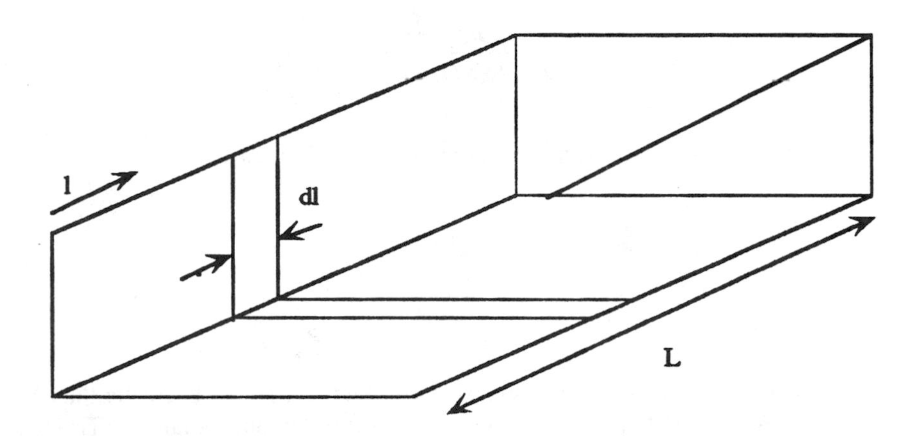

Figure 3.7 One channel of a catalyst support

Define the length of a mass-transfer unit, L_m, as: $L_m = u/k_c a$. The number of transfer units, N, and the concentration at the reactor outlet are given by

$$N = \frac{L}{L_m}, \qquad [A]_L = [A]_0 \, e^{-N}, \qquad N = -\ln\frac{[A]_L}{[A]_0} \tag{3.55}$$

To calculate the length of a mass-transfer unit, an estimate of the mass-transfer coefficient is needed. The latter can be expressed in terms of the Sherwood number, Sh, defined, in this situation, as:

$$\mathrm{Sh} = k_c d/D_A$$

where

d is the effective channel diameter ($d = 4/a$)

D_A is the diffusivity of the reactant

There are different empirical correlations for estimating the Sherwood number, depending on the flow regime. Many of these take advantage of the analogy between Sh in mass-transfer and Nu in heat-transfer.

Consider, for example, laminar flow in a circular channel. According to Holman (1990), for this situation Nu = 4.364. Assuming perfect analogy between heat and mass-transfer, Sh = 4.364. Therefore,

$$L_m = \frac{u}{k_c a} = \frac{ud^2}{4\,\mathrm{Sh}\,D_A} = \frac{ud^2}{17.46 D_A} \tag{3.56}$$

For laminar flow in a channel with square cross section, Sh = Nu = 3.61. Then,

$$L_m = \frac{ud^2}{14.43 D_A} \tag{3.57}$$

where d is the height of the channel.

When the Reynolds number for flow inside the reactor channels is greater than 2,000, the flow is considered turbulent. For that case, use the Chilton-Colburn analogy (Treybal 1980), which establishes an analogy between mass transfer and momentum transfer:

$$\mathrm{St}_D\,\mathrm{Sc}^{2/3} = \frac{k_c}{u}\,\mathrm{Sc}^{2/3} = \frac{f}{2} \tag{3.58}$$

where

- Sc = Schmidt number = $\mu/\rho D_A$
- St_D = Stanton number for mass-transfer = Sh/(Re Sc)
- f = Fanning friction factor

Solving Eq. (3.58) for kc, and substituting in the definition of Lm,

$$L_m = \frac{2}{fa}\,\mathrm{Sc}^{2/3} \tag{3.59}$$

To calculate the Fanning friction factor, assume the channel is "hydraulically smooth" (Bennett and Myers 1982),

$$\frac{1}{\sqrt{f}} = 1.7631\ln\left(\mathrm{Re}\sqrt{f}\right) - 0.6 \tag{3.60}$$

It is evident that the size of the channels is a very important design parameter in catalytic incineration. Smaller channels have greater surface area per unit volume of reactor, resulting in a smaller incinerator. However, the pressure drop increases as the size of the channels decreases. Another difficulty with small channels could be plugging caused by entrained particulate matter.

Retallick (1981) suggests that the reactor length calculated by this method should be considered a minimum effective length for achieving the desired degree of VOC destruction. Doubling the calculated length is an adequate safety factor in most cases.

Example 3.8 Design of Catalytic Incinerator

The waste gas of Example 3.5 is preheated in a heat exchanger to 600 K and incinerated in a catalytic bed at that temperature to obtain 99.9% destruction of the toluene. At that temperature, and for this catalyst, mass transfer is the rate limiting step.
(a) The catalyst has circular channels with an effective diameter of 1.17 mm, and a fractional open area for flow of 65%. To avoid excessive pressure drop, the gas velocity (based on the total cross-sectional area of the incinerator) should not exceed 10 m/s. Estimate the dimensions of the catalytic unit required. Assume that the transport properties of the combustion gases are similar to those of nitrogen.

Solution

The flow rate of waste gas in Example 3.5 was 600 m^3/min at 300 K and 1 atm. Since it is preheated in a heat exchanger to 600 K, the flow rate entering the catalytic bed is 1,200 m^3/min. The gas velocity through the channels is u = 10/0.65 = 15.4 m/s. From the data given by the problem, d = 1.17 mm. Assuming the combustion gases behave like nitrogen at 600 K, $\rho = 0.5687$ kg/m^3, $\mu = 2.91 \times 10^{-5}$ kg/m-s. Calculate the Reynolds number to determine the flow regime: Re = (0.5687)(15.4)(0.00117)/(2.91×10^{-5}) = 352 (laminar).

The value of L_m is from Eq. (3.56). The diffusivity in that equation is approximately that of toluene in air (Treybal 1980), corrected for temperature assuming a $T^{3/2}$ dependence: DA = 2.23×10^{-5} m^2/s, L_m = 0.054 m. Use Eq. (3.55) to calculate the number of transfer units required: N = 6.91. Therefore, $L = NL_m$ = 0.37 m. Following Retallick's suggestion, the total length is 0.74 m. Calculate the diameter with Eq. (3.41) based on the volumetric flow rate and the specified velocity of 10 m/s, D = 1.60 m. Therefore, a cylindrical catalyst bed with a diameter of 1.6 m and a packed depth of 0.74 m should destroy, at least, 99.9% of the toluene. Recall from Example 3.6 that the same performance in a thermal reactor requires a cylindrical combustion chamber with a diameter of 2.12 m and a length of 5.0 m.
(b) Assume that the waste gas of part a carries a substantial amount of particulate matter. To avoid plugging of the catalyst bed, a different catalyst support structure is considered. It has circular channels with an effective diameter of 3.34 mm , and fractional open area of 50%.

The gas velocity (based on total cross-section area) should not exceed 25 m/s. Estimate the dimensions of the catalyst bed required.

Solution

Calculate the Reynolds number. The new value of velocity is $u = 25/0.5 = 50$ m/s; Re = (0.5687)(50)(0.00334)/(2.91×10^{-5}) = 3,266 (turbulent). Therefore, the Chilton-Colburn analogy gives an estimate of L_m. Use Eq. (3.60) to estimate the friction factor. Define $x = f^{-1/2}$. Then,

$$x = 1.763\ln\left(\frac{3{,}266}{x}\right) - 0.6 \tag{3.61}$$

Rearranging Eq. (3.61),

$$g(x) = x + 1.763\ln x - 13.665 = 0 \tag{3.62}$$

$$g'(x) = 1 + \frac{1.763}{x} \tag{3.63}$$

The root of Eq. (3.62) can be found using the root-finding program presented in Example 2.8, with the help of Eq. (3.63). Only the SUBROUTINE FUNCD needs to be modified. The answer is $x = 9.6655$, $f = 0.0107$. Calculate $a = 4/d = 1198$ m^{-1}. Calculate Schmidt number: Sc = 2.29. Substituting in Eq. (3.59), $L_m = 0.271$ m. From part a, $N = 6.91$; therefore, $L = 1.88$ m. Following Retallick's suggestion, the total length is 3.76 m. Calculate the diameter based on the volumetric flow rate and the specified velocity of 25 m/s, from Eq. (3.41): $D = 1.0$ m.

3.5.2 Design of Catalytic Incinerators Based on Space Velocity

A simplified method to determine the total volume of catalyst bed for a given incineration problem is to specify the space velocity. Space velocity is defined as the waste gas flow rate per unit volume of catalyst; its inverse is the residence time. Katari et al. (1987) suggest the following values for space velocity during catalytic incineration:

Table 3.4 Space Velocities for Catalytic Incineration, h^{-1}

Catalyst	90% Conversion	95% Conversion
Precious metals	40,000	30,000
Base metals	15,000	10,000

Example 3.9 Space Velocity Calculations

Calculate the space velocity for the catalytic incinerator of Example 3.8.

Solution

From Example 3.8, the volumetric flow rate of the waste gas is 1,200 m^3/min. For part a of that example, calculate the volume of the catalytic bed, a cylinder with a diameter of 1.6 m and a length of 0.74 m . The volume is 1.488 m^3. Therefore:

$$\text{Space velocity} = \frac{1{,}200\ m^3/min}{1.486\ m^3} = 806\ min^{-1} = 48{,}400\ h^{-1} \qquad (3.64)$$

A similar calculation for part b results in a space velocity of 24,400 h^{-1}, which is closer to the specifications of Table 3.4, assuming that the catalyst is a precious metal.

3.6 PRESSURE DROP CALCULATIONS

Your objectives in studying this section are to

1. Understand the importance of pressure drop calculations in air pollution control equipment design.
2. Estimate pressure drops resulting from either thermal or catalytic incinerators including recuperative heat exchangers.

The energy required to operate an air pollution control device is mainly of two kinds: (1) work to move the waste gas stream through the equipment; (2) additional energy of a form unique to the type of device (e.g., auxiliary fuel during incineration). The first kind is due to friction. The second kind may be completely absent in some devices, such as cyclones, and even when present may be small relative to the first kind.

The first kind is always present. Hence it is important to be able to estimate the pressure drop in the gas stream as it moves through the device, and to calculate the corresponding power loss that must be replaced by the fan or blower that moves the stream. The total pressure drop owing to frictional losses is made up of losses in the duct work plus loss in the collector itself. The first may be calculated by the standard methods of fluid mechanics, in which it is usually assumed that the gas is noncompressible. The latter requires special treatment which may be different for each control device.

Once the total pressure drop, ΔP_{TOT}, is calculated, the power required to operate the fan or blower, W, is given by:

$$W = \frac{Q\,\Delta P_{TOT}}{E_m} \tag{3.65}$$

where

Q is the volumetric flow rate of the gas at the blower location
E_m is the mechanical efficiency of the blower-motor system.s

The total pressure drop across an incinerator system depends on the number and types of equipment in the system and on design considerations. The estimations of actual pressure drops are from complex calculations based on the specific system's waste gas and flue gas conditions and the equipment. For the purpose of preliminary estimates, Katari et al. (1987) suggest approximate values for thermal incinerators and recuperative heat exchangers, which are summarized in Table 3.5.

For catalytic incinerators designed with space velocity specifications, Katari et al. suggest a pressure drop of 1.5 kPa. For those designed with mass transfer calculations, a more accurate estimate can be made. By definition

$$\Delta P = \frac{2 f L u^2 \rho}{d} \tag{3.66}$$

For turbulent flow in a "hydraulically smooth" channel, the friction factor is calculated from Eq. (3.60). For laminar flow, $f = 16/\text{Re}$.

Equation (3.66) gives the pressure drop for flow across the catalyst bed. To account for inlet expansion and outlet contraction losses, add approximately 50 Pa to the pressure drop calculated from Eq. (3.66).

Table 3.5 Approximate Pressure Drops across Thermal Incinerators and Recuperative Heat Exchangers

Device	Pressure Drop, kPa
Thermal incinerator	1.00
Heat exchanger, 35% efficiency	1.00
Heat exchanger, 50% efficiency	2.00
Heat exchanger, 70% efficiency	3.74

For regenerative thermal oxidizers, the pressure drop depends on the thermal effectiveness. Vatavuk (1990) suggests 4.0 kPa for an RTO with a thermal effectiveness of 85%, and 5.0 kPa for 95%.

Example 3.10 Pressure Drop for Thermal Incinerator with Heat Recovery

Estimate the total pressure drop for the thermal incinerator of Example 3.5. Neglect friction losses owing to the ductwork and stack. Calculate the power required to operate the blower. Assume the blower is located at the entrance of the system (point 1 on Figure 3.4). The mechanical efficiency of the motor-blower system is 60%.

Solution

Estimate the total pressure drop from Table 3.5. The thermal incinerator contributes 1.0 kPa, and the 35% heat exchanger an additional 1.0 kPa. Therefore, neglecting ductwork and stack pressure drops, $\Delta P_{TOT} = 2.0$ kPa. From Example 3.5, the volume flow rate of waste gas at the blower location is 600 m^3/min, or 10 m^3/s. Substituting in Eq. (3.65), W = (10)(2)/0.6 = 33.3 kW.

Example 3.11 Pressure Drop Calculations for Catalytic Incinerator

Estimate the total pressure drop for the catalytic incinerator of Example 3.8, part a. Assume that the preheating occurs in a recuperative heat exchanger with a thermal effectiveness of 70%. Calculate the power required to operate the blower. Assume that the blower is located at the entrance of the system and has a mechanical efficiency of 60%.

Solution

Use Eq. (3.66) to calculate the pressure drop through the catalytic bed. Since the flow is laminar, $f = 16/\text{Re} = 16/352 = 0.045$. Substituting in Eq. (3.66),

$$\Delta P = \frac{2(0.045)(0.74)(15.4)^2(0.5687)}{0.00117} = 7{,}680 \text{ Pa} = 7.68 \text{ kPa}$$

Add 0.05 kPa to account for entrance effects, and 3.74 kPa owing to the recuperative heat exchanger. Therefore, $\Delta P_{TOT} = 11.47$ kPa.

Use Eq. (3.65) to calculate the power required to operate the blower.

$$W = 10(11.47)/0.6 = 191 \text{ kW}.$$

Comments

The combination of a catalytic incinerator and a recuperative heat exchanger with 70% effectiveness virtually eliminates the need for auxiliary fuel in this case, but results in a much higher pressure drop than a thermal incinerator. Chemical energy requirements are reduced, but the electricity demand increases substantially.

3.7 CAPITAL AND ANNUAL OPERATING COSTS OF INCINERATORS

Your objectives in studying this section are to

1. Estimate purchased equipment cost for thermal and catalytic incinerators, and RTOs.
2. Estimate total capital investment and total annual cost for incinerators.
3. Choose between viable alternative incinerator designs.

The capital costs for VOC incinerators can be very high because of the high temperatures they must be able to withstand. The major contributor to the annual operating cost is usually the auxiliary fuel cost. Heat recovery and catalytic units reduce the net fuel cost considerably.

Most thermal and catalytic incinerators handling waste gas flowrates up to 15 m3/s (at "standard" conditions of 295 K and 101.3 kPa) are often sold as packaged modular units. Each unit consists of (1) the combustion unit and burner; (2) recuperative heat exchanger; (3) blower and motor; (4) instrumentation and controls; (5) 3-m stack; and 6) for catalytic units only, a filter-mixer to distribute the gas flow and remove particulate matter. Larger incinerators, handling flows up to 50 standard m3/s, have been built. These are "custom" units, part vendor fabricated and part field assembled.

3.7.1 Thermal Incinerator Equipment Costs

Katari et al. (1987) present cost data from a number of equipment manufacturers. The cost data on Table 3.6, taken from that source and updated to June 1990, relate thermal incinerator equipment costs (EC) to the waste gas volume flow rate (Q) at "standard" conditions of 295 K and 101.3 kPa, expressed in m^3/s, and to the effectiveness of the recuperative heat exchanger. The cost data apply to dilute VOC content waste gases containing up to 25% LEL to which no outside air is added (Category 1 of Figure 3.1) and incinerated at a combustion temperature of 1,100 K. They can also be applied to other cases requiring slightly different combustion temperatures without introducing significant errors to the costs. For example, if a combustion chamber were sized for a combustion temperature of 1,200 K instead of 1,100 K, the equipment cost increase would be less than 5%.

The data on Table 3.6 apply to flowrates in the range of 2.36 to 23.6 standard m^3/s.

Table 3.6. Thermal Incinerators Equipment Costs (June 1990)

$$\ln \mathrm{EC} = \left[a + b \ln Q + c(\ln Q)^2\right](10^{-3}) \tag{3.67}$$

Effectiveness (%)	*a*	*b*	*c*
0	11,061	80.00	70.00
35	11,646	39.84	85.00
50	11,746	75.58	81.00
75	12,214	–112.63	131.00

EC is the equipment cost, in $ of June 1990. Q is the volumetric flow rate of waste gases, in m^3/s @ 295 K and 1 atm.

Example 3.12 Thermal Incinerator Equipment Cost

Estimate the equipment cost for the thermal incinerator of Example 3.5.

Solution

The waste gas flow rate in Example 3.5 is 600 m^3/min at 300 K and 1 atm. Therefore, $Q =$ (600)(295)/(60)(300) = 9.83 standard m^3/s. Substitute this value of Q in Eq. (3.67). The values of a, b, and c are from Table 3.6, for a recuperative heat exchanger with an effectiveness of 35%. The answer is: EC = $195,000 of June 1990.

3.7.2 Catalytic incinerators equipment costs

Katari et al. (1987) present catalytic incinerator equipment costs developed from cost information received from equipment manufacturers for flowrates between 2.36 and 23.6 m^3/s at standard conditions. Table 3.7 summarizes them, updated to June 1990. These cost data apply to dilute VOC waste gases requiring a temperature of approximately 600 K at the catalytic bed inlet.

Table 3.7. Catalytic Incinerators Equipment Costs (June 1990)[a]

Effectiveness	a	b	c
0%	11,355	-112.26	205.00
35%	11,852	-251.05	231.00
50%	11,891	-203.95	225.00
70%	11,800	21.95	174.00

[a]Constants to be used with Eq. (3.67), in Table 3.6.

Example 3.13 Catalytic Incinerator Equipment Cost

Estimate the equipment cost for the catalytic incinerator of Example 3.11.

Solution

The waste gas flow rate is the same as that for Example 3.12, 9.83 m3/s @ 295 K and 101.3 kPa. A recuperative heat exchanger with an effectiveness of 70% is used. Substituting the appropriate constants from Table 3.7 in Eq. (3.67), the answer is EC = $347,700 of June 1990.

3.7.3 Total Capital Investment for Incinerators

The total incinerator capital investment depends on whether the unit is a skid-mounted modular one, or custom made. Systems handling waste gas flowrates of more than 15 m3/s at standard conditions are custom made. For estimating purposes, the total capital investment (TCI) of a skid-mounted modular unit can be calculated as 125% of the total purchased equipment cost (*B*). Conversely, Table 2.2 (Section 2.2) presents typical capital costs factors applicable to custom installations of both thermal and catalytic systems. The TCI also includes costs for land, working capital, and offsite facilities, which are not included in the direct-indirect installation factors. However, these items are rarely required with incinerator systems. No factor is provided for site preparation or buildings, because these costs are highly site specific and depend very little on the purchased equipment cost.

Example 3.14 Total Capital Investment for Thermal Incinerator

Estimate the TCI required for the incinerator of Example 3.12. Assume that the duct work and other auxiliaries add $15,000 to the equipment cost. Assume, also, that no site preparation, buildings, working capital, land, or offsite facilities are required.

Solution

The waste gas flow rate for this example is 9.83 m^3/s, therefore the incinerator is a skid-mounted modular unit. The purchased cost of the incinerator and auxiliaries is: A = $195,000 + $15,000 = $210,000. According to Table 2.2, the total purchased equipment cost, B, is given (for both modular and custom units) by: $B = 1.18A$ = $247,800. Because this is a modular unit and no working capital, and so on, is required: TCI = $1.25B$ = $309,750 of June 1990.

3.7.4 Estimating Total Annual Cost for Thermal and Catalytic Incinerators

Table 3.8 presents suggested (Katari et al. 1987) factors for estimating incinerator annual costs. For thermal incinerators, the cost of auxiliary fuel is, usually, the major direct annual cost. For catalytic incinerators, the cost of replacing the catalyst load can be a major portion of the total annual cost. The life of a given load may be 2 to 10 yr. A conservative estimate of the catalyst replacement cost can be based on the lower life time of 2 yr. The initial costs of precious metal and base metal (manganese dioxide) catalysts, taken from Katari et al. (1987) and updated to June 1990, are $127,000/m^3$ and $25,200/m^3$ respectively. The catalyst replacement labor is minimal compared to the catalyst cost. Taxes and freight on the replacement catalyst are estimated as 8% of the initial catalyst cost.

Example 3.15 Total Annual Cost of Thermal Incinerator

Estimate the TAC for the incinerator of Example 3.14. Assume the system operates 350 days per year, 24 hours per day. The cost of auxiliary fuel is $2/10^6$ kJ, the cost of electricity

Table 3.8 Factors for Estimating Incinerator Annual Costs

Item	Suggested Factor
Direct operating costs	
Operating labor	0.5 h/shift
Supervisory labor	15% of operating labor
Maintenance labor	0.5 h/shift
Maintenance. materials	100% of maintenance labor
Replacement parts	Periodic replacement of catalyst
Utilities	
Fuel	As required
Electricity	As required
Indirect operating costs	
Overhead	60% of sum of operating, supervisory, maintenance labor and maintenance materials
Administrative charges	2% × TCI
Property tax	1% × TCI
Insurance	1% × TCI
Capital recovery cost	CRF × (TCI – 1.08 × Initial catalyst cost)

Source: Katari et al. (1987).

is \$0.10/kW-h. The useful life of the incinerator is 10 yr with no salvage value. The minimum attractive rate of return for the company is 12%. The operating labor rate is \$10/h; the maintenance labor rate is \$15/h.

Solution

Follow the suggestions of Table 3.8 to estimate annual costs. From Example 3.14, TCI = \$309,750

I Direct operating costs:

1. Operating labor: (0.5)(3)(350)(10)	= \$5,250
2. Supervisory labor = (0.15)(5,250)	= 790
3. Maintenance labor:(0.5)(3)(350)(15)	= 7,900
4. Maintenance materials:(1.0)(7,900)	= 7,900
5. Utilities:	
a. Fuel (See Example 3.5)	
(0.427)(16)(50,150)(60)(24)(350)(2)/106	= 345,400
b) Electricity (See Example 3.10)	
(33.3)(24)(350)(0.10)	= 28,000
Total direct operating costs	= 395,240
II. Indirect operating costs:	
1. Overhead: 0.6(5250 + 790 + 7900 +7900)	= 13,100
2. Administrative charges, property tax and insurance: 0.04(309,750)	= 12,390
3. Capital recovery cost = CRF × TCI:	
CRF = 0.177 (corresponding to n =10, i =0.12)	
Capital recovery cost	= 54,800
Total indirect operating costs	= 80,290
Total annual cost (TAC)	= \$475,530/yr

Example 3.16 Capital Investment and Annual Cost for Catalytic Incinerator

Estimate the total capital investment and total annual cost for the catalytic incinerator of Example 3.13. Assume economic conditions are the same as those in Examples 3.14 and 3.15. The waste gases are preheated to 545 K in a 70% effective heat exchanger. The temperature rises to 650 K through the catalytic bed, for an average combustion temperature of 600 K. No auxiliary fuel is required.

Solution

From Example 3.13, the equipment cost is \$347,700. Assume an additional expense of \$15,000 for duct work and other auxiliaries. Then A = \$362,700. The total purchased equipment cost is: $B = 1.18A$ = \$428,000. Since this is a modular unit ($Q = 9.83$ m^3/s), TCI = $1.25B$ = \$535,000.

Follow the suggestions of Table 3.8 to estimate annual costs. The direct operating costs are the same as those of the previous example, except for the replacement parts and utilities costs. To calculate the replacement parts cost, assume that the catalyst, a precious metal, must be replaced every 2 yr. The volume of the catalytic bed is taken from Example 3.9 as 1.49 m^3. The initial cost of the catalyst is, then: 1.49(127,000) = \$189,000. The total replacement cost, including freight and taxes, and neglecting labor, is 1.08(189,000) = \$204,120. The capital recovery factor for a life of 2 yr and interest rate of 12% is 0.592. Therefore, the annualized cost of replacement parts is: (0.592)(204,120) = \$120,800/yr.

Because no auxiliary fuel is required, electricity is the only utility. The power to operate the blower is, from Example 3.11, 191 kW. Therefore, the cost of electricity is (191)(24)(350)(0.10) = \$160,440/yr.

The total direct operating costs are \$303,080/year. The indirect operating costs follow Table 3.8. The capital recovery factor used to calculate the capital recovery cost corresponds to a useful life of 10 yr and an interest rate, or rate of return, of 12%, CRF = 0.177. The capital recovery cost is (0.177)(533,950 – 204,120) = \$58,400/yr. The total indirect operating costs are \$92,900/yr. Then, TAC = \$395,980/yr.

3.7.5 Total Capital Investment and Annual Costs For RTOs

Vatavuk (1990) presents equations to relate the installed cost of RTOs, C_{RTO}, as function of the waste gas volumetric flow rate at standard conditions, Q, and the effectiveness of thermal recovery. The equations, updated to June 1990, are

1. For 85% effectiveness,

$$C_{RTO} = 392{,}000 + 33{,}117Q \tag{3.68}$$

2. For 95% effectiveness,

$$C_{RTO} = 474{,}400 + 41{,}342Q \tag{3.69}$$

The same reference suggests multiplying the installed cost given by Eqs. (3.68) or (3.69) by a factor of 1.2 to estimate the total capital investment.

RTOs incur the same kind of direct and indirect annual costs as thermal and catalytic incinerators. Auxiliary fuel requirements are usually quite low. The system lifetime is 20 yr as opposed to 10 yr for thermal and catalytic units.

3.7.6 Choice between Alternative Incinerator Designs

It is evident from the discussion, so far, that there are many options open in incineration problems: thermal, or catalytic units with different effectiveness of heat recovery, and RTOs with different degrees of heat recovery. The measures of merit discussed in Chapter 2 determine the best alternative. The EUAC method is the simplest one to use. Remember that, when the total annual cost includes a capital recovery cost, EUAC = –TAC.

Example 3.17 Choice between Incineration Alternatives

The thermal incinerator of Example 3.15 (Alternative A) and the catalytic unit of Example 3.16 (Alternative B) are two viable alternatives to destroy 99.9% of the toluene present in a stream of waste gas.
(a) Based on the EUAC measure of merit, choose the best alternative.
(b) Calculate the profitability index when B is chosen over A.

Solution

(a) The following table summarizes pertinent results for the alternatives taken from Examples 3.15 and 3.16:

	Alternative A	Alternative B
TCI	$309,750	$535,000
CRC	54,800	58,400
EUAC = –TAC	–475,530	–395,980

Because the total annual cost includes the capital recovery cost, the EUAC for each alternative is equal to the negative of the corresponding TAC. Therefore, alternative B is to be chosen because it has the highest EUAC.

(b) To determine the profitability index (PI) when choosing plan B over plan A, calculate the annual net cash flow (NCF) for each alternative. Remember that when cash flows are calculated no capital recovery costs are included. This applies for the initial capital investment and for the costs of replacing the catalyst every 2 yr. The following table presents the results of the NCF calculations:

Year	Plan A	Plan B	B – A
0	–309,750	–535,000	–225,250
1	–420,730	–216,780	203,950
2	–420,730	–420,900	–170
3	–420,730	–216,780	203,950
4	–420,730	–420,900	–170
5	–420,730	–216,780	203,950
6	–420,730	–420,900	–170
7	–420,730	–216,780	203,950
8	–420,730	–420,900	–170
9	–420,730	–216,780	203,950
10	–420,730	–216,780	203,950

Discount all cash flows to the initial time and set the net present worth to zero. Define $x = (1 + \text{PI})^{-1}$. Then,

$$203{,}950\left(x^{10} + x^{9} + x^{7} + x^{5} + x^{3} + x\right) - 170\left(x^{8} + x^{6} + x^{4} + x^{2}\right) - 225{,}250 = 0$$

The root of this equation is $x = 0.6453$, which corresponds to PI = 0.54. Therefore, over the useful life of the equipment, there is an annual 54% return on the additional investment of \$225,250 required to implement the catalytic incinerator as compared to the thermal one, an attractive investment indeed.

3.8 CONCLUSION

Incineration is an effective method to control emissions of VOCs. Very high destruction efficiencies can be achieved, however the process can be quite expensive, depending on the amount of auxiliary fuel required to maintain the operating temperature. Other air pollutants—such as carbon dioxide, carbon monoxide, nitrogen oxides, and particulate matter—are formed during incineration. Another drawback of the process is that it is a destructive technique; any potential value of the VOCs "goes up in smoke."

Other techniques able to recover the VOCs in a useful way may be more benign to the atmosphere, and even more attractive economically. The next chapter treats one of those processes: adsorption.

REFERENCES

Bennett, C.O., and Myers, J. E. *Momentum, Heat, and Mass Transfer,* 3rd. ed., McGraw-Hill, New York (1982).

Cooper, C. D., and Alley, F.C. *Air Pollution Control; A Design Approach*, Prindle, Weber & Schmidt, Boston, MA (1986).

Cooper, C.D., Alley, F. C., and Overcamp, T. J. *Environmental Progress*, 1:2 (May 1982).

Danielson, J. A. (Ed.) *Air Pollution Engineering Manual,* 2nd. ed., AP-40, Washington, D C: U.S. Environmental Protection Agency (1973).

Hemsath, K.H., and Susey, P. E. "Fume Incineration Kinetics and Its Applications", *American Institute of Chemical Engineers Symposium Series No. 137,* **70**:439 (1974).

Himmelblau, D. M. *Basic Principles and Calculations in Chemical Engineering,* 5th ed., Prentice Hall, Englewood Cliffs, NJ (1989).

Holman, J. P. *Heat Transfer*, 7th ed. McGraw-Hill, New York (1990).

Katari, V. S., Vatavuk, W. M., and Wehe, A. H. *JAPCA*, **37**:91 (1987).

Retallick, W. B. *Chemical Engineering*, 77:123 (January 12, 1981).

Ross, R. D. "Thermal Incineration," Chapter 17 in *Air Pollution Control and Design Handbook*, P. N. Cheremisinoff and R. A. Young (Eds.), Marcel Dekker, New York (1977).

Treybal, R. E. *Mass-Transfer Operations*, 3rd. ed., McGraw-Hill, New York (1980).

Vatavuk, W. M. *Estimating Costs of Air Pollution Control,* Lewis, Chelsea, MI (1990).

PROBLEMS

The problems at the end of each chapter have been grouped into four classes (designated by a superscript after the problem number)

Class a: Illustrates direct numerical applications of the formulas in the text.
Class b: Requires elementary analysis of physical situations, based on the subject material in the chapter.
Class c: Requires somewhat more mature analysis.
Class d: Requires computer solution.

3.1^{b}. Categorization of waste gases for incineration

Categorize the following waste gas streams for incineration, according to the scheme of Figure 3.1

(a) 1.5% acetone, 6% oxygen, and 92.5% nitrogen (the lower heating value of acetone is 29.12 MJ/kg)
(b) 5,200 ppm cyclohexane, 10% oxygen at 300 K and 1 atm. To avoid having to use LEL monitors and controls, use dilution air, if necessary, available at 300 K and 1 atm.

3.2^{b}. A method to estimate LELs for hydrocarbons

The LEL has been measured for many substances and is available in the literature. Occasionally, it is necessary to estimate the LEL for a gas. One simple method is based on the observation that, for many hydrocarbons, the LEL is a fixed fraction of the concentration required for stoichiometric combustion of the gas in air, y_{st}.

a) Using the data in Table 3.1 for ethane, toluene, and cyclohexane, develop a rule of thumb for estimating LELs for hydrocarbons in terms of y_{st}.

Answer: LEL = 0.54 y_{st}

(b) Estimate the LEL for n-pentane, and compare it to the observed value of 1.5%.

3.3b. Consecutive, first-order, irreversible reactions

Consider a set of consecutive, first-order, irreversible reactions as represented by Eq. (3.12) of the text. The reaction rate constants are: $k_1 = 1\ s^{-1}$, $k_2 = 5\ s^{-1}$.

(a) Determine at what time the concentration of the intermediate product reaches its maximum value.

Answer: 0.40 s

(b) Estimate the maximum concentration of the intermediate product, as a fraction of the original concentration of reactant A.

Answer: 13.4%

(c) Calculate the dimensionless concentration of the intermediate product when the concentration of reactant A has dropped to 0.1% of its original value.

Answer: 0.025%

3.4a. Temperature required for thermal destruction of methane

Estimate the temperature required in an isothermal plug flow thermal incinerator with a residence time of 0.5 s to give 99% destruction of methane. Assume an oxygen mole fraction of 0.15 and a pressure of 1 atm.

Answer: 976 K

3.5c. CO formation during thermal incineration

A kinetic model has been proposed (Dryer, F.L., and Glassman, I. *High-Temperature Oxidation of CO and CH_4,* 14th International Symposium on Combustion, The Combustion Institute, Pittsburgh, PA, 1973) for the oxidation of carbon monoxide of the form:

$$r_{CO} = 1.3 \times 10^{10} \exp\left(-\frac{20{,}130}{T}\right)[CO][H_2O]^{0.5}[O_2]^{0.25}$$

where [ ] is concentration in mol/m³, and r_{CO} is the rate of oxidation, in mol/m³-s. Such an equation can be combined with the VOC kinetic model presented in the text (with CO as an intermediate product) to build a complete global model for the processes occurring in a thermal incinerator.

Consider the incineration of methane as described in Problem 3.4. Estimate the concentration of CO at the incinerator outlet. Assume that the molar fractions of oxygen and water remain relatively constant during the combustion at 0.15 and 0.05, respectively. The initial concentration of methane is 1,000 ppm.

Answer: 19.7 ppm

3.6a. Maximum recuperative heat-transfer effectiveness

A thermal incinerator used for processing a stream of air polluted with chlorobenzene operates at 1,200 K. The waste gas enters the system at 400 K. Determine the maximum recuperative heat-transfer effectiveness possible.

Answer: 64.1%

3.7a. Residence time required in a thermal incinerator.

A waste stream containing 2,000 ppm ethylene and 10% oxygen will be incinerated at 1,000 K and 1 atm. If the outlet ethylene concentration is not to exceed 10 ppm, calculate the residence time required.

Answer: 0.74 s

3.8a. Enthalpy change of combustion gases

From the data on Table 3.3, calculate the heat liberated when 1 kmol of CO_2 is cooled from 1,200 K to 500 K at a constant pressure of 1 atm.

Answer: 35.79 MJ

3.9c. Material and energy balances

The gaseous effluent from a printing press flows at the rate of 20 m^3/s, at 350 K and 1 atm. Its composition is: 3,000 ppm n-hexane, 20% O_2, and 79.7% N_2.

(a) Calculate the temperature required to reduce the hexane concentration to 10 ppm in a thermal incinerator with a residence time of 0.5 s. Assume an average oxygen concentration of 15%

Answer: 992 K

(b) Calculate—through combined material and energy balances—the auxiliary fuel (CH_4 at 298 K) required to maintain the operation temperature using a recuperative heat exchanger with an effectiveness of 20% . Assume that the heat losses are 10 percent of the standard heat liberated during combustion.

3.10b. Dimensions of a thermal incinerator

(a) Calculate the volumetric flow rate and the composition of the flue gas in Problem 3.9.

Answer: 57.7 m^3/s, 2.5% CO_2

(b) If the velocity of the flue gas is 10 m/s, calculate the dimensions of the combustion chamber.

Answer: $D = 2.7$ m

(c) Show that the flow through the combustion chamber is fully turbulent.

3.11a. Total capital investment for a thermal incinerator

Estimate the total capital investment required for the incinerator of Problem 3.9. Assume no special buildings or site preparation are required. The duct work and other auxiliary equipment add $25,000 to the equipment cost.

Answer: TCI = $412,260 (June 1990)

3.12c. Catalytic incinerator design

Consider catalytic incineration of the waste gas of Problem 3.9. It will be preheated to 550 K in a chamber fitted with a methane-fired burner. The methane is available at 298 K. Heat losses in the preheater are approximately 5% of the standard heat of combustion. Although the preheater outlet temperature is higher than the auto ignition temperature of the hexane, assume no oxidation of the VOC takes place. The gases will then flow through a catalytic bed where the VOC concentration will be reduced to 10 ppm. The catalyst is a precious metal; mass transfer is the rate limiting step. Its support consists of circular channels, 2 mm in diameter, with a fractional open area of 70%. The average temperature of the bed will be 650 K. The velocity of the gases, based on total cross-sectional area, should not exceed 10 m/s.

(a) Calculate the mass flow rate of methane required.

Answer: 5.4 kg/min

(b) Calculate the dimensions of the catalytic bed.

Answer: L = *1.12 m;* D = *2.18 m*

3.13[a]. Pressure drop through catalytic incinerator

(a) Estimate the pressure drop through the catalytic incinerator of Problem 3.12.

Answer: 4.07 kPa

(b) Assume that the pressure drop through the preheater and duct work is 1.0 kPa. A blower, located at the preheater entrance and with a mechanical efficiency of 60%, forces the gases through the system. Calculate the power to operate the blower.

Answer: 169 kW

3.14[a]. Catalytic incinerator equipment cost

Estimate the equipment cost for the catalytic incinerator of Problem 3.12. The cost of the preheater burner is included in the cost data for catalytic incinerators given in the text.

Answer: $319,280 (June 1990)

3.15[b]. Total annual cost for an RTO

Refer to Example 3.17. Consider as a third alternative the use of an RTO with a thermal effectiveness of 85%. Assume that no auxiliary fuel is required. Calculate the TAC for this alternative (the useful life for RTOs is 20 yr) and compare it to the other two alternatives, based on the EUAC measure of merit.

Answer: TAC = $240,680

3.16[c]. Incineration design problem

A chemical process emits 25 m^3/s at 310 K and 1 atm, containing a 50–50 mixture of hexane and heptane in air. The total VOC concentration is 2,500 ppm. Recent regulations require that solvent emissions from this plant be reduced by 95%. Design an incineration system for this purpose. Specify type of incinerator (thermal, catalytic, or RTO), heat-recovery effectiveness, dimensions (except for RTOs), total capital investment required, and total annual cost. The following data apply:

8,280 h/yr
Utilities costs
Fuel (methane), \$5.00/GJ
Electricity, \$0.08/kW-h
Labor rates
Operating, \$15.00/h
Maintenance, \$20.00/h
Minimum attractive rate of return, 20%/yr
No working capital, land, site preparation, or buildings required.

Give your results in dollars of June 1990. Justify your design selection based on the EUAC method.

3.17b. Flare design

Flaring is a VOC combustion process in which the VOCs are piped to a remote, usually elevated, location and burned in an open flame in the open air using a specially designed burner tip, auxiliary fuel, and steam or air to promote mixing for nearly complete destruction (Stone, D. K., et al. *J. Air Waste Manage. Assoc.*, **42**:333, 1992).

Steam-assisted flares are single burner tips, elevated above ground, that burn the vented gases in essentially a diffusion flame. This type of flare system injects steam into the combustion zone to promote turbulence and to induce air into the flame. They account for the majority of the flares installed. The EPA design and operation requirements for steam-assisted elevated flares are summarized in Table 3.9. It is standard practice to size the flare tip so that the design velocity for the total gas flow rate—including auxiliary fuel, if needed—is 80% of u_{max}. The flare tip diameter, D, is calculated and rounded to the next commercially available size. The minimum tip diameter is 25 mm; larger sizes are available in 50-mm increments from 50 to 600 mm, and in 150-mm increments above 600 mm up to a maximum of 2,300 mm.

Table 3.9 EPA Parameters for Steam-Assisted Flares

Net Heating Value of Waste Stream, H_v (MJ/scm)	Maximum Velocity at Flare Tip, u_{max} (m/s)
11.2[a]	18.3
11.2–37.2	$\ln (u_{max}) = 0.0727H_v + 2.0927$
>37.2	122

[a]For $H_v < 11.2$ MJ/scm, auxiliary fuel is required to raise the net heating value of the mixture to, at least, 11.2 MJ/scm.
Source: From Stone et al. (1992).

The height of a flare is usually determined based on the ground level limitations of thermal radiation intensity. If operating personnel are required to remain in the unit area performing their duties, the recommended design flare radiation level (excluding solar radiation) is 1.58 kW/m^2. The minimum flare height usually is 10 m. The following equation may be used to determine the minimum flare height, L, where thermal radiation must be limited.

$$L = \sqrt{\frac{\tau f R}{4\pi K}}$$

where:

τ = fraction of heat intensity transmitted (assume 1.0)

f = fraction of heat radiated (design value of 0.2)

R = net heat release, kW

K = allowable radiation intensity (1.58 kW/m^2)

A municipal landfill generates 2.0 m^3/s of a waste gas (86% CH_4, 4% N_2, and 10% CO_2) at 298 K and 101.3 kPa. Design a steam-assisted flare to combust this waste gas.

Answer: D = *250 mm*

3.18[a]. Capital costs of flares

Flare equipment costs are a function of stack height, tip diameter, and support type as follows (Stone, D. K., et al. *J. Air Waste Manage. Assoc.*, **42**:488, 1992):

Self Support ($L < 30$ m)

$$EC = (78.0 + 0.36D + 2.46L)^2$$

Guy Support (30 m $< L <$ 60 m)

$$EC = (103 + 0.341D + 1.54L)^2$$

Derrick Support ($L > 60$ m)

$$EC = (76.4 + 0.107D + 5.38L)^2$$

where:

EC = equipment cost, \$ (March 1990)

L = flare height, m (10 m minimum)

D = tip diameter, mm

Auxiliary equipment—including a knock-out drum to remove any liquids that may be present in the waste gas, and piping—contribute an additional 10% to the equipment cost. When instrumentation, sales tax, and freight are added, the purchased equipment cost, $B = 1.28EC$. Furthermore, TCI = $1.92B$.

Estimate the TCI and CRC for the flare of Problem 3.17. Assume that the useful life of the system is 15 yr, and a minimum attractive rate of return of 12%.

Answer: CRC = \$18,900/yr

4

Adsorption Devices

4.1 INTRODUCTION

Adsorption is a separation process based on the ability of certain solids to remove gaseous components preferentially from a flow stream. In air pollution control applications, the pollutant gas or vapor molecules present in a waste gas stream collect on the surface of the solid material. Adsorption is useful in removing objectionable odors and pollutants from industrial gases as well as recovering valuable solvent vapors from air and other gases. It is a particularly useful technique when the gaseous emissions present in the waste gas are valuable enough to recover for recycling or resale.

Before reviewing the basic theory and design of adsorption systems, we will describe the components and operation of a typical fixed-bed adsorber system as shown on Figure 4.1. The system shown in that figure uses two vessels in which beds of adsorbent are located. The waste gas from a process enters the main blower and passes through a cooler. The reason for cooling is that the VOC-adsorbing capacity of the solid increases as the temperature decreases. The cooled gas stream passes through one of the adsorbent beds, where most of the VOC is removed. The "clean air" is either vented to the atmosphere or returned to the source process. Eventually, a substantial portion of the bed becomes saturated with the VOC, and the pollutant starts to "breakthrough" in the effluent. When that happens, the waste gas stream is switched to the idle bed, which has been regenerated and cooled.

The expended bed is then regenerated by direct contact with steam. The adsorbed VOC is displaced from the adsorbent, and the mixture of steam and organic vapor is condensed and collected in a decanter for initial separation. If the solubility of the condensed VOC in water is low enough, decanting is sufficient; otherwise an additional separation unit, such as distillation, is required. The VOC is recovered for reuse or sale, and the waste water, usually, must be further purified for reuse or discharge.

The adsorbents used for air pollution control include activated carbon, alumina, bauxite, and silica gel. Activated carbon is, by far, the most frequently used adsorbent, and has virtually displaced all other materials in solvent recovery systems.

The term activated as applied to adsorbent materials refers to the increased internal and external surface area imparted by special treatment processes. Any carbonaceous material can be converted to activated carbon. Coconut shells, bones, wood, coal, petroleum coke, lignin, and lignite all serve as raw materials for activated carbon. However, most industrial grade carbon is made from bituminous coal (Cooper and Alley 1986).

Activated carbon is manufactured by first dehydrating and carbonizing the raw material. Activation is completed during a controlled oxidation step in which the carbonized material is heated in the presence of an oxidizing gas. For certain carbons the dehydration can be accomplished by using chemical agents. The ideal raw material has a porous structure that provides a uniform pore distribution and high adsorptive capacity when activated. Activated carbon is tailored for a specific end use by both raw material selection and control of the activation process. Carbons for gas-phase applications have a specific surface area in the range of 800 to 1,200 m^2/g, and a porosity in the range of 35% to 40%. Most of the pore volume is distributed over a narrow range of pore diameters, usually ranging from 0.4 to 3.0 nm. Bulk density is of the order of 500 kg/m^3.

To minimize pressure drop in fixed beds, granular or pelletized carbon is used. Typically, the particle size of granular carbon is about 1.0 mm, which corresponds to a screen size between 10 and 20 mesh on the U.S. sieve scale (see Table 4.1).

Table 4.1 U.S. Sieve Series; Sieve Opening

Mesh	3	6	10	12	16	20	32
Opening (mm)	6.73	3.36	1.68	1.41	1.00	0.84	0.50

4.2 ADSORPTION EQUILIBRIA

Your objectives in studying this section are to

1. Understand the nature of physical, or van der Waals, adsorption.
2. Calculate equilibrium adsorption concentrations using Langmuir or Freundlich adsorption isotherms.

In many respects the equilibrium adsorption characteristics of a vapor or gas on a solid resemble the equilibrium solubility of a gas in a liquid (Treybal 1980). Figure 4.2 shows several equilibrium adsorption isotherms for a particular activated carbon as adsorbent. The concentration of adsorbed gas (adsorbate) on the solid is plotted against the equilibrium partial pressure, p^*, of the vapor or gas at constant temperature. Examples of such isotherms are shown on Figure 4.2. At 373 K, for example, pure acetone vapor at a partial pressure of 25.33 kPa is in equilibrium with an adsorbate concentration of 0.2 kg adsorbed acetone/kg carbon. Increasing the partial pressure of acetone will cause more to be adsorbed, and decreasing it will cause acetone to desorb from the carbon.

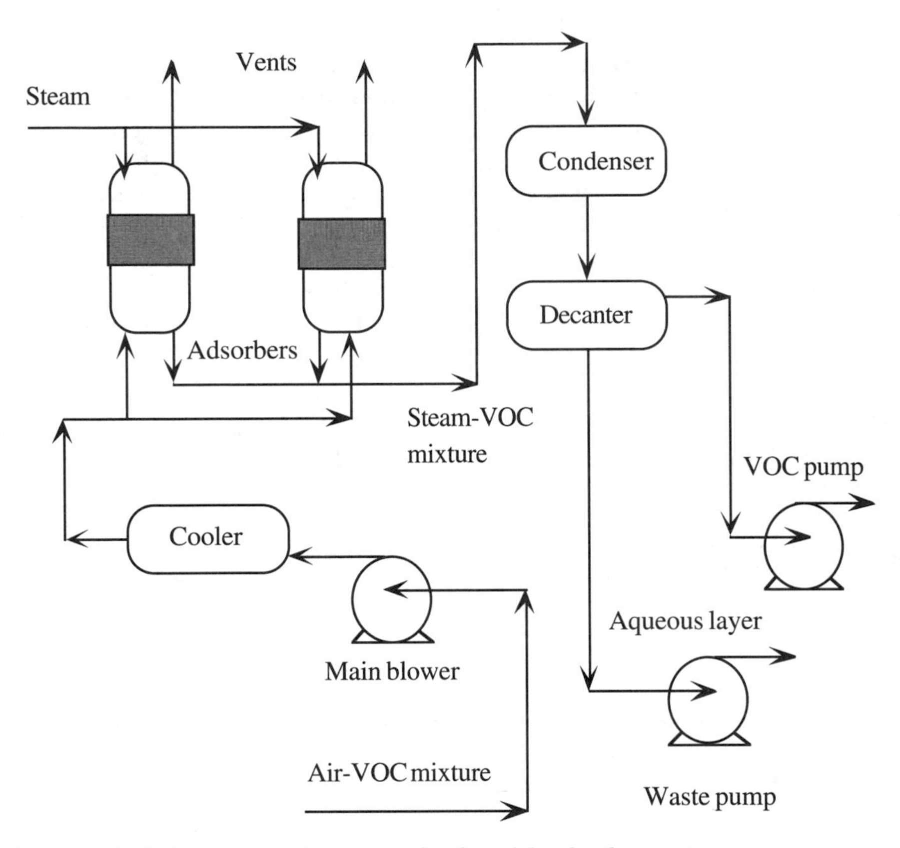

Figure 4.1. Schematic diagram of a fixed-bed adsorption system

Because adsorption is an exothermic process, the concentration of adsorbed gas decreases with increased temperature at a given equilibrium pressure, as the several acetone isotherms of Figure 4.2 illustrate. Different gases and vapors are adsorbed to different extents under comparable conditions. Figure 4.2 shows that benzene is more readily adsorbed than acetone at the same temperature and gives a higher adsorbate concentration for a given equilibrium partial pressure. As a general rule, gases and vapors are more readily adsorbed the higher their molecular weight.The adsorption equilibria is highly dependent on the solid used as adsorbent. For example, the equilibrium curves for acetone and benzene on silica gel would be entirely different from those of Figure 4.2. Even differences in the origin and method of preparation of a given adsorbent will result in significant differences in the equilibrium adsorption.

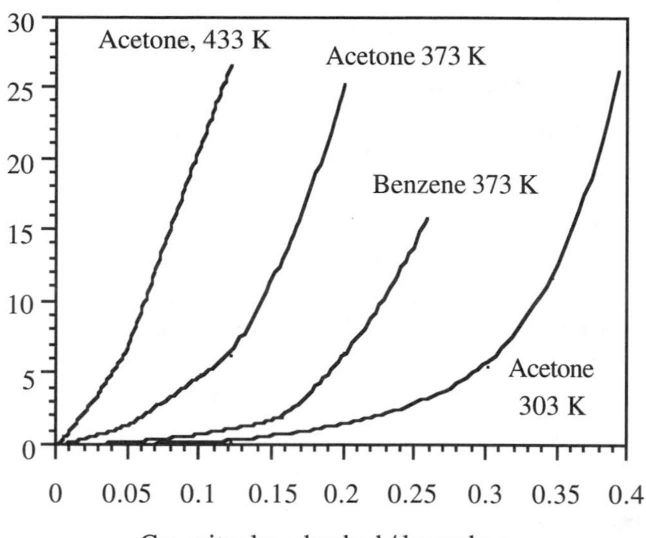

Figure 4.2 Equilibrium adsorption of vapors on activated carbon (From Treybal, R. E. *Mass-Transfer Operations*, 3rd ed. McGraw-Hill, New York, 1980; reprinted with permission).

There are two distinct adsorption mechanisms: physical adsorption and chemisorption. Physical adsorption, also referred to as van der Waals adsorption, involves a weak bonding of gas molecules to the solid. The bond energy is similar to the attraction forces between molecules in a liquid. The adsorption process is exothermic, and the heat of adsorption is slightly higher than the heat of vaporization of the adsorbed material. The forces holding the gas molecules to the solid are easily overcome by either the application of heat or the reduction of pressure. Most of adsorption applications to air pollution control involve physical adsorption.

Chemisorption involves an actual chemical bonding by reaction of the adsorbate with the adsorbing solid. It is virtually irreversible. Except in some very specialized applications, recovery of a substance through chemisorption is not feasible.

One of the most useful mathematical models to describe adsorption equilibria is the Langmuir isotherm. Its theoretical development is based on the following assumptions: (1) the adsorbed phase is a unimolecular layer, and (2) at equilibrium, the rate of adsorption is equal to the rate of desorption from the surface. Define f as the fraction of the total solid surface occupied by adsorbate molecules. The rate of adsorption, r_a, is proportional to the partial pressure of the adsorbate, p, and to the fraction of the solid surface area available for adsorption, $(1 - f)$. Therefore,

$$r_a = C_a p(1 - f) \tag{4.1}$$

where C_a is a constant. Conversely, the rate of desorption, r_d, is proportional to the fraction of the surface area occupied by the adsorbate:

$$r_d = C_d f \tag{4.2}$$

where C_d is a constant. At equilibrium, the rate of adsorption is equal to the rate of desorption. The fraction of the surface covered is, then, given by

$$f = \frac{C_a p^*}{C_d + C_a p^*} \tag{4.3}$$

Because the adsorbed phase is a unimolecular layer, the mass of adsorbate per unit mass of adsorbent, m, is also proportional to the surface covered:

$$m = C_m f \tag{4.4}$$

where C_m is a constant. Combining Eqs. (4.3) and (4.4),

$$m = \frac{k_1 p^*}{k_2 p^* + 1} \tag{4.5}$$

where $k_1 = C_a C_m / C_d$, and $k_2 = C_a / C_d$. Equation (4.5) is known as Langmuir isotherm. At very low adsorbate equilibrium partial pressure $k_2 p^*$ is approximately equal to zero, and Eq. (4.5) becomes

$$m = k_1 p^* \tag{4.6}$$

Conversely, at high equilibrium partial pressure,

$$m = \frac{k_1}{k_2} \tag{4.7}$$

Hence, over an intermediate range of partial pressures:

$$m = k(p^*)^n \tag{4.8}$$

where:

k = constant

n = constant with a value between 0 and 1

Equation (4.8) is known as Freundlich isotherm. Even though it is not based on a rigorous theoretical background, the Freundlich isotherm gives an adequate description of adsorption equilibrium in many air pollution control applications. The values of k and n are obtained from experimental data. Table 4.2 gives values of the Freundlich isotherm parameters for some adsorbates on Calgon type "BPL" activated carbon (4 × 10 mesh). Note that these isotherms may not be extrapolated outside of the partial pressure ranges shown. Data for other compounds, adsorbents, partial pressures, and temperatures are available from adsorbent vendors and the literature (e.g., *Handbook of Chemistry and Physics*).

Table 4.2 Freundlich Isotherm Parameters for Some Adsorbates

Adsorbate	Temperature(K)	$k \times 100$	n	Partial Pressure (Pa)
Acetone	311	1.324	0.389	0.69–345
Acrylonitrile	311	2.205	0.424	0.69–103
Benzene	298	12.602	0.176	0.69–345
Chlorobenzene	298	19.934	0.188	0.69–69
Cyclohexane	311	7.940	0.210	0.69–345
Dichloroethane	298	8.145	0.281	0.69–276
Phenol	313	22.116	0.153	0.69–207
Toluene	298	20.842	0.110	0.69–345
Trichloroethane	298	25.547	0.161	0.69–276
m-Xylene	298	26.080	0.113	0.69–6.9
m-Xylene	298	28.313	0.0703	6.9–345

Source: *EAB Control Cost Manual*, 3rd. ed. U.S. Environmental Protection Agency, Research Triangle Park, NC, 1987.

The amount adsorbed is expressed in kg adsorbate/kg adsorbent.

The equilibrium partial pressure is expressed in Pa.

Data are for the adsorption on Calgon type "BPL" activated carbon (4 × 10 mesh).

Should not be extrapolated outside of the partial pressure ranges shown.

Example 4.1 Equilibrium Adsorption on Activated Carbon.

Calculate the equilibrium adsorptivity of toluene (in air) on Calgon "BPL" activated carbon (4 × 10 mesh). The temperature is 298 K and the total pressure is 101.3 kPa. The equilibrium concentration of toluene is 1,000 ppm.

Solution

Calculate the equilibrium partial pressure of toluene: $p^* = 1{,}000\,(101.3)/106 = 0.1013$ kPa = 101.3 Pa. According to Table 4.2, Freundlich isotherm applies for this partial pressure. Substituting the appropriate parameters from that table in Eq. (4.8),

$$m = 0.20842(101.3)^{0.110} = 0.346\ \frac{\text{kg toluene}}{\text{kg carbon}}$$

Example 4.2 Adsorption Equilibrium; Material Balances

A waste gas contains 0.3% toluene in air at 298 K and 101.3 kPa. One hundred kilomoles of this mixture are put in a closed container with 82 kg of Calgon "BPL" activated carbon. Originally, there is no toluene adsorbed on the carbon. The system is allowed to reach equilibrium at constant temperature and pressure. Calculate the equilibrium concentration of toluene in the gaseous phase, and the amount adsorbed by the carbon. Assume that the air does not adsorb on the carbon.

Solution

Basis: 100 kmol of initial gaseous mixture

Let x = kmol of toluene in the gaseous phase in equilibrium. Then, the mass of toluene adsorbed is $(0.3 - x)(92)$ kg. The equilibrium adsorptivity is

$$m = 92(0.3 - x)/82 = 1.122(0.3 - x)\ \text{kg toluene/kg carbon}$$

Conversely,

$$p^* = 101{,}300x\,/(99.7 + x)\ \text{Pa}.$$

These two quantities are related through the Freundlich isotherm.

$$1.122(0.3 - x) = 0.20842\left[\frac{101{,}300}{99.7 + x}\right]^{0.11}$$

This equation is solved iteratively for x. The answer is $x = 0.03$ kmol. Therefore, the equilibrium toluene concentration in the gaseous phase is $0.03/(99.7 + 0.03) = 0.0003 = 300$ ppm. The mass of toluene adsorbed by the carbon is $(0.3 - 0.03)(92) = 24.84$ kg.

4.3 DYNAMICS OF FIXED-BED ADSORPTION

Your objectives in studying this section are to

1. Generate the breakthrough curve for a fixed-bed adsorption system.
2. Estimate the time required to reach the breakpoint in a fixed-bed adsorption system.
3. Estimate the mass of adsorbent required for a given adsorption application.

Steady-state adsorption requires continuous movement of both fluid and adsorbent through the equipment at constant rate, with no change in composition at any point in the system with passage of time. The inconvenience and relatively high cost of continuously transporting solid particles as required for steady-state operation make it more economical to pass the fluid mixture to be treated through a stationary bed of adsorbent. As increasing amounts of fluid pass through such a fixed-bed, the solid adsorbs increasing amounts of adsorbate, and an unsteady state prevails. The dynamic behavior of such an operation is the subject of this section.

Consider a binary gaseous mixture with an adsorbate concentration of C_0. The gas passes continuously down through a relatively deep fixed-bed of adsorbent initially free of adsorbate. The uppermost layer of solid at first adsorbs rapidly and effectively, and what little adsorbate is left in the solution is substantially all removed by subsequent layers of solid in the lower part of the bed. The bulk of the adsorption occurs over a relatively narrow adsorption zone in which the concentration changes rapidly. The effluent from the bottom of the bed is practically adsorbate free. As the gas continues to flow, the uppermost layer of the bed becomes saturated, and the adsorption zone moves downward as a wave at a rate ordinarily much slower than the linear velocity of the fluid through the packed bed. When the lower portion of the adsorption zone reaches the bottom of the bed, the concentration of adsorbate suddenly rises to an appreciable value for the first time. The system is said to have reached the breakpoint. The adsorbate concentration in the effluent now rises rapidly as the adsorption zone passes through the bottom of the bed. Soon the bed is completely exhausted and the outlet composition is exactly equal to the inlet composition. The portion of the effluent concentration curve between the breakpoint and exhaustion is termed the breakthrough curve. Figure 4.3 shows a typical breakthrough curve.

The shape and time of appearance of the breakthrough curve greatly influence the design and the method of operating a fixed-bed adsorber. The curves generally have an S shape, but they may be steep or relatively flat. The breakpoint is very sharply defined in some cases and in others poorly defined. The rate of the adsorption process, the nature of the adsorption equilibrium, the fluid velocity, the feed concentration, and the length of the bed all contribute to the shape of the curve for a given system.

Consider the idealized breakthrough curve shown in Figure 4.3 resulting from flow of an inert gas through an adsorbent bed with a rate G' kg/m2-s containing an inlet solute concentration of Y_0 kg solute/kg inert gas. The total amount of solute-free gas that has passed through the bed up to any time is w kg/m2 of bed cross section. The gas breakpoint and exhaustion concentrations are denoted by Y_B and Y_E, respectively. The total amount of solute-free gas that has passed through the bed at the breakpoint is w_B, at exhaustion it is w_E. The adsorption zone, taken to be of constant height z_a, is that part of the bed in which the concentration profile from Y_B to Y_E exists at any time.

If θ_a and θ_E are the times required for the adsorption zone to move its own length and down the entire bed, respectively, then,

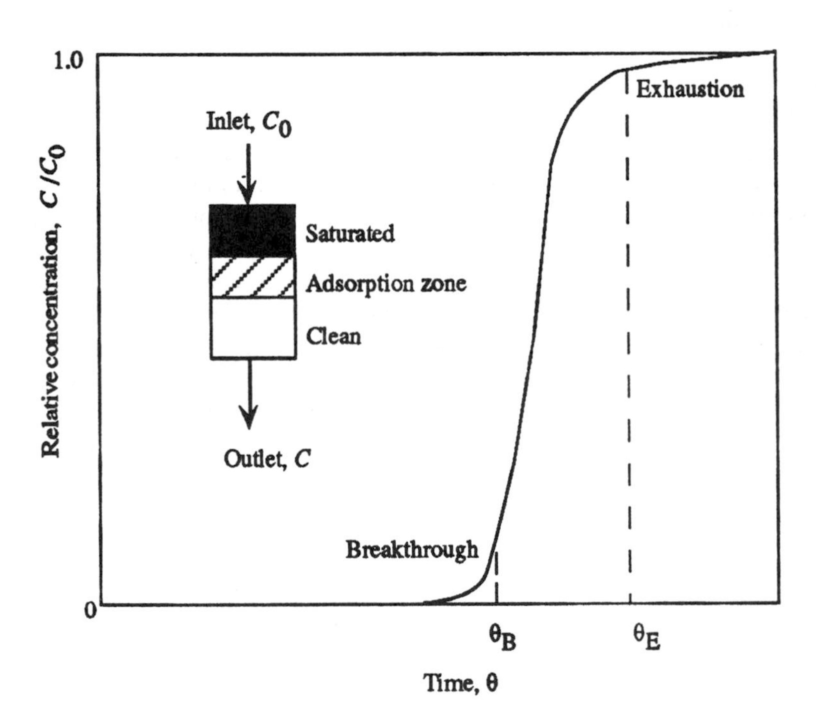

Figure 4.3. Typical breakthrough curve for adsorption of a gas on a solid

$$\theta_a = \frac{w_a}{G'} = \frac{w_E - w_B}{G'} \tag{4.9}$$

and

$$\theta_E = \frac{w_E}{G'} \tag{4.10}$$

Some time is required to form the adsorption zone at the beginning of the bed when the gas is first introduced. If θ_F is the time required for the adsorption zone to form, then it follows that $\theta_E - \theta_F$ is the time available for the zone to move through the bed once it is formed. In this interval, the adsorption zone travels a distance z_a every θ_a of elapsed time. Then, if Z is the length of the bed, and the shape of the adsorption zone does not change after it is established,

$$z_a = Z\frac{\theta_a}{\theta_E - \theta_F} \tag{4.11}$$

It is now necessary to find an expression for the time required to form the adsorption zone for use in the preceding equation.

The last part of the adsorption occurs as the zone moves out of the bed. The solute removed from the gas in the adsorption zone is U kg/m^2 of bed cross section. It is given by

$$U = \int_{w_B}^{w_E} (Y_0 - Y)\,dw \tag{4.12}$$

If all the adsorbent in the zone were saturated, the solid in the zone would contain $Y_0 w_a$ kg of adsorbate/m^2. Consequently, at the breakpoint, when the zone is still inside the column, the fractional ability of the zone to still adsorb solute is

$$\phi = \frac{U}{Y_0 w_a} = \frac{\int_{w_B}^{w_E} (Y_0 - Y)\,dw}{Y_0 w_a} = \int_0^1 \left(1 - \frac{Y}{Y_0}\right) d\frac{w - w_B}{w_a} \tag{4.13}$$

The formation of the adsorption zone at the beginning of the bed may be assumed to follow the same sort of pattern as has just been described for the departure of the zone; therefore Eq. (4.13) can be used to describe the formation of the adsorption zone.

The fraction ϕ is obviously some function of the shape and slope of the breakthrough curve, and it normally approaches a value of 50%. It is useful in establishing a relation between the time required to establish the adsorption zone and that required for it to advance through the bed a distance equal to its thickness. If $\phi = 0$, so that the adsorbent in the zone is essentially saturated, θ_F at the top of the bed should be substantially the same as θ_a. Conversely, if $\phi = 1.0$, so that the solid in the zone contains essentially no adsorbate, the zone-formation time should be very short, essentially zero. These limiting conditions, at least, are described by:

$$\theta_F = (1-\phi)\theta_a \tag{4.14}$$

Substituting Eq. (4.14) into Eq.(4.11),

$$z_a = Z\frac{\theta_a}{\theta_E - (1-\phi)\theta_a} = Z\frac{w_a}{w_E - (1-\phi)w_a} \tag{4.15}$$

If the column contains $ZA_c\rho_s$ kg of adsorbent, where A_c is the cross-sectional area of the bed and ρ_s is the apparent solid density in the bed, at complete saturation the bed would contain $ZA_c\rho_s m^*$ kg of adsorbate, where m^* is the adsorbate concentration on the solid in equilibrium with the gaseous feed. At the breakpoint, $Z - z_a$ of the bed is saturated, and z_a of the bed is saturated to the extent of $1 - \phi$. The degree of overall bed saturation at the breakpoint, α, is thus:

$$\alpha = \frac{(Z - z_a)\rho_s m^* A_c + z_a \rho_s (1-\phi) m^* A_c}{Z\rho_s m^* A_c} = \frac{Z - \phi z_a}{Z} \tag{4.16}$$

In the fixed bed, the adsorption zone really moves downward through the solid, as we have seen. Imagine, instead, that the solid moves upward through the column countercurrent to the fluid fast enough for the adsorption zone to remain stationary within the column as in Figure (4.4a). Here, the solid leaving at the top of the column shows in equilibrium with the entering gas, and all the adsorbate is removed from the effluent gas. As shown in Figure (4.4b), the operating line for the entire adsorber passes through the origin, and intersects the equilibrium curve at the point (Y_0, m^*).

Over the differential height dz in the adsorption zone, the rate of adsorption is given by

$$G' \, dY = K_Y a (Y - Y^*) dz \tag{4.17}$$

where K_Y is the overall mass-transfer coefficient for transfer from gas to solid phase. For the adsorption zone, therefore,

$$z_a = \frac{G'}{K_Y a} \int_{Y_B}^{Y_E} \frac{dY}{Y - Y^*} = H_{tOG} N_{tOG} \tag{4.18}$$

For any value of z less than z_a, but within the adsorption zone,

$$\frac{z}{z_a} = \frac{\int_{Y_B}^{Y} \frac{dY}{Y - Y^*}}{\int_{Y_B}^{Y_E} \frac{dY}{Y - Y^*}} = \frac{w - w_B}{w_a} \tag{4.19}$$

The breakthrough curve can be plotted from Eqs. (4.18) and (4.19).

The effective rate of adsorption is determined by one or more of several diffusional steps. Individual steps in the transport mechanism follow.

1. Diffusion in the sorbed state (in a uniform liquidlike or solid phase or a pore-surface layer). This is designated as particle-phase diffusion.
2. Reaction at the phase boundary, usually very fast.
3. Pore diffusion in the fluid phase, within the particles.
4. Mass transfer from the fluid phase to the external surfaces of the adsorbent particles.
5. Mixing, or lack of it, between different parts of the contacting equipment. For instance, in column operation with slow flow rates, the breakthrough curve may be broadened by axial dispersion.

For most air pollution control applications, steps 3 and 4 are the rate limiting steps. The following equation (Vermeulen et al. 1973) can be used to evaluate the volumetric mass-transfer

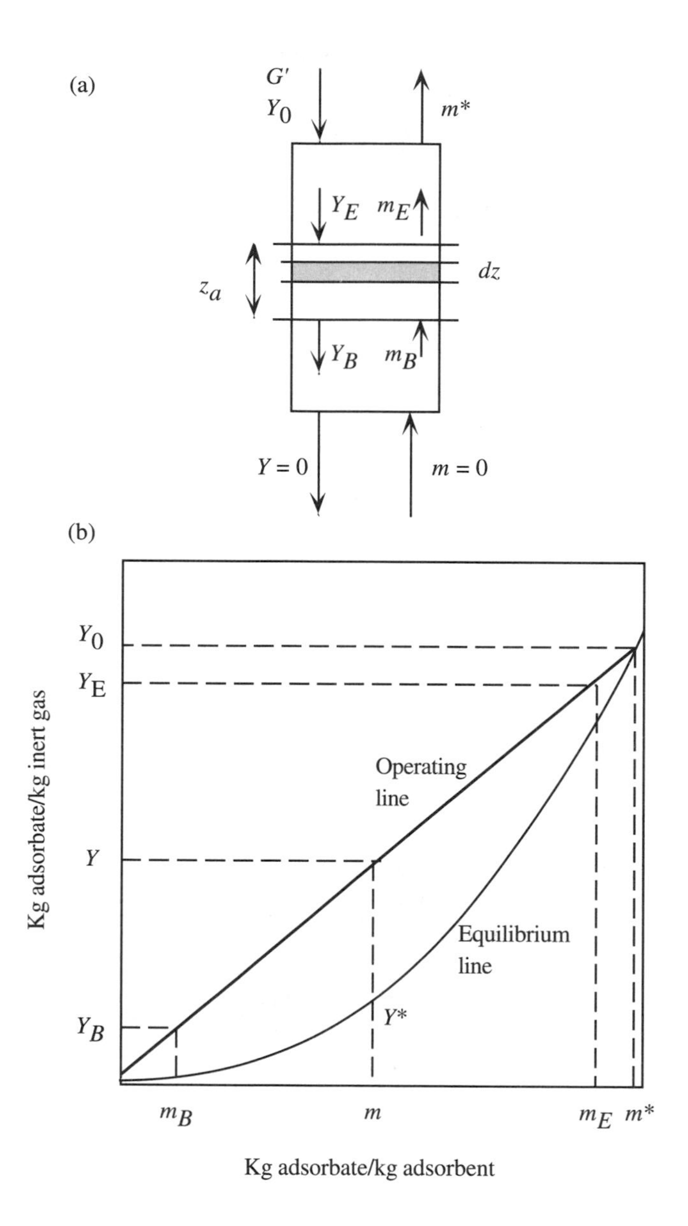

Figure 4.4 Adsorption zone

coefficient corresponding to step 4 (for laminar-flow gas-solid contact)

$$k_Y a = \frac{10.9Q(1-\varepsilon)p_{B,M}M_B}{d_p A_c RT}\left[\frac{D_f A_c}{d_p Q}\right]^{0.51}\left[\frac{D_f \rho_f}{\mu}\right]^{0.16} \tag{4.20}$$

where Q is the volumetric flow rate of gas, d_p is the average particle diameter, ε is the bed porosity, $p_{B,M}$ is the logarithmic mean partial pressure of the inert gas (approximately equal to the total pressure for most air pollution control applications), M_B is the molecular weight of the inert gas, and ρ_f and D_f are the fluid density and difussivity respectively. This equation is valid for laminar flow, which occurs if

$$\text{Re}_p = \frac{Q d_p \rho_f}{A_c \mu} \le 10 \tag{4.21}$$

For turbulent-flow, gas-solid contact, Treybal (1980) suggests the following correlation, valid for Re_p = 90-4,000:

$$k_Y a = \frac{12.36Q(1-\varepsilon)p_{B,M}M_B}{\varepsilon d_p A_c RT}\left[\frac{D_f A_c}{d_p Q}\right]^{0.575}\left[\frac{D_f \rho_f}{\mu}\right]^{0.092} \tag{4.22}$$

Vermeulen et. al. (1973) recommend estimating the volumetric mass-transfer coefficient corresponding to step 3 using:

$$k_S a = \frac{60 D_p \rho_s}{d_p^2} \tag{4.23}$$

where D_p is the diffusion coefficient inside the particle. From work on diffusion inside catalyst particles, the diffusion coefficient in the gas-filled pores inside the particle can be estimated from (Satterfield and Sherwood 1963)

$$\frac{1}{D_p} = \frac{\tau}{\chi}\left[\frac{1}{D_K} + \frac{1}{D_f}\right], \qquad D_K = \frac{0.194\chi}{S_g \rho_p}\sqrt{\frac{T}{M}} \tag{4.24}$$

where

D_K = Knudsen diffusion coefficient, m^2/s

τ = tortuosity factor ≈ 4.0

S_g = surface area per gram of solid, m^2/g

ρ_p = particle density, kg/m^3

χ = internal porosity

The overall mass-transfer coefficient is related to the individual coefficients in the usual way (Treybal 1980):

$$\frac{1}{K_Y a} = \frac{1}{k_Y a} + \frac{Y_0}{m^* k_S a} \tag{4.25}$$

The following examples illustrate the use of the previous equations to (1) generate the breakthrough curve for a fixed-bed adsorption system, (2) estimate the time required to reach the breakpoint in a fixed-bed of given dimensions, and (3) estimate the mass of adsorbent required for a given operation time before regeneration.

Example 4.3 Breakthrough Calculations

A waste gas consists of 5% acetone in air at 300 K and 1 atm. It flows at the rate of 2.3 kg/s. The acetone vapors are to be removed by passing the gas mixture downward through a bed of activated carbon with a total cross-sectional area of 5 m^2 and a packed depth of 0.3 m. The porosity of the bed is 40%, the bulk density of the carbon is 630 kg/m^3, and the average particle size is 6 mm. The equilibrium data for this system are as follows:

Y kg acetone/kg air	0	0.0053	0.032	0.063	0.115
m kg acetone/kg carbon	0	0.10	0.20	0.25	0.30

The breakpoint will be considered as that time when the acetone content of the effluent is 5% of the acetone content of the feed. The bed will be considered exhausted when the acetone content of the effluent is 95% of the acetone content of the feed.

(a) Generate the breakthrough curve for these conditions.

(b) Estimate the time required to reach the breakpoint.

Solution

(a) Calculate the feed, breakpoint, and exhaustion gas concentrations: $Y_0 = (0.05)(58)/(.95)(29) = 0.105$ kg acetone/kg air; $Y_B = 0.05Y_0 = 0.0053$; $Y_E = 0.95Y_0 = 0.0998$. Draw the adsorption isotherm as shown in Figure (4.4b). The operating line is drawn through the origin, and intersects the equilibrium curve at $Y_0 = 0.105$, corresponding to a solid equilibrium concentration $m^* = 0.295$ kg acetone/kg solid.

To generate the breakthrough curve, the integrals in Eq. (4.19) must be evaluated numerically. Define:

$$I(Y) = \int_{Y_B}^{Y} \frac{dY}{Y - Y^*}, \qquad Y_B \le Y \le Y_E \tag{4.26}$$

Then,

$$\frac{w - w_B}{w_a} = \frac{I(Y)}{N_{tOG}} \tag{4.27}$$

Table 4.3 presents the results of the numerical integrals . Figure 4.5 shows the shape of the breakthrough curve in dimensionless terms.

(b) To determine the height of the adsorption zone, use Eq. (4.18). From Table 4.3, $N_{tOG} =$ 4.62 units. H_{tOG} must be evaluated. $G' = (2.3)/[5\ (1 + .105)] = 0.416$ kg air/m2-s. $Q =$ $(2.3)(8.314)(300)/\{[(58)(.05) + (29)(.95)](101.3)\} = 1.86$ m3/s. The superficial velocity is $Q/A_c = 1.86/5 = 0.372$ m/s. The density and viscosity of the fluid at 300 K and 1 atm are 1.24 kg/m3 and 1.8×10^{-5} kg/m-s, respectively. The average particle size is 6 mm. Calculate the Reynolds number: $Re_p = (.372)(1.24)(0.006)/(1.8 \times 10^{-5}) = 154$. Therefore, use Eq. (4.22) to estimate the fluid side mass-transfer coefficient. Use the Wilke-Lee equation (Treybal 1980) to estimate the diffusivity of acetone in air at the given temperature and pressure: $D_f = 1.12 \times 10^{-5}$ m2/s. Assume that $p_{B,M} = P$ (total pressure). From Eq. (4.22), $k_Y a$ = 65.8 kg/m3-s.

Use Eq. (4.23) to estimate the mass-transfer coefficient inside the particle. Equation (4.24) must be used first to estimate the corresponding diffusivity. Assume an internal porosity of 60% and a surface area of 1,200 m2/g. The particle density is (630)/(1 – 0.40) = 1,050 kg/m3. The Knudsen diffusion coefficient is $D_K = (0.194)(0.6)(300/58)^{0.5}/(1{,}200)(1{,}050) =$ 0.021×10^{-5} m2/s. Therefore, $D_p = 0.309 \times 10^{-7}$ m2/s. Substituting in Eq. (4.23), $k_S\,a =$ $(60)(0.309 \times 10^{-7})(630)/(6.0 \times 10^{-3})^2 = 32.43$ kg/m3-s.

Table 4.3 Numerical Evaluation of Integrals for Example 4.3

Y	Y^*	$(Y-Y^*)^{-1}$	I	$(w-w_B)/w_a$	Y/Y_0
0.0053	~0.000	188.7	0	0	0.05
0.02	0.002	55.6	1.8	0.3896	0.191
0.03	0.004	38.5	2.27	0.4913	0.2857
0.04	0.006	29.4	2.61	0.5650	0.3810
0.05	0.011	25.6	2.89	0.6255	0.4762
0.06	0.019	24.4	3.14	0.6797	0.5714
0.07	0.032	26.3	3.39	0.7338	0.6667
0.08	0.046	29.4	3.67	0.7944	0.7619
0.09	0.065	40.0	4.02	0.8701	0.8571
0.0998	0.088	84.7	N_{tOG} =4.62	1.0000	0.9500

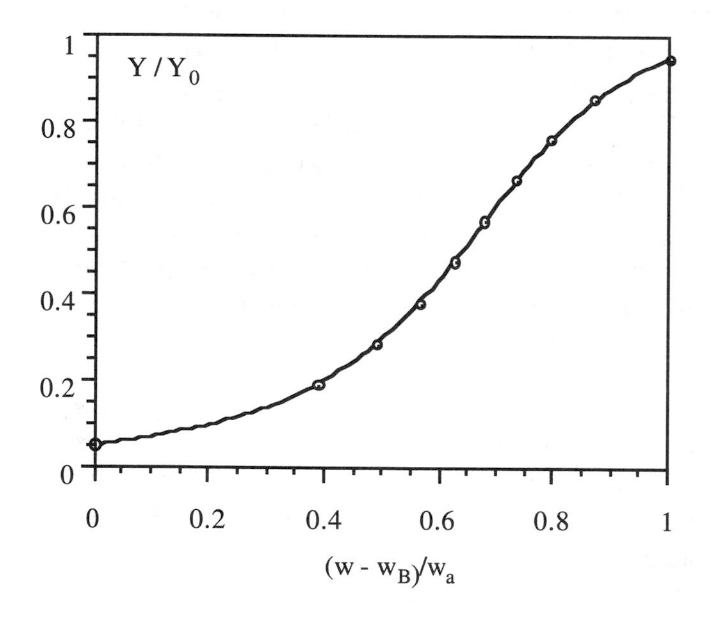

Figure 4.5 Breakthrough curve for Example 4.3

Use Eq. (4.25) to calculate the overall mass-transfer coefficient. This equation neglects any mass-transfer resistance due to the adsorbed layer: $K_Y a = [(65.8)^{-1} + (0.105)/(0.295)(32.43)]^{-1} = 38.2$ kg/m^3-s. Therefore, $H_{tOG} = 0.416/38.2 = 0.0109$ m. The height of the adsorption zone is $z_a = (4.62)(0.0109) = 0.050$ m.

According to Eq. (4.13), $\phi = 1 -$ (area under the curve in Figure 4.5). That area is evaluated numerically. The result is $\phi = 0.602$. Calculate the degree of overall bed saturation at the breakpoint using Eq. (4.16): $\alpha = 0.90$. The total mass of adsorbent in the bed is (5)(0.3)(630) = 945 kg of carbon. At the breakpoint, the amount of acetone adsorbed is (945)(0.295)(0.90) = 251 kg. All that acetone entered the bed with the waste gas during the period of time from the start of operation to the breakpoint. Therefore, the breakpoint is given by $\theta_B = (251)(1.105)/[(0.105)(2.3)] = 1{,}148$ s = 19.1 min.

Example 4.4 Breakthrough Calculations; Freundlich Isotherm

It can be shown (see Problem 4.5) that when a Freundlich isotherm adequately describes the adsorption equilibrium:

$$N_{tOG} = \ln\frac{\eta_E}{\eta_B} + \frac{1}{b-1}\ln\left[\frac{1-\eta_B^{b-1}}{1-\eta_E^{b-1}}\right] \tag{4.28}$$

$$I(Y) = I(\eta) = \ln\frac{\eta}{\eta_B} + \frac{1}{b-1}\ln\left[\frac{1-\eta_B^{b-1}}{1-\eta^{b-1}}\right] \tag{4.29}$$

where $b = n^{-1}$, and $\eta = Y/Y_0$.

The adsorption equilibrium of benzene on a silica gel at 298 K is described by a Freundlich isotherm with parameters $k = 0.00185$ and $n = 0.667$, with the equilibrium partial pressure of the adsorbate in Pa. A waste gas containing 9,290 ppm of benzene in air at 298 K and 2 atm is to stripped of the benzene vapors by passing the gas mixture downward through a bed of the aforementioned silica gel. It is desired to operate for 90 min, at a superficial gas velocity of 0.5 m/s. The waste gas flows at a rate of 1 m^3/s. The breakpoint will be considered as that time when the effluent air contains 150 ppm of benzene, and the bed will be considered exhausted when the effluent air contains 9,000 ppm of benzene. Silica gel has a bulk density of 625 kg/m^3, a porosity of 40%, and an average particle size of 6 mm. Determine the dimensions of the packed bed, and the total amount of adsorbent required.

Assume that the mass-transfer resistance owing to the fluid inside the pores is negligible ($K_Y = k_Y$).

Solution

Calculate the cross-sectional area of the adsorber: $Q/A_c = 0.5$ m/s. Therefore, $A_c = Q/0.5 = 2$ m^2. Calculate the feed, breakpoint, and exhaust concentrations: $Y_0 = 9{,}290(78)/[(10^6)(29)] = 0.025$ kg benzene/kg air (benzene-free basis), $Y_B = 150(78)/[(10^6)(29)] = 0.000403$ kg benzene/kg air, $Y_E = 9{,}000(78)/[(10^6)(29)] = 0.0242$ kg benzene/kg air. Calculate $\eta_B = Y_B/Y_0 = 0.01612$, $\eta_E = Y_E/Y_0 = 0.9683$, $b = 1/n = 1.5$. Substituting in Eq. (4.28), $N_{tOG} = 12.1$ units. The next step is to calculate H_{tOG}. Calculate the fluid density with the ideal gas law: $\rho_f = 2.408$ kg/m^3. The viscosity of air at 298 K is 1.8×10^{-5} kg/m-s; $G' = (1)(2.408)/(2) = 1.204$ kg/m^2-s; $Re_p = G'd_p/\mu = 401$. Use Eq. (4.22) to calculate k_Ya. The diffusivity of benzene in air at 298 K and 2 atm is 0.5×10^{-5} m^2/s; $K_Ya = k_Ya = 144$ kg/m^3-s. Therefore, $H_{tOG} = G'/K_Ya = 0.0084$ m. The height of the adsorption zone is $z_a = (12.1)(0.0084) = 0.101$ m.

To generate the dimensionless breakthrough curve, use Eq. (4.29) as shown in the following table.

Y/Y_0	0.0162	0.10	0.20	0.30	0.40	0.500
$(w-w_B)/w_a$	0	0.1912	0.2837	0.3503	0.4084	0.464
Y/Y_0	0.60	0.70	0.80	0.90	0.95	0.968
$(w-w_B)/w_a$	0.5227	0.5887	0.6719	0.8008	0.9221	1.0

Plot the breakthrough curve and, by numerical integration, obtain $\phi = 0.405$.
If Z is the bed depth, the degree of overall bed saturation at the breakpoint is, from Eq. (4.16), $\alpha = [Z - (.405)(.101)]/Z = (Z - 0.041)/Z$. The total mass of adsorbent in the bed is (2Z)(625) kg. At saturation, the adsorbate concentration on the solid is, from Freundlich's isotherm, $m^* = 0.284$ kg benzene/kg carbon. Therefore, the total amount of benzene adsorbed at breakthrough is $(2Z)(625)(0.284)(Z - 0.041)/Z$. The mass of benzene that must be removed from the air over a 90-min period is $(2.408)(0.025)(60)(90)/(1.025) = 317$ kg. Equating this mass removed to that adsorbed on the packing at the breakpoint and solving, $Z = 0.934$ m. Therefore, the total mass of adsorbent required is $(2)(0.934)(625) = 1{,}168$ kg.

4.3.1 Thomas Solution

One of the most useful treatments of the adsorption design problem is that of Thomas (1944). He solved the set of partial differential equations that describes the dynamics of fixed-bed adsorption systems having Langmuir-type phase equilibrium by introducing a transformation of the dependent variables. Consider an operation in which a bed of adsorbent, initially free of adsorbate, that is, $m(z, 0) = 0$, is fed at $\theta = 0$ with a gas containing Y_0 kg adsorbate/kg inert gas. Eventually, the whole bed will come to equilibrium with the gaseous feed. Then the solid will contain m^* kg adsorbate/kg solid. The gradual breakthrough of the adsorbate in the gaseous effluent, i.e., the function $Y(z, \theta)/Y_0$, and the gradual accumulation of the adsorbate on the solid, that is, $m(z, \theta)/m^*$, are the aims of the Thomas solution. It can be expressed as

$$\frac{Y}{Y_0} = \frac{J(N/K, NT)}{J(N/K, NT) + \left[1 - J(N, NT/K)\right]\exp\left[\left(1 - K^{-1}\right)(N - NT)\right]} \tag{4.30}$$

$$\frac{m}{m^*} = \frac{1 - J(NT, N/K)}{J(N/K, NT) + \left[1 - J(N, NT/K)\right]\exp\left[\left(1 - K^{-1}\right)(N - NT)\right]} \tag{4.31}$$

where

$$N = \frac{2Kz}{(K+1)H_{tOG}} \tag{4.32}$$

$$K = 1 + k_2 p_0 \tag{4.33}$$

$$T = \frac{Y_0 G' t}{m^* \rho_s z}, \qquad t = \theta - \frac{\varepsilon A_c z}{Q} \tag{4.34}$$

$$J(\alpha, \beta) = 1 - e^{-\beta} \int_0^{\alpha} e^{-x} I_0\left(2\sqrt{\beta x}\right) dx \tag{4.35}$$

and $I_0(x)$ is the modified Bessel function of the first kind and order zero.

For large values of N, if K is much greater than one, Gilliland and Baddour (1953) have shown that $J(N/K, NT)$ approaches one and $J(N, NT/K)$ approaches zero. For those conditions, Eq. (4.30) simplifies to

$$\frac{Y}{Y_0} = \frac{1}{1+\exp\left[\left(1-K^{-1}\right)N(1-T)\right]} \tag{4.36}$$

For desorption of adsorbed material from a solid bed, containing m_0 kg of adsorbate/kg of solid, with a fluid that initially contains no adsorbate:

$$\frac{Y}{Y_0{}^*} = \frac{1-J(N/K, NT)}{1-J(N/K, NT)+\left[J(N, NT/K)\right]\exp\left[\left(K^{-1}-1\right)N(T-1)\right]} \tag{4.37}$$

where $Y_0{}^*$ is the fluid concentration in equilibrium with m_0. For relatively large values of N when K is much greater than one, Eq. (4.37) simplifies to (Gilliland and Badour 1953):

$$\frac{Y}{Y_0} = \frac{\sqrt{\dfrac{K}{T}}-1}{K-1}, \qquad \frac{1}{K} \le T \le K \tag{4.38}$$

Numerical values of the J function have been listed by Schumann (1929), by Furnas (1930), and by Hougen and Marshall (1947). Following, you will find a function subprogram to evaluate J. It uses Gauss-Legendre quadrature (Constantinides 1987) to estimate numerically the integral. It incorporates a subroutine to evaluate the roots and weight factors needed for the Gauss-Legendre quadrature. Approximations to the J function for large values of the arguments suggested by Thomas (1944) and Klinkenberg (1948) are used.

```
      FUNCTION FUNCJ (ALPHA, BETA)
      PARAMETER (X1 = 0.0, N = 16)
      DIMENSION X(20), W(20)
      IF (ALPHA*BETA .LE. 100.0) THEN
        CALL GAULEG (X1, ALPHA, X, W, N)
        SUM = 0.0
        DO 10 I = 1, N
          SUM = SUM+W(I)*EXP(-X(I))*BESSI0(2.0*SQRT(BETA*X(I)))
10      CONTINUE
```

```
    FUNCJ = 1.0 - EXP(-BETA)*SUM
  ELSE
    IF (ALPHA*BETA .LE. 3600.) THEN
      FUNCJ = 0.5*(1.-ERF((SQRT(ALPHA)-SQRT(BETA))))+(EXP(
 *     -(SQRT(ALPHA)-SQRT(BETA))**2))/(3.5449*((ALPHA*BETA)
 *     **(0.25)+BETA**(0.5)))
    ELSE
       FUNCJ=0.5-0.5*ERF((SQRT(ALPHA)-SQRT(BETA)))
    END IF
  END IF
  END

     FUNCTION BESSI0(X)
  REAL*8 Y,P1,P2,P3,P4,P5,P6,P7,
 *   Q1,Q2,Q3,Q4,Q5,Q6,Q7,Q8,Q9
  DATA P1,P2,P3,P4,P5,P6,P7/1.0D0,3.5156229D0,3.0899424D0,1.2067492D
  *0,
 *   0.2659732D0,0.360768D-1,0.45813D-2/
  DATA Q1,Q2,Q3,Q4,Q5,Q6,Q7,Q8,Q9/0.39894228D0,0.1328592D-1,
 *   0.225319D-2,-0.157565D-2,0.916281D-2,-0.2057706D-1,
 *   0.2635537D-1,-0.1647633D-1,0.392377D-2/
  IF (ABS(X).LT.3.75) THEN
   Y=(X/3.75)**2
   BESSI0=P1+Y*(P2+Y*(P3+Y*(P4+Y*(P5+Y*(P6+Y*P7)))))
  ELSE
   AX=ABS(X)
   Y=3.75/AX
   BESSI0=(EXP(AX)/SQRT(AX))*(Q1+Y*(Q2+Y*(Q3+Y*(Q4
 *     +Y*(Q5+Y*(Q6+Y*(Q7+Y*(Q8+Y*Q9))))))))
  ENDIF
  RETURN
  END

FUNCTION ERF(X)
  IF (X . GT. 0) THEN
    ERF = 1.-  ERFCC(X)
  ELSE
    ERF = ERFCC(X) -1.
  END IF
  END

  FUNCTION ERFCC(X)
  Z=ABS(X)
  T=1./(1.+0.5*Z)
  ERFCC=T*EXP(-Z*Z-1.26551223+T*(1.00002368+T*(.37409196+
 *   T*(.09678418+T*(-.18628806+T*(.27886807+T*(-1.13520398+
```

```
     *  T*(1.48851587+T*(-.82215223+T*.17087277)))))))))
      IF (X.LT.0.) ERFCC=2.-ERFCC
      RETURN
      END

      SUBROUTINE GAULEG(X1,X2,X,W,N)
C     THIS SUBROUTINE CALCULATES THE ROOTS,X, AND WEIGHT FACTORS,W,
C     FOR GAUSS-LEGENDRE QUADRATURE WITH N POINTS IN THE INTERVAL
C     BETWEEN X1 AND X2
```



```
      IMPLICIT REAL*8 (A-H,O-Z)
      REAL*4 X1,X2,X(N),W(N)
      PARAMETER (EPS=3.D-14)
      M=(N+1)/2
      XM=0.5D0*(X2+X1)
      XL=0.5D0*(X2-X1)
      DO 12 I=1,M
          Z=COS(3.141592654D0*(I-.25D0)/(N+.5D0))
1       CONTINUE
            P1=1.D0
            P2=0.D0
            DO 11 J=1,N
                P3=P2
                P2=P1
                P1=((2.D0*J-1.D0)*Z*P2-(J-1.D0)*P3)/J
11        CONTINUE
            PP=N*(Z*P1-P2)/(Z*Z-1.D0)
            Z1=Z
            Z=Z1-P1/PP
          IF(ABS(Z-Z1).GT.EPS)GO TO 1
          X(I)=XM-XL*Z
          X(N+1-I)=XM+XL*Z
      W(I)=2.D0*XL/((1.D0-Z*Z)*PP*PP)
      W(N+1-I)=W(I)
12    CONTINUE
      RETURN
      END
```

Example 4.5 Breakthrough Calculations using Thomas Solution

Repeat Example 4.3 using Thomas solution. The equilibrium data given in that example can be accurately described by a Langmuir isotherm with parameters $k_1 = 0.000328$ kg acetone/kg carbon-Pa, and $k_2 = 0.000867$ Pa^{-1}.

Solution

Calculate the initial partial pressure of acetone in the mixture: $p_0 = 0.05(101.3) = 5.065$ kPa. Use Eq. (4.33) to calculate $K = 5.391$. At the bed exit $z = Z = 0.3$ m. From Example 4.3, $H_{tOG} = 0.0109$ m. Therefore, the number of transfer units at the bed exit is $N = 2(5.391)(0.3)/[(6.391)(0.0109)] = 46.43$. Because the number of transfer units is large and K is greater than one, use Eq. (4.36) to calculate the dimensionless breakthrough time, T_B, corresponding to $Y/Y_0 = 0.05$. The answer is $T_B = 0.922$. Now, use Eq. (4.34) to calculate $t_B = (0.922)(0.295)(630)(0.3)/[(0.105)(0.416)] = 1{,}177$ s. Use the second part of Eq. (3.34) to calculate $\theta_B = 1{,}177.3$ s $= 19.62$ min.

Comments

There is good agreement between the results obtained using the Thomas solution and those obtained by the graphical method used in Example 4.3. The advantage of the Thomas solution is that it requires no graphical construction, and it can be easily programmed for computer application, even when the complete form of Eq. (4.30) must be used.

Example 4.6 Regeneration of Fixed-Bed Adsorber

An activated carbon fixed-bed adsorber is used to recover a VOC (molecular weight of 44) from air. The bed has a cross-sectional area of 10 m^2 and a depth of 0.5 m. The carbon has a bulk density of 500 kg/m^3 and a porosity of 40 percent. At the break point the solid contains 0.10 kg VOC/kg carbon. The bed is to be regenerated using saturated steam at 373 K. The regeneration time should not exceed 15 min. The bed will be considered fully regenerated when the VOC concentration in the exit steam is within 5% of the concentration of the gaseous phase in equilibrium with the initial VOC content of the solid ($Y/Y_0^* = 0.05$). The adsorption equilibrium is described by a Langmuir isotherm with parameters $k_1 = 0.0056$ kg VOC/kg carbon-kPa, and $k_2 = 0.022$ kPa^{-1}. Mass-transfer conditions are such that $H_{tOG} = 0.50(G')^{0.51}$. Estimate the rate at which the steam must be supplied, in kg/s. Calculate, also, the ratio of kg steam used per kg of VOC recovered.

Solution

Use the equation for the Langmuir isotherm to calculate the partial pressure of the VOC in equilibrium with the solid: $p_0^* = 0.10/[0.0056 - (0.10)(0.022)] = 29.4$ kPa. Therefore, $Y_0^* =$

(29.4)(44)/[(101.3 – 29.4)(18)] = 1.00 kg VOC/kg steam. From Eq. (3.34), assuming that t is approximately equal to θ, G' = (0.10)(500) (0.5)T/[(1.00)(15)(60)] = 0.027T kg steam/m2-s. Use Eq. (4.33) to calculate K = 1 + (0.022)(29.4) = 1.667. At the bed exit, the number of transfer units is N = 2(1.667)(0.5)/[(2.667)H_{tOG}] = 0.625/H_{tOG}. Next, assume values of G', calculate T, H_{tOG}, and N. Use Eq. (4.37) to calculate the value of Y/Y_0* corresponding to each value of G'. The results are summarized in the following table:

G' (kg/m2-s)	T	H_{tOG} (m)	N	Y/Y_0
0.03	1.11	0.0836	7.48	0.350
0.04	1.48	0.0970	6.44	0.233
0.06	2.22	0.1200	5.21	0.180
0.08	2.96	0.1380	4.53	0.053

A steam mass velocity of about 0.08 kg/m2-s will regenerate the bed according to specifications in 15 min. The steam must be supplied at a rate of (.08)(10) = 0.8 kg/s. The total amount of steam required during the regeneration step is 720 kg. The total amount of VOC recovered is (0.10)(10)(0.5)(500) = 250 kg. Therefore, the ratio of steam required to VOC recovered is 720 / 250 = 2.88 kg steam/kg VOC.

4.4 HEAT AND MOMENTUM TRANSFER IN FIXED-BED ADSORBERS

Your objectives in studying this section are to

1. Estimate the time required to cool a thermally regenerated fixed bed adsorber to its operating temperature.
2. Estimate the pressure drop through a fixed bed adsorber.

Heat- and momentum-transfer considerations are very important in the design and operation of fixed-bed adsorbers. Most adsorbing devices are regenerated through intimate contact with a hot fluid. Once the regeneration step is completed, the adsorbing solid must be cooled to its operating temperature. This is accomplished by blowing cold air through the bed. The heat-transfer dynamics for this process resembles the adsorption dynamics discussed in the previous section.

The pressure drop that results when a waste gas stream flows through an adsorption device is a very important design parameter. The energy expenditure required to overcome the pressure drop through a poorly designed fixed-bed could account for a significant fraction of the annual operating cost for the device.

4.4.1 Cooling of Thermally Regenerated Fixed-Bed

Consider a fixed-bed that has been thermally regenerated and is initially at a uniform temperature Θ_{s0}. A cold fluid, at an initial temperature Θ_{f0}, is forced through the bed at a mass velocity of G_c' kg/m2-s. If the fluid inside the pores of the adsorbent is at all times in thermal equilibrium with the solid (Ruthven 1984), the bed temperature as a function of time and longitudinal position, $\Theta_s(\theta, z)$, is given by

$$\frac{\Theta_{s0} - \Theta_s}{\Theta_{s0} - \Theta_{f0}} = 1 - J(N_{th} T_{th}, N_{th}) \tag{4.39}$$

where

$$N_{th} = \frac{ha\,z}{G_c' C_f}, \qquad N_{th} T_{th} = \frac{hat}{\rho_s C_s}, \qquad t = \theta - \frac{z A_c \varepsilon}{Q} \tag{4.40}$$

ha = volumetric heat-transfer coefficient

C_f and C_s = specific heat of the fluid and adsorbent, respectively

The volumetric heat-transfer coefficient for packed beds can be estimated by a correlation suggested by Bradshaw (1963).

$$\frac{ha}{G_c' C_f} = \frac{15(1-\varepsilon)}{d_p}\left[\frac{(1-\varepsilon)}{\mathrm{Re}_p}\right]^{0.5} \mathrm{Pr}^{-2/3} \tag{4.41}$$

where

$\mathrm{Pr} = C_f\,\mu / k_f$

k_f= thermal conductivity of the fluid.

Equation (4.41) is valid for $\mathrm{Re}_p/(1 - \varepsilon)$ from 400 to 10,000. The fluid properties are evaluated at $\Theta_{av} = (\Theta_{f0} + \Theta_{s0})/2$.

Example 4.7 Cooling of Fixed-Bed Adsorber

The adsorber of Example 4.6, regenerated using saturated steam at 373 K, must be cooled down to a temperature not exceeding 300 K. Air at 298 K and 101.3 kPa is forced through the bed, at the rate of 8.9 m3/s, for that purpose. Estimate the time required to achieve the desired temperature. The specific heat of the adsorbing medium is 850 J/kg-K. The average particle size is 6 mm.

Solution

The initial solid and fluid temperatures are given, as well as the solid temperature at the bed outlet. From Eq. (4.39), (373 – 300)/(373 – 298) = 0.973 = 1 – $J(N_{th}T_{th}, N_{th})$. Therefore, $J(N_{th}T_{th}, N_{th})$ = 0.0267. At 298 K and 101.3 kPa the density of air is 1.18 kg/m3. The air mass velocity is Gc' = (1.18)(8.9)/10 = 1.05 kg/m2-s. Calculate Θ_{av} = (373 + 298)/2 = 336 K. At this temperature, the viscosity of air is 2×10^{-5} kg/m-s. The Reynolds number is Re_p = (1.05) (0.006)/2 × 10^{-5} = 315; $\mathrm{Re}_p/(1 - \varepsilon)$ = 525. The Prandtl number for air at this temperature is Pr = 0.714. At the average temperature, Q/A_c = (8.9)(336)/ [(298)(10)] = 1.0 m/s. The specific heat of air at this temperature is C_f = 1.0 kJ/kg-K, its density is ρ_f = 1.05 kg/m3. Substituting in Eq. (4.41), $ha/G_c'C_f$ = (15)(1 – 0.4)(1/525)0.5(0.714)-2/3/(0.006) = 81.9 m-1. Calculate the number of heat-transfer units at the bed exit using Eq. (4.40): N_{th} = (81.9)(0.5) = 41.0. Also, $N_{th}T_{th}$ = (81.9)(1.05)t/[(500)(0.85)] = 0.202t. Therefore, $J(0.202t, 41)$ = 0.0267. This last equation is solved iteratively for t. The answer is t = 300 s = 5 min. The total time elapsed is θ = 300 + (0.3)(0.4)/1.0 = 300.12 s = 5 min.

4.4.2 Pressure Drop Across Fixed-Beds.

The pressure drop through a packed bed is a standard problem in chemical engineering, and considerable attention has been devoted to it. It can be estimated with the following equation (Ergun 1952):

$$\frac{\Delta P \varepsilon^3 d_p \rho_f}{Z(1-\varepsilon)(G')^2} = \frac{150(1-\varepsilon)}{\mathrm{Re}_p} + 1.75 \qquad (4.42)$$

The first term on the right side of Eq. (4.42) corresponds to a friction factor in laminar flow; the second one corresponds to turbulent flow. Unlike in flow inside a pipe, there is a smooth transition from the laminar to the turbulent regime for flow through a packed bed.

Example 4.8 Pressure Drop through Fixed-Bed Adsorber

Estimate the pressure drop during the cooling process described in Example 4.6.

Solution

All the information needed to calculate the pressure drop is available from Example 4.7. Substituting in Eq. (4.42),

ΔP = (150/525 + 1.75)(0.5)(1 – 0.4)(1.05)2/[(0.4)3(0.006)(1.05)] = 1,670 Pa.

4.5 PRACTICAL DESIGN CONSIDERATIONS FOR FIXED-BED ADSORBERS

Your objectives in studying this section are to

1. Select an appropriate adsorption cycle time for a given design problem.
2. Determine the total adsorbent requirement for it.
3. Determine the number, and size of adsorbing vessels required.

The primary sizing parameter for adsorbers is the total adsorbent requirement. This parameter directly determines the equipment cost, and indirectly determines the size and number of the adsorber vessels, and the auxiliaries such as the system fan. The adsorbent requirement, in turn, incorporates several system variables, such as (1) adsorption, regeneration, and cooling times, (2) waste gas volumetric flow rate, (3) allowable pressure drop, and (4) working capacity of each bed.

For continuously operated fixed-bed systems, an extra adsorbent bed (or beds) is included to accommodate the inlet waste gas while the other bed is being regenerated. If the desorption plus cooling time is considerably shorter than the adsorption time, it may be more economical to have two or more beds adsorbing, with only one desorbing. This arrangement can reduce the total adsorbent requirement. The amount of adsorbent in one bed, W_{ci}, is given by:

$$W_{ci} = \frac{Y_0 G' A_c \theta_B}{m_w} \tag{4.43}$$

where m_w is the working capacity of the bed, given by $m_w = m^*a$. The total carbon requirement for the system, W_c, is given by:

$$W_c = W_{ci} N_a \left(1 + \frac{N_d}{N_a}\right) \tag{4.44}$$

where N_a, N_d = number of beds adsorbing and desorbing, respectively, at any time. The following expression relates N_a and N_d to the adsorption time, θ_B, the regeneration time, θ_R, and the cooling time, θ_C:

$$N_a = \frac{\theta_B N_d}{\theta_R + \theta_C} \tag{4.45}$$

Example 4.9 Total adsorbent requirement for Fixed-Bed Adsorber

A fixed-bed adsorbing system has adsorption, desorption, and cooling times of 4, 1, and 0.5 h. At any given time, there will be five beds adsorbing, each one loaded with 2,000 kg of adsorbent. Calculate the number of beds desorbing at any given time, and the total adsorbent requirement for the system.

Solution

Use Eq. (4.45) to determine the number of beds desorbing at any time, Nd = (5)(1.0 + 0.5)/(4.0) = 1.875 or 2. Hence, there are "five vessels on, two off". Use Eq. (4.44) to calculate the total adsorbent requirement, W_c = (5)(2,000)(1 + 2/5) = 14,000 kg.

Comments

Notice that if only one bed with 10,000 kg was used instead of five beds with 2,000 kg of adsorbent each, a total of two beds would be required ("one on, one off"). The total adsorbent requirement for this arrangement would be 20,000 kg, and the regenerated bed would be idle for 2.5 h of each cycle.

The number of vessels in the system depends on the waste gas flowrate, the superficial velocity across the bed, and the maximum allowable vessel diameter. According to Vatavuk (1990), because of transportation constraints, the largest vessel diameter that can be shop fabricated is approximately 4 m. Superficial gas velocities are, usually, limited to 0.4 to 0.5 m/s to avoid excessive pressure drops. Therefore, the largest flowrate that a 4-m-diameter, vertically erected vessel can handle would be $Q_{max} = (\pi/4)(16)(0.5) = 6.3$ m^3/s.

The adsorption cycle time is a very important parameter in sizing the system. Shorter adsorption times reduce the total adsorbent requirements, reducing equipment costs. However, shorter adsorption times require the bed to be regenerated more often. This increases utility costs (steam, cooling water, and electricity). Too-frequent desorption tends to shorten the carbon life. For any given application, the optimum adsorption cycle time can be determined from a detailed economic evaluation of alternatives. Usually, it will be in the range of 8 to 12 h (Vatavuk 1990).

Example 4.10 Design of Carbon Adsorber for Control of VOC Emissions

A waste gas from a chemical process flows at the rate of 36.2 m^3/s at 310 K and 102 kPa, and consists of 315 ppm n-hexane in air. It will be processed in a carbon adsorber to recover the n-hexane. The carbon particle size is 6 × 10 mesh, the bed porosity is 40%, the bulk density is 500 kg/m^3. An 8-h adsorption cycle is suggested: 6 h of adsorption, 1.5 h desorption, and 0.5 h cooling. Saturated steam at 383 K will be used for regeneration. From calculations based on the dynamics of the process, the following results are available:

1. The working capacity of the bed is 0.085 kg VOC/kg carbon, with a very steep breakthrough curve.
2 For the desorption step, use 3 kg steam/kg VOC recovered.
3 Ambient air will be used for cooling at the rate of 3.6 m^3/s-metric ton of carbon.

The mixture of steam and n-hexane will be condensed and cooled to 303 K using cooling water, which will enter the condenser at 293 K, and leave it at 308 K. The solubility of n-hexane in water at 303 K is 0.003 mole percent. Calculate the number of adsorbing vessels to be used and their dimensions, and the total carbon requirement. Calculate the utility requirements for an 8-h cycle.

Solution

Calculate the number of adsorbing vessels at any time, $N_a = Q/Q_{max} = 36.2/6.3 = 5.75$ or 6 beds. Use Eq. (4.45) to calculate the number of desorbing vessels, $N_d = N_a(2/6) = 2$ beds. Calculate the carbon requirement per bed using Eq. (4.43). The waste gas flowrate per bed is 6 m^3/s. Therefore, Y_0A_cG' = kg hexane/s = (6)(315)(102)(86)/[(106)(8.314)(310)] = 0.00641 kg/s. The carbon requirement per bed is W_{ci} = (0.00641)(6)(3,600)/0.085 = 1,630 kg. The total carbon requirement is W_c = 1,630(6)(1 + 2/6) = 13,037 kg carbon. The volume of each bed is calculated from the individual carbon requirement and the bulk density of the bed: 1,630/500 = 3.259 m^3. For a 4-m diameter bed, the depth is $Z = 3.259/ [(\pi/4)(4)2]$ = 0.259 m. Summarizing, a total of 8 beds is required for continuous operation; at any instant 6 beds will be adsorbing and 2 will be regenerating and/ or cooling. Each bed will have a diameter of 4 m and a depth of 0.259 m.

Calculate the utility requirements for an 8-hour-cycle:

1. Electricity: Use Eq. (4.42) to calculate the pressure drop during adsorption, ΔP = 0.75 kPa. Assuming a mechanical efficiency of 60%, the power required is (36.2)(0.75)/0.6 = 45.3 kW. Since the adsorption part of the cycle is continuous, the electric energy requirement for the adsorption step is (45.3)(8) = 362.4 kW-h/cycle.

For the cooling step, the volumetric flow rate of air per bed is (3.6)(1.63) = 5.9 m^3/s. The pressure drop, calculated with Eq. (4.44) is virtually the same as during adsorption, namely 0.75 kPa. Because there will be two beds cooling simultaneously, the power required is (2)(5.9)(0.75)/0.6 = 14.75 kW. The cooling time is only 25% of the regeneration/cooling period, therefore the electric energy required for the cooling portion of the cycle is (14.75)(8)(0.25) = 29.5 kW-h/cycle. Hence, the total electric energy requirement is 391.9 kW-h/cycle.

2. Steam: During each cycle, each bed will adsorb 138.6 kg of hexane. Therefore, the total amount of steam required to regenerate two beds will be (138.6)(3)(2) = 831.3 kg steam/1.5-h = 554.2 kg/h. The regeneration time is 75% of the regeneration/cooling period, therefore the steam requirement is (554.2)(8)(0.75) = 3,325 kg/cycle.

3. Cooling water: During the desorption step, a gaseous mixture consisting of 831.3 kg of steam and 277.2 kg of hexane at 383 K will flow through the condenser where it will be condensed and cooled to 303 K. Calculate the enthalpy change for this stream, $\Delta H = -2.276 \times 10^6$ kJ. Because the cooling water must absorb this enthalpy while experiencing a temperature rise of 15 K, the cooling water required is $(2.276 \times 10^6)/[(4.18)(15)]$ = 36,300 kg/1.5-h = 24,200 kg/h. The condenser will operate for 75% of the regeneration/ cooling period, therefore the cooling water required during an 8-h cycle is (24,200)(8)(0.75) = 145,200 kg/cycle.

4. Waste water treatment: During each 8-h cycle, 3,325 kg of steam will be condensed while in intimate contact with hexane. The solubility of hexane in water at 303 K is

0.003 molar percent. The amount of hexane dissolved in the condensed steam is calculated to be 5.5 mol/cycle. For complete oxidation of hexane to carbon dioxide and water the stoichiometric oxygen requirement is 9.5 mol O_2/mol hexane. Therefore, the *chemical oxygen demand* (COD) of the waste water is (5.5)(9.5)(32)/1,000 = 1.68 kg O_2/cycle.

4.6 COSTING PROCEDURE FOR FIXED-BED ADSORBERS

Your objectives in studying this section are to

1. Estimate the total capital investment required for a fixed-bed adsorber.
2. Estimate the corresponding total annual cost.

4.6.1 Total Capital Investment

Like most other types of pollution control devices, fixed-bed adsorbers can be sold either as packaged or custom units, depending on the size of the adsorber as measured by the total adsorbent requirement, Wc. Vatavuk (1990) obtained vendor quotes for carbon steel, steam-regenerated packaged activated carbon adsorbers. These prices (updated to June 1990 dollars) can be fitted with the following regression equation:

$$EC = 257.4(W_c)^{0.848}, \qquad 160 \le W_c \le 6{,}400\,\text{kg} \tag{4.46}$$

These prices include the adsorber vessels, carbon, condenser, decanter, system fan and motor, bed cooling fan and motor, instruments and controls, and internal ducting. The cost of a boiler to provide regeneration steam is not included.

For custom-built fixed-bed carbon adsorbers, the same source suggests the following correlation (updated to June 1990 dollars):

$$EC = 68.13(W_c)^{0.86}, \qquad 6{,}400 \le W_c \le 100{,}000\,\text{kg} \tag{4.47}$$

These prices include the same auxiliary equipment as do the packaged adsorber prices.

For stainless steel construction, the equipment cost is about twice that for carbon steel construction.

Example 4.11 Equipment Cost for Fixed-Bed Carbon Adsorber

Estimate the equipment cost for the fixed-bed adsorber of Example 4.10.

Solution

From Example 4.10, the total carbon requirement is 13,037 kg. A custom-built system is required. The equipment cost is estimated from Eq. (4.47): $EC = 68.13(13{,}037)0.86 =$ \$235,700 of June 1990.

Because packaged adsorbers come equipped to be virtually "plugged into" emission sources, the costs for installing them are minimal. A reasonable installation cost estimate would be 25% of the total purchased equipment cost. Table 4.4 presents average installed-cost factors for custom-built carbon adsorbers.

Example 4.12 TCI for Carbon Adsorber

Estimate the total capital investment required for the carbon adsorber of Example 4.11. Assume no site preparation, buildings, land, or working capital are required.

Solution

The equipment cost calculated in Example 4.11 includes instrumentation and controls. Therefore, according to Table 4.4, TCI = 1.75(1.08)(235,700) = \$445,500.

4.6.2 Annual Costs

Table 4.5 presents suggested (Vatavuk 1990) factors for estimating fixed-bed adsorber annual costs. The adsorbent has a shorter life than the rest of the unit—typically 3 to 5 years compared to 10 years for the equipment. Hence, the adsorbent replacement cost must be calculated

Table 4.4 Average Cost Factors for Carbon Adsorbers

Cost Item	Cost Factor
Direct costs	
1) Purchased equipment cost	
Adsorber and auxiliary equipment	1.00 *A*
Instrumentation and controls	0.10 *A*
Taxes and freight	0.08 *A*
Total purchased equipment cost	*B* = 1.18 *A*
2) Direct installation costs	
Foundations and supports	0.08 *B*
Erection and handling	0.20 *B*
Electrical	0.08 *B*
Piping	0.05 *B*
Insulation and painting	0.03 *B*
Site preparation (SP), building (Bldg)	As required
Total direct installation costs	0.44 *B* + SP + Bldg
Indirect installation costs	
Engineering and supervision	0.10 *B*
Construction, fee, and field	0.15 *B*
Start-up and performance tests	0.03 *B*
Contingency	0.03 *B*
Total direct plus indirect costs	1.75 B + *B* + Bldg

From Neveril, R.B., et al. *JAPCA*, 28: , (1978). Reprinted with permission from *JAPCA*.

separately. A typical activated carbon price is $5.00/kg. Sales tax and freight must be added to this price. The carbon replacement labor cost is usually negligible.

The recovery credits apply to those adsorbates captured by the system, condensed, and separated that have value either as recyclable or resalable commodities. Not all adsorbates can be reused, regardless of their purity. Others may be reusable, but the costs to separate them from the steam condensate may exceed their value.

Example 4.13 TAC for Fixed-Bed Carbon Adsorber.

Estimate the TAC for the adsorber of Example 4.12. The system will operate for 24 h/day, 350 d/yr. Labor rate are $10/h and $15/h for operation and maintenance, respectively. Rates

Table 4.5 Factors for Estimating Adsorbers Annual Costs

Item	Suggested Factor
Direct operating costs	
Operating labor	0.5 h/shift
Supervisory labor	15% of operating labor
Maintenance labor	0.5 h/shift
Maint. materials	100% of maintenance labor
Replacement parts	Periodic replacement of adsorbent
Utilities	
Electricity	As required
Steam	As required
Cooling water	As required
Waste water treatment	As required
Indirect operating costs	
Overhead	60% of the sum of all labor and maintenance materials
Administrative charges	2% × TCI
Property tax	1% × TCI
Insurance	1%× TCI
Capital recovery cost	CRF × (TCI – 1.08 × Initial adsorbent cost)
Recovery credits	
Recovered adsorbate	As applicable

Source: Vatavuk (1990).

for utilities are (1) \$0.10/kW-h for electricity, (2) \$13/metric ton of steam, (3) \$0.05/m3 of cooling water, and (4) \$1/kg of COD for waste water treatment. The value of the recovered hexane, as it comes from the decanter, is \$0.20/L. The carbon service life is 3 yr, the equipment useful life is 12 yr. The initial cost of the carbon is \$5/kg, and the sales tax and freight is approximately 8% of this figure. The minimum attractive return on investment is 15%. Assume that all the conditions of Example 4.10 apply to this example.

Solution

Use information from Examples 4.10 and 4.12, and Table 4.5 to estimate the TAC. Some of the calculations follow; the results are presented in tabular form. To estimate the annual cost of replacing the carbon, the CRF corresponding to $n = 3$ yr and $i = 0.15$ is calculated: CRF = 0.438. Hence, the annualized cost of replacing the carbon is (13,037)(5)(1.08)(0.438) = \$30,835/yr. To estimate the capital recovery cost, the CRF corresponding to $n = 12$, $i = 0.15$ is calculated: CRF = 0.184. The capital recovery cost is equal to 0.184[445,500 –

13,037(5)(1.08)] = $69,020/yr. The recovery credit is calculated assuming that 95% of all the hexane will be recovered, a conservative estimate given the characteristics of the breakthrough curve and the low solubility of hexane in water. The density of liquid hexane is 660 kg/m^3. The recovery credit is: (138.6)(24)(350)(0.95)(1,000)(0.2)/660 = $335,160/yr.

Summary of Annual Costs

Total labor	$13,920
Maintenance materials	7,880
Replacement parts	30,835
Total utilities	95,920
Indirect costs	99,930
Total	248,485
Recovery credit	335,160
TAC	– 86,675

The use of this system results in a yearly profit of $86,675, a very attractive investment, both from the financial and environmental viewpoints.

4.7 CONCLUSION

Adsorption is an effective method for the control of VOC emissions. It is a particularly attractive option, when applicable, because it can turn a potential air pollution problem into a valuable resource. Most applications of adsorption to air pollution control problems involve the use of fixed-bed systems. Because of its very nature, fixed-bed adsorption is a transient process. Various methods of analyzing the dynamics of the process were presented. Sizing and costing procedures were illustrated. An example of a typical application to a chemical process showed that solvent recovery through activated carbon can be a profitable investment, besides helping to maintain a clean environment.

REFERENCES

Bradshaw, R. D. *AIChE J.*, **9**:590 (1963).

Constantinides, A. *Applied Numerical Methods With Personal Computers*, McGraw-Hill, New York (1987).

Cooper, C. D., and Alley, F.C. *Air Pollution Control; A Design Approach*, Prindle, Weber & Schmidt, Boston, MA (1986).

EAB *Control Cost Manual,* 3rd. ed., U.S. E.P.A., Research Triangle Park, NC (1987).

Ergun, S. *Chem. Eng. Progr.*, **48**:89 (1952).

Furnas, C. *C. Trans. AIChE*, **24**:142 (1930).

Gilliland, E. R., and Baddour, R. F. *Ind. Eng. Chem.*, **45**: 330 (1953).

Hougen, O. A., and Marshall, W. R., Jr. *Chem. Eng. Progr.*, **43**:197 (1947).

Klinkenberg, A. *Ind. Eng. Chem.*, **40**:1970 (1948).

Neveril, R. B., Price, J. U., and Engdahl, K. L. *JAPCA,* **28**:1269 (1978).

Press, W. H., Flannery, B. P., Teukolsky, S. A., and Vetterling, W. T. *Numerical Recipes; The Art of Scientific Computing,* Cambridge University Press, New York,, (1989).

Ruthven, D. M. *Principles of Adsorption and Adsorption Processes,* Wiley, New York (1984).

Satterfield, C. N., and Sherwood, T. K. *The Role of Diffusion in Catalysis*, Addison-Wesley, Reading, MA (1963).

Schumann, T. E. W. *J. Franklin Inst.,* **208**:405 (1929).

Thomas, H. *J. Amer. Chem. Soc.*, **66:**1664 (1944).

Treybal, R. E. *Mass-Transfer Operations*, 3rd ed., McGraw-Hill, New York (1980).

Vatavuk, W. M. *Estimating Costs of Air Pollution Control,* Lewis, Chelsea, MI (1990).

Vermeulen, T., Klein, G., and Hiester, N. K. Sec. 16 in J. H. Perry (ed.), *Chemical Engineers' Handbook,* McGraw-Hill, New York (1973).

PROBLEMS

The problems at the end of each chapter have been grouped into four classes (designated by a superscript after the problem number)

Class a: Illustrates direct numerical applications of the formulas in the text.
Class b: Requires elementary analysis of physical situations, based on the subject material in the chapter.
Class c: Requires somewhat more mature analysis.
Class d: Requires computer solution.

4.1[a]. Equilibrium adsorption of m-xylene on activated carbon

Calculate the equilibrium adsorptivity of m-xylene (in air) on Calgon "BPL" activated carbon (4×10 mesh). The temperature is 298 K and the total pressure is 200 kPa. The equilibrium concentration of m-xylene in the gas phase is 100 ppm.

Answer: 0.035 kg/kg carbon.

4.2[b]. Equilibrium adsorption of methane on activated carbon

The equilibrium adsorption of methane on a given activated carbon was studied by Grant et al. (Grant, R. J., Manes, M., and Smith, S. B. *AIChE J*, **8**:403, 1962). They proposed the following Langmuir-type adsorption isotherm:

$$q = 3.0 \times 10^{-3} \frac{K_A p}{1 + K_A p} \quad \text{kmol } CH_4/\text{kg of solid}$$

where the partial pressure of methane is expressed in atmospheres, and

$$K_A = 0.346 \exp\left[2{,}200\left(\frac{1}{T} - \frac{1}{298.1}\right)\right]$$

and T is the equilibrium temperature, in K. If the equilibrium partial pressure of methane is 0.1 atm, at what temperature will the adsorptivity of this carbon be 1.0 g/kg?

Answer: 319 K

4.3d. Adsorption equilibrium isotherms

The following data has been reported for equilibrium adsorption of benzene on certain activated carbon at 306 K:

Benzene Partial Pressure (Pa)	Benzene Adsorbed (kg/kg solid)
0	0
15.33	0.124
33.46	0.206
133.29	0.278
374.54	0.313
1,042.30	0.348

(a) Fit a Langmuir-type isotherm to this data. Linearize the model by rewriting it as

$$\frac{p^*}{m} = \frac{1}{k_1} + \frac{k_2}{k_1} p^*$$

Estimate k_1 and k_2, and the correlation coefficient, r.

Answer: $k_1 = 0.0104$ *kg/kg-Pa*

b) Fit a Freundlich-type isotherm to the same data. Linearize by rewriting it as

$$\ln m = \ln k + n \ln p^*$$

Estimate the values of k and n, and the correlation coefficient. Comment on which model best describes the observed data.

Answer: $n = 0.2274$

4.4a. Adsorption equilibrium; material balances

A waste gas contains 1,974 ppm of phenol in air at 313 K and 101.3 kPa. One hundred kilomoles of this mixture are put in a closed container with 50 kg of Calgon "BPL" activated carbon completely free of phenol. The system is allowed to reach equilibrium at constant temperature and pressure. Calculate the equilibrium concentration of phenol in the gaseous phase and the amount of phenol adsorbed on the solid. Assume that the air does not adsorb on the carbon.

Answer: 165 ppm

4.5b. Breakthrough calculations; Freundlich isotherm

Derive Eqs. (4.28) and (4.29) for an adsorption system whose equilibrium characteristics are adequately described by a Freundlich-type isotherm.

4.6b. Breakthrough calculations; graphical method

A waste gas flows at the rate of 10 m^3/s, and contains 3,700 ppm of benzene in air at 306 K and 101.3 kPa. The activated carbon of Problem 4.3 will be used to recover the benzene vapors in this gas. The carbon particle size is 3×6 mesh, its bulk density is 550 kg/m^3, the bed porosity is 40%, the internal particle porosity is 65%, and the specific surface area is 1,200 m^2/g. Calculate the dimensions of a cylindrical fixed-bed with a breakthrough time of 4 hr. The breakpoint will occur when the outlet benzene concentration is 5% of the inlet concentration. The bed will be considered exhausted when the outlet concentration is 95% of the inlet concentration. Design for a superficial velocity of 0.25 m/s. Use the graphical method illustrated in Example 4.3.

Answer: Z = *0.25 m*

4.7a. Number of overall gas transfer units; Freundlich isotherm

Use the results of Problem 4.3b to estimate N_{tOG} for Problem 4.6, assuming that the Freundlich-type isotherm adequately describes the adsorption equilibrium for this case. Compare it with the result obtained in Problem 4.6.

Answer: 3.48

4.8[b]. Breakthrough calculations; Thomas solution

Solve Problem 4.6 using the Thomas solution, assuming that the adsorption equilibrium can be described by a Langmuir-type isotherm (see Problem 4.3a).

Answer: $Z = 0.247$ *m*

4.9[d]. Breakthrough calculations; Thomas solution

A waste gas flows at the rate of 5 m^3/s at 300 K and 100 kPa. It is mostly air, with traces of a VOC. The partial pressure of the VOC is 500 Pa. The gas will be processed in a fixed-bed adsorber to recover the VOC vapors. The equilibrium distribution of this VOC between the gas phase and the solid adsorbent at 300 K is described by a Langmuir-type isotherm with $k_2 = 0.001$ Pa^{-1}. The VOC content of the feed, Y_0, is 0.01 kg VOC/kg air. The bulk density of the solid is 500 kg/m^3 and its external porosity is 45%. The solid concentration in equilibrium with the feed is 0.10 kg VOC/kg solid. Design a cylindrical fixed-bed adsorber for this process with a breakthrough time of 8 hr. The breakpoint will occur when the outlet VOC concentration is 1.5% of that in the feed. Design for a mass velocity of 0.25 kg/m^2-s. The mass-transfer conditions are such that $H_{tOG} = 0.10$ m.

Answer: $Z = 2.3$ *m.*

4.10[a]. Pressure drop through a fixed-bed adsorber

Estimate the pressure drop for flow through the fixed-bed adsorber of Problem 4.9. The average particle size corresponds to 3 mesh.

Answer: 296 Pa

4.11[c]. Thermal regeneration of fixed-bed adsorber

The fixed-bed adsorber of Example 4.3 will be regenerated using saturated steam at 373 K at the rate of 0.25 m^3/s. The bed will be considered completely regenerated when the acetone concentration in the exit steam is 1% of the gas-phase concentration in equilibrium with the initial solid concentration ($Y/Y_0^* = 0.01$). At this temperature, the adsorption equilibrium is described by a Langmuir-type isotherm with $k_1 = 0.028$ kg acetone/kg carbon-kPa, and $k_2 = 0.074$ kPa^{-1}. Calculate the time required for regeneration of the bed.

Answer: 45.3 min

Hint: During thermal regeneration of a fixed-bed, the gaseous phase is no longer a dilute solution of adsorbate in the inert gas. On a mass basis, the adsorbate can even be the most abundant species. Be careful when calculating the fluid properties required to estimate the mass-transfer coefficients. They should be calculated at a fluid composition that is the arithmetic average between the composition of the fluid immediately adjacent to the solid (in equilibrium with it) and that of the bulk fluid (basically pure steam). The properties of the mixture at the average composition are calculated from the properties of the pure gases. The viscosity of the mixture is evaluated by a method based on the kinetic-theory of gases (Reid, R. C. and Sherwood, T. K. *The Properties of Gases and Liquids, Their Estimation and Correlation*, 2nd. ed., McGraw-Hill, New York, 1966):

$$\mu = \sum_{i=1}^{n} \frac{\mu_i}{1 + \sum\limits_{\substack{j=1 \\ j \neq i}}^{n} \phi_{ij} \dfrac{y_j}{y_i}}$$

where y_i is the mol fraction of component i, and

$$\phi_{ij} = \frac{\left[1 + \sqrt{\mu_i / \mu_j}\left(M_j / M_i\right)^{1/4}\right]^2}{\sqrt{8\left(1 + M_i / M_j\right)}}$$

4.12d. Cooling of fixed-bed adsorber

After regeneration with steam, the fixed-bed adsorber of Problem 4.11 will be cooled down to 300 K using air at 293 K and 101.3 kPa at a rate of 4 m^3/s. Estimate the time required for the cooling process and the resulting pressure drop.

Answer: 1.76 min

4.13c. Toxic exposure control and personal protective equipment (Problem No. 65, *Safety, Health, and Loss Prevention in Chemical Processes: Problems for Undergraduate Engineering Curricula*, © 1990 by the American Institute of Chemical Engineers, reprinted by permission of The Center for Chemical Process Safety of AIChE.)

Although respirators, which are devices that are worn over the face to prevent inhaling harmful materials, are the "last line of defense" against exposure to airborne contaminants, they are nevertheless very important safety equipment items, and it is vital that they be used appropriately, in recognition of their limitations.

Chemical cartridge respirators provide protection against vapors and gases being inhaled. One type of device uses an adsorbent, such as activated carbon, to adsorb organic vapors and thus to purify the air that the wearer inhales. The analysis of the performance of an air purifying respirator can be done by the method of analysis of any fixed-bed adsorber. This problem illustrates the effect that the concentration of contaminant has on the service life of an adsorption canister.

A generalized correlation of adsorption potential shows that the logarithm of the amount adsorbed is linear with the function $(T/V)[\log 10(f_s/f)]$ over a wide range of values, where

T = temperature, K

V = molar volume of liquid at the normal boiling point, cm^3/mole

f_s = fugacity of saturated liquid (approximate as vapor pressure)

f = fugacity of the vapor (approximate as partial pressure)

A particular activated carbon used in a respirator canister exhibits the following adsorption characteristics for dichloropropane (DCP):

Amount Adsorbed cm^3 of Liquid/100 g Solid	$(T/V)[\log_{10}(f_s/f)]$ K-mol/cm^3
1.0	21
10.0	11

The canister contains 75 g of this carbon, and tests have shown that at the break point 82% of the adsorbent is saturated.

Regulations permit carbon canister, full face mask respirators at DCP concentrations of up to 750 ppm. If a worker were using this respirator in a DCP concentration of 750 ppm when the temperature is 300 K, how long will it take for breakthrough to occur? Assume that the worker breathes at the rate of 45 L/min. If, owing to an accident, a worker is caught in a DCP concentration of 2,000 ppm, how long might he have before breakthrough? The density of liquid DCP at 300 K is 1.16 g/cm^3, and its molar volume at the normal boiling point is 100 cm^3/mol.

Answer: 147 min, 74 min

4.14[a]. Total adsorbent requirement

A fixed-bed carbon adsorber will be used to recover the VOC vapors in a waste gas which flows at a rate of 25.2 m^3/s. The density of the gas is 1.1 kg/m^3, and it contains 0.005 kg VOC/kg of air. The working capacity of the adsorbent is 0.20 kg VOC/kg solid. Its bulk density is 650 kg/m^3. The adsorption cycle consists of 8 h of adsorption, 1 h of regeneration, and 30 min of cooling. Calculate the number of vessels adsorbing at any moment, the amount of carbon and dimensions of each vessel, and the total adsorbent requirement for the system.

Answer: 24,950 kg

4.15[a]. Design of a carbon adsorber for control of VOC emissions

Repeat Example 4.10, but assuming that the VOC concentration in the waste gas is 1,000 ppm. Calculate the number of adsorbing vessels needed and their dimensions, the total carbon requirement, and the utilities required for an 8-h cycle.

Answer: W_c = *41,368 kg*

4.16[a]. Total capital investment for a carbon adsorber

Estimate the total capital investment required for the carbon adsorber of Problem 4.15. Assume that the working capital required is equal to 10% of the TCI.

Answer: TCI = $1,336,200

4.17b. Total annual cost for a fixed-bed carbon adsorber.

White Rubber Company of Ravenna, Ohio, is a major manufacturer of rubber insulating gloves and sleeves that protect linemen who work on high transmission lines. Operating 24 h/d, 6 d/wk, the plant uses a tremendous volume of a naphtha solvent in their manufacturing process (*Chemical Processing*, January 1983). In 1979 they installed an activated carbon solvent recovery system to prevent costly air pollution problems and to protect the firm from the escalating price of petroleum-based products (the price of the solvent went from $0.05/L in 1973 to $0.37/L in 1983). The system consists of three fixed-bed vessels, each one holding 4,540 kg of carbon. The three stainless steel vessels, set up in parallel configuration, use the activated carbon in 2-h adsorption and 1-h desorption/cooling cycles.

The fully automated system operates at 316 K and handles an air flow of 13.51 m^3/s containing an average of 0.0024 kg of solvent/kg of air with 95% recovery. The density of the liquid solvent is 800 kg/m^3. When installed in 1979, the system had an expected life of 20 yr, with the carbon having a life expectancy of 5 yr. The initial cost of the carbon was $3.00/kg. Because the vessels were made of stainless steel, the equipment cost was twice the cost of a similar carbon steel system. In 1983, the cost of recovering the solvent (excluding the capital recovery cost and the carbon replacement cost) was $0.10/L. Estimate the total annual cost for the system in 1983. The minimum attractive return on investment for White Rubber Co. was 20%. The M&S Index was 599.4 in 1979.

Answer: TAC = – $182,980/yr

4.18b. Profit from an activated carbon solvent recovery system.

Adhesive coated plastic film is manufactured by Bertek Inc. in St. Albans, Vermont, for use in the pharmaceutical field. During the manufacturing process, Freon-113 (1,1,2-trichloroethane) is used. The volatile Freon was lost to the atmosphere through evaporation. In 1983, following EPA's request, Bertek installed an activated carbon solvent recovery system to "contain and control" the Freon vapors (*Chemical Processing*, November 1986).

The system was designed to handle 1.42 m^3/s of air at 300 K and 101.3 kPa, containing 0.01 kg Freon/kg air, with 96% recovery of the vapors. The plant operates 24 h/d,

350 d/yr. The recovery system consists of two stainless steel vessels, each one loaded with 612 kg of 4 × 10 mesh activated carbon. The bulk density of the adsorbent is 550 kg/m^3, and the external porosity of the beds is 40%. The adsorption cycle consists of a 4-h adsorption step, followed by a 4-h desorption/cooling step. The process requires 0.3 kg of steam/kg of carbon during regeneration. The condenser uses 30 L of cooling water/kg of steam. The liquid Freon is completely insoluble in water, therefore no waste water treatment is required.

The useful life of the equipment is expected to be 15 yr, that of the carbon 5 yr. Despite the fact that the recovered solvent cannot be reused in the pharmaceutical field, its quality is sufficient enough to draw reasonable prices on the solvent resale market with recycling firms. Bertek claimed (in 1986) that it was making a profit of $100,000/yr with the system. Assume that a similar system was installed in June 1990 with similar operating conditions. Estimate what the selling price of the recovered solvent needs to be in order to make a profit similar to the one claimed by Bertek. Use the cost data presented in Example 4.13. Assume that the minimum attractive return on investment for Bertek is 20%, and that the equipment cost for stainless steel is twice the cost for carbon steel.

Answer: $0.59/L

4.19c. Fixed-bed adsorber design problem

Problem 3.16 called for the design of an incineration system for the control of VOC emissions from a chemical process. Design an activated carbon fixed-bed adsorber for the same purpose. The adsorbent properties are: coal-based, granular activated carbon; 10 × 30 mesh size; internal porosity, 70%; external porosity, 40%, bulk density; 480 kg/m^3; surface area, 1,100 m^2/g.

A generalized correlation of adsorption potential shows that the logarithm of the amount adsorbed is linear with the function $(T/V)[\log 10(f_s/f)]$ over a wide range of values, where

T = temperature, K

V = molar volume of liquid at the normal boiling point, cm3/mol

f_s = fugacity of saturated liquid (approximate as vapor pressure)

f = fugacity of the vapor (approximate as partial pressure)

The particular activated carbon under consideration exhibits the following adsorption characteristics for hydrocarbons:

Amount Adsorbed cm^3 of Liquid/100 g Solid	$(T/V)[\log_{10}(f_s/f)]$ K-mol/cm^3
1.0	23
10.0	12

Consider two alternatives: a 2-h cycle, and an 8-h cycle. For the 2-h cycle stainless steel construction is required. The useful life of the carbon will be 18 mo for the shorter cycle and 36 mo for the longer cycle. Assume that the cost data presented in Example 4.13 is valid. The market value of the recovered mixed solvent is \$0.25/L. Compare your results with those of Problem 3.16.

4.20b. Moving bed adsorption system for control of VOCs

The development of a moving adsorbent bed VOC control system could help lead to smaller designs by allowing higher superficial gas velocities than in fixed-bed units (Larsen, E. S. and Pilat, M. J. *J. Air Waste Manage. Assoc.*, **41**:1199, 1991). Figure 4.6 is a schematic diagram of a proposed cross-flow moving bed adsorption system to destroy 99.9% of toluene emissions from a chemical process.

(a) Estimate the adsorbent recirculation rate, in kg/h.

Answer: 17.56 kg/h

(b) Estimate the flow rate of purge air required.

Answer: 1.71 m^3/min

(c) Estimate the amount of methane required as auxiliary fuel in the incinerator.

4.21a. Fluidized bed adsorbers

In some applications, fluidized bed adsorbers are preferred over fixed-bed units. The pressure drop is lower for the fluidized adsorber, and the possibility of bed clogging by aerosols in the waste gas is greatly reduced. Hori et al. (*JAPCA*, **38**:269, 1988) proposed the following semi-empirical equation to estimate the 10% breakthrough time for a fluidized adsorption bed:

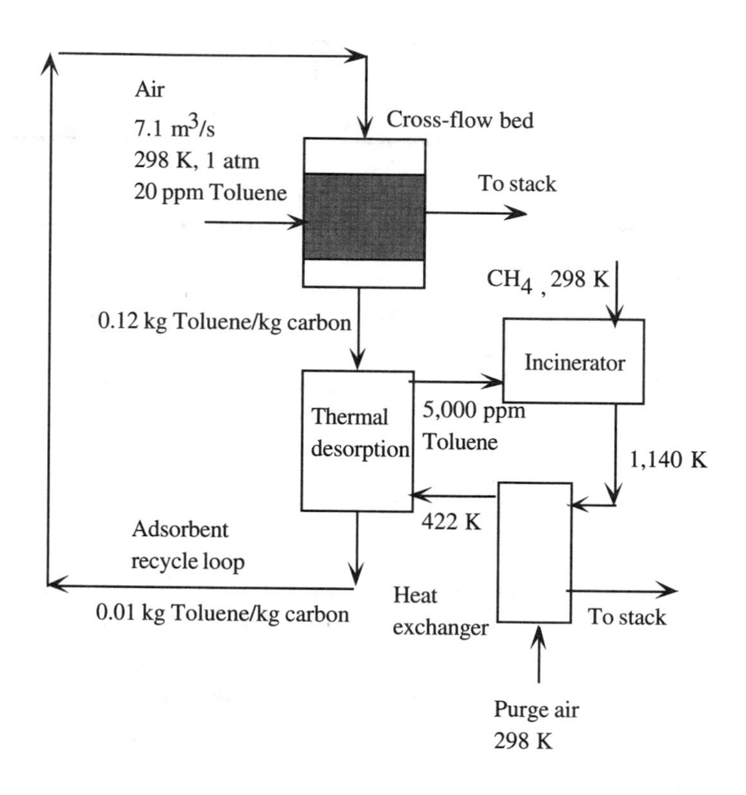

Figure 4.6 Schematic diagram of a cross-flow adsorption system.

$$\theta_B = \frac{\rho_0 Z_0 A_c m^*}{Q C_0}\left(1 - \frac{Z_0}{2Z_f}\right)$$

where

Z_0 is the static bed height

Z_f is the bed height in the fluidized state

ρ_0 is the static bed density

Q is the volumetric flow rate

C_0 is the inlet concentration

Consider the fixed-bed of Example 4.3. If the adsorbent particle size is reduced to 1 mm, the gas velocity is high enough to fluidize the adsorbent to a depth 50% higher than that of the static bed. Estimate the 10% breakthrough time under those conditions. Assume that the equilibrium data and static bed density given in Example 4.3 apply.

Answer: 14.1 min

4.22b. Pressure swing adsorption (PSA)

In PSA systems, heatless regeneration of the adsorbent during the desorption cycle is achieved by reducing the total pressure and purging the bed at low pressure with a small fraction of the product stream (Ruthven, 1984, p. 361; Treybal, 1980, p.628). Figure 4.7 shows the method of operation. The waste gas under pressure flows through four-way valve 3 into adsorber bed 1, and the clean gas leaves at the top to the stack. A portion of the clean gas passes through valve 4 and at lower pressure into bed 2 to desorb the adsorbate accumulated there from a previous cycle. It leaves through valve 3. Check valves 5 and 6 allow flow to the right and left, respectively. After a short interval (30 to 300 s), beds 1 and 2 switch roles. Proper selection of the extremes of the pressure swing results in a purge stream adsorbate concentration many times higher that in the feed to the PSA process. The purge can be further processed to recover the adsorbate. The design of PSA systems was covered in detail by White et al. (*Chem. Engng. Prog.*, **85**:25, 1989).

The BOC Group developed a new process involving PSA which improves the economics of partial-oxidation and ammoxidation reactions (*Chem. Eng.*, January 1991, p. 23). Conventional processes incinerate unreacted hydrocarbons in the gas stream as wastes. BOC's PSA process recovers and recycles hydrocarbons, most notably propylene. Additional capital required for a 200,000-metric ton/yr acrylonitrile plant is about $8 million, which will generate savings of $4.7 million/yr for 10 years. Estimate the profitability index for this investment.

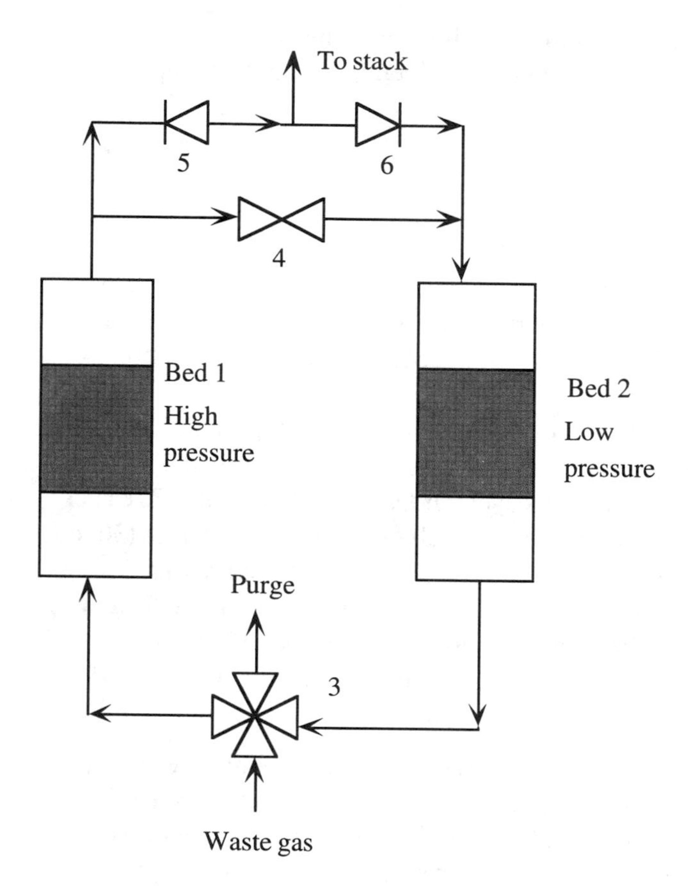

Figure 4.7 Pressure swing adsorption system

5

Flue Gas Desulfurization

5.1 INTRODUCTION

SO_2 is emitted from coal-fired power plants (about two thirds of U.S. emissions), from industrial fuel combustion, sulfuric acid manufacturing, and smelting of nonferrous metals. The two approaches to SO_2 emission control are (1) to remove the sulfur from the fuel before it is burned, or (2) to remove SO_2 from the exhaust gases. There has been a significant amount of effort expended worldwide on the development of processes in both categories. This chapter considers only methods for the removal of SO_2 from exhaust gases.

There are two types of effluent gas desulfurization tasks. The first is to remove SO_2 from power plant flue gases. These gases generally contain low concentrations of SO_2 (less than 0.5% by volume), but are emitted at tremendous volumetric flow rates. For example, a coal-fired power plant burning 2% sulfur coal (by weight) produces 40,000 kg of SO_2 for every 10^6 kg of coal burned. The second desulfurization problem is the removal of SO_2 from streams containing relatively high concentrations at low flow rates. Streams of this type are typical of those emitted from smelter operations. A smelter emission gas typically contains SO_2 at a concentration of about 10% by volume.

This chapter concentrates largely on the problem of sulfur dioxide removal from power plant flue gases, so-called flue gas desulfurization (FGD), because it represents a more prevalent and, in many respects, a more difficult problem than that of SO_2 removal from smelting and other industrial operations. Elliot et al. (1982) reviewed a number of processes for the cleaning of smelter gases, and the interested reader is referred to this source.

5.2 MATERIAL AND ENERGY BALANCES

Your objective in studying this section is to

Apply material and and energy balances to calculate the volumetric flow rate of flue gases from combustion sources and calculate their sulfur dioxide content.

This section illustrates the use of material and energy balances to calculate the volumetric flow rate and composition of exhaust gases from combustion sources. This is the preliminary step in the design of any FGD system.

Example 5.1 Exhaust Gases from Coal-Fired Power Plant

Cogentrix, Inc., a cogeneration outfit, plans to build a coal-fired, 300-MW power plant in Mayagüez, a city on the west coast of Puerto Rico (Cogentrix 1990). The analysis of a typical coal is 68.95% C, 2.25% H, 6.0% O, 1.4% N, 1.5% S, 0.1% Cl, 7.8% H_2O, and 12.0% ash. The heating value of the coal is 27,000 kJ/kg. The power cycle has a thermal efficiency of 35%. To reduce the formation of nitrogen oxides, the excess air will be limited to 20%. Assume complete combustion. Enough CO is formed to cause air pollution problems but not enough to alter significantly the material balance calculations. The flue gases from the boiler flow through a baghouse for particulate removal. They flow out of the baghouse at 530 K. The fabric filters remove more than 99% of the particulate matter. The effluent is quenched to 350 K with seawater before entering the FGD system. Calculate the volumetric flow rate and composition of the flue gases entering the desulfurization system.

Solution

Figure 5.1 is a schematic diagram of the system. The first step is to calculate the rate at which the coal is needed.

Basis: 1 minute

The amount of coal needed is $(300 \times 10^6)(60)/[(0.35)(27 \times 10^3)] = 1{,}905$ kg/min. From the stoichiometry of the combustion reactions and the amount of coal fired, calculate the rate of production of all the combustion gases. The reactions are:

$$C + O_2 \rightarrow CO_2 \tag{5.1}$$

$$2\,H_2 + O_2 \rightarrow 2\,H_2O \tag{5.2}$$

$$S + O_2 \rightarrow SO_2 \tag{5.3}$$

$$Cl + H \rightarrow HCl \tag{5.4}$$

Theoretical oxygen required = 1,905 [(0.6895)(1)/(12) + (0.0225)(0.5)/(2) + (0.015)(1)/(32) – (0.06)/(32)] = 117.49 kmol of O_2. Notice that only the first three reactions are considered for this calculation, and the oxygen supplied by the coal is subtracted. For 20% excess air, the oxygen supplied with the air is (1.20)(117.49) = 141.27 kmol. The nitrogen supplied

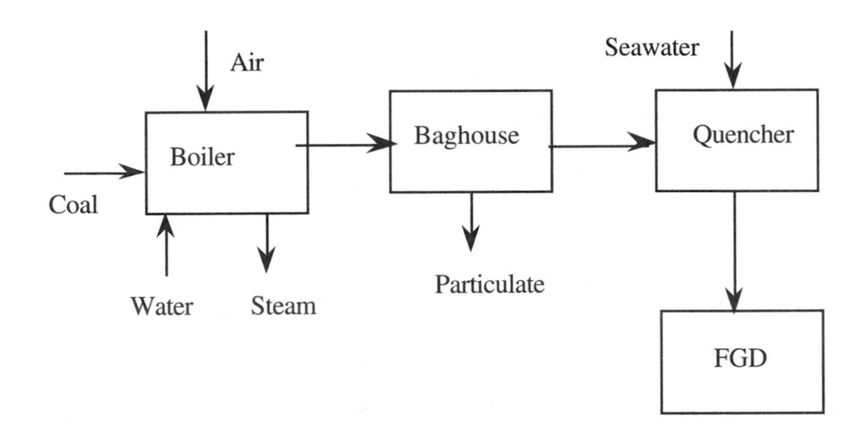

Figure 5.1. Schematic of the air pollution control equipment for the coal-fired power plant of Example 5.1

with the air is (79)(141.27)/(21) = 531.2 kmol. The following table summarizes the material balance calculations to determine the molar flow rate of the gases leaving the boiler and their composition:

Component	In	Formed	Consumed	Out	Fraction
CO_2	0.0	109.45	0.0	109.45	0.1573
SO_2	0.0	0.89	0.0	0.89	0.0013
HCl	0.0	0.05	0.0	0.05	7.18×10^{-5}
H_2O	8.26	21.42	0.0	29.68	0.0426
O_2	144.84	0.0	121.06	23.78	0.0342
N_2	532.13	0.0	0.0	532.13	0.7646
			Total	696.0 kmol/min	

The volumetric flow rate of the gases entering the baghouse is from the ideal gas law: $V = nRT/P$ = (696.0)(8.314)(530)/[(101.3)(60)] = 504.7 m³/s.

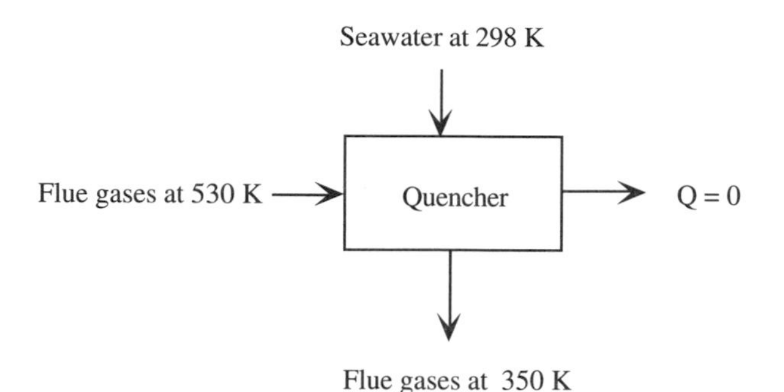

Figure 5.2. Quenching of flue gases with seawater.

To calculate the particulate loading to the baghouse, assume that 90% of the ash in the coal leaves the boiler with the flue gases in the form of "fly ash." The particulate loading is (1,905)(0.12)(0.90) = 205.7 kg/min. Most of this ash is removed in the baghouse. To calculate the total flow rate and composition of the gases entering the FGD system, combine material and energy balances around the quencher. Assuming adiabatic quenching, $\Delta H = 0$.

$$\Delta H_{Gases} + \Delta H_{Water} = 0 \tag{5.5}$$

From Table 3.3, $\Delta H_{Gases} =$ 29.68 (1,742 – 7,961) + 109.45 (1,862 – 9,878) + 23.78(1,479 – 7,085) + 532.13 (1,410 – 6,645) = –3,982,000 kJ. Notice that the particulate matter, SO_2, and HCl were not included in the previous calculation since they are present in such small amount at the baghouse outlet. From Eq. (5.5), ΔH_{Water} = 3,982,000 kJ = $m_w \Delta h_w$, where m_w is the mass of seawater added and Δh_w is the enthalpy change when 1 kg of water goes from the liquid state at 298 K to the vapor state at 350 K. From the Steam Tables (Haar et al. 1984), Δh_W = 2,533 kJ/kg. Therefore, m_W = 3,982,000/2,533 = 1,572 kg. The moles of seawater added are 87.3 kmol. The following table summarizes the molar flows and concentrations of all the gases leaving the quencher (or entering the FGD system).

Component	Moles	Molar Fraction
CO_2	109.45	0.1397
SO_2	0.89	1,136 ppm
HCl	0.05	64 ppm
H_2O	116.98	0.1494
O_2	23.78	0.0304
N_2	532.13	0.6794
Total	783.20	1.0000

Use the ideal gas law to calculate the volumetric flow rate at the entrance of the FGD system: $V = (783.20)(8.314)(350)/[(101.3)(60)] = 375\ m^3/s$.

Comments

The volumetric flow rate to be processed by the FGD system is truly enormous, whereas the SO_2 concentration is extremely low. This combination makes the design of FGD systems unique among mass-transfer processes.

5.3 GAS ABSORPTION APPLIED TO FGD

Your objective in studying this section is to

Apply the concepts of gas absorption with and without chemical reaction to the design of FGD units.

Most flue gas desulfurization systems depend on the selective absorption of the SO_2 by a liquid. Absorption by pure water is rarely used since the solubility of SO_2 in water is only moderate. Enormous amounts of pure water are required for removal efficiencies of the order of 95%, typical of the new and proposed performance standards. An alkali agent is added to the water, which reacts in the liquid phase with the absorbing SO_2, increasing its solubility.

5.3.1 Gas absorption without chemical reaction

The application of the concepts of gas absorption without chemical reaction to FGD systems is simple because the simplifications for dilute solutions are always valid. The operation and equilibrium lines are straight. The determination of the minimum liquid flow rate and the number of transfer units are straightforward.

Example 5.2 Liquid Flow Rate for Sulfur Dioxide Absorber

The gases leaving the quencher of Example 5.1 enter a packed-bed absorber where they come into intimate contact with pure water at 300 K. Calculate the flow rate of water for 93% removal of the SO_2. Assume that the design water flow rate is 50% higher than the minimum. The equilibrium line at this temperature is $y^* = 11.71x$, where x and y^* are the SO_2 mole fractions in the water and gas, respectively. Assume that the gases cool down to 300 K immediately on contact with the water.

Solution

From Example 5.1, the initial SO_2 concentration in the gas is 1,136 ppm. For 93% removal, the final concentration in the gas is 80 ppm. Because the operation and equilibrium lines are straight, the minimum liquid rate condition is when the lines cross at the point (y_1, x_{max}) which lies on the equilibrium line as shown on Figure 5.3. From the equilibrium relation, x_{max} = 1,136/11.71 = 97 ppm. The slope of the operation line is

$$\frac{L_{min}}{G} = \frac{y_1 - y_2}{x_{max}} = \frac{1,136 - 80}{97} = 10.89 \qquad (5.6)$$

From Example 5.1, G = 783.20 kmolmin. Therefore, L_{min} = 8,529 kg/min; L =1.5 L_{min} = 12,794 kmol/ min. Hence, the water flow rate is 230,283 kg/min (3.32×10^8 kg/d).

Comments

This example shows that absorption with pure water is not a practical method for desulfurization of power plant effluents because it requires enormous amounts of water.

5.3.2 Absorption with chemical reaction

The equilibrium vapor pressure of SO_2 over the liquid depends on the concentration of dissolved SO_2. It can be reduced almost to zero by adding a reagent to the absorbing liquid that reacts with the dissolved solute, effectively "pulling" more of the solute into solution.

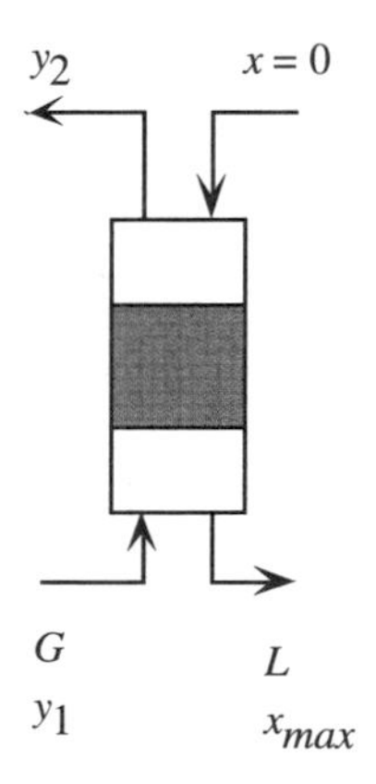

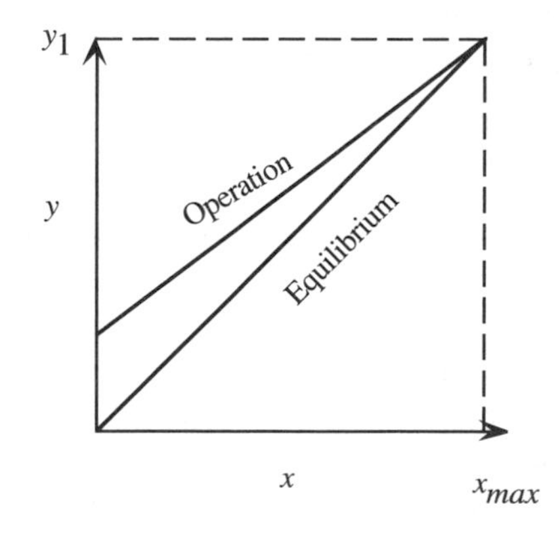

Figure 5.3 Minimum liquid rate for gas absorption.

Consider an absorption tower in which an alkaline reagent is added to the fresh liquid feed at the top of the column. It is necessary to account for the chemical state of the dissolved SO_2 to compute its equilibrium partial pressure and, therefore, the gas-phase driving force for absorption.

Let

G = total molar flow rate of entering gas, per unit cross-sectional area

y_0 = mole fraction of SO_2 in the entering gas

y_1 = desired mole fraction of SO_2 in the exiting gas

z = distance measured from the top of the tower

W = volumetric flow rate of liquid fed at the top of the column, per unit cross-sectional area

Performing a balance on gas-phase SO_2 over a section of the packed-bed of depth dz,

$$\frac{d}{dz}\left(\frac{y}{1-y}\right)=\frac{k_y a(y-y_i)}{G\,(1-y_0)} \tag{5.7}$$

which is to be solved subject to

$$y\,(Z_T)=y_0 \tag{5.8}$$

where Z_T is the packed height of the tower and $k_y a$ is the volumetric mass-transfer coefficient. All the quantities in Eq. (5.7) are known except y_i, the interfacial SO_2 mole fraction. To determine y_i consider the behavior of the liquid phase as a function of position in the tower. The absorption of SO_2 by water leads to the following equilibria:

$$SO_2\,(g) + H_2O \leftrightarrow SO_2 \cdot H_2O \tag{5.9}$$

$$SO_2 \cdot H_2O \leftrightarrow H^+ + HSO_3^- \tag{5.10}$$

$$HSO_3^- \leftrightarrow H^+ + SO_3^{2-} \tag{5.11}$$

$$H_2O \leftrightarrow H^+ + OH^- \tag{5.12}$$

The concentrations of the dissolved sulfur species in the liquid, given the equilibrium partial pressure of SO_2 ($p^*_{SO_2}$), are found from the equilibrium constant expressions in Table 5.1

to be

$$[SO_2 \cdot H_2O] = K_{hs}\, p^*_{SO_2} \tag{5.13}$$

$$[HSO_3^-] = \frac{K_{hs} K_{s1}\, p^*_{SO_2}}{[H^+]} \tag{5.14}$$

$$[SO_3^{2-}] = \frac{K_{hs} K_{s1} K_{s2}\, p^*_{SO_2}}{[H^+]^2} \tag{5.15}$$

Table 5.1 Equilibrium Constants for Aqueous Absorption of SO_2

Reaction No.	Equilibrium Constant Expression	Equilibrium Constant [a,b]
(5.9)	$K_{hs} = \frac{[SO_2 \cdot H_2O]}{p^*_{SO_2}}$	$\log K_{hs} = \frac{1{,}376.1}{T} - 4.521$
(5.10)	$K_{s1} = \frac{[H^+][HSO_3^-]}{[SO_2 \cdot H_2O]}$	$\log K_{s1} = \frac{853}{T} - 4.74$
(5.11)	$K_{s2} = \frac{[H^+][SO_3^{2-}]}{[HSO_3^-]}$	$\log K_{s2} = \frac{621.9}{T} - 9.278$
(5.12)	$K_w = [H^+][OH^-]$	$pK_w = \frac{4{,}424.1}{T} + 0.01656T - 5.782$

[a] Values of K_{hs}, K_{s1}, and K_{s2} from Maahs (1982).

[b] Value of K_w derived from data in Harned and Owen (1958).

Concentration units are kmol m^{-3}; pressure units are atm.

As we noted, to enhance the solubility of SO_2 it is customary to raise the pH of the feed over that of pure water through the addition of an alkaline substance. Let us presume that an amount of nonvolatile salt MOH that dissociates in solution into M^+ and OH^- is added to the feed water such that the initial pH is pH_0. Electroneutrality must always be maintained locally in the liquid, so the concentration of the ion M+ is found from the specified initial pH,

$$[H^+]_0 + [M^+]_0 = [OH^-]_0 \tag{5.16}$$

and because MOH is nonvolatile,

$$[M^+] = [M^+]_0 = \frac{K_w}{[H^+]_0} - [H^+]_0 \tag{5.17}$$

Electroneutrality at any time is expressed as

$$[M^+] + [H^+] = [OH^-] + [HSO_3^-] + 2[SO_3^{2-}] \tag{5.18}$$

which can be written in terms of $[H^+]$ as

$$[M^+] + [H^+] = \frac{K_w}{[H^+]} + \frac{K_{hs} K_{s1} p^*_{SO_2}}{[H^+]} + \frac{2 K_{hs} K_{s1} K_{s2} p^*_{SO_2}}{[H^+]^2} \tag{5.19}$$

or

$$[H^+]^3 + [M^+][H^+]^2 - (K_w + K_{hs} K_{s1} p^*_{SO_2})[H^+] - 2 K_{hs} K_{s1} K_{s2} p^*_{SO_2} = 0 \tag{5.20}$$

The local hydrogen ion concentration in the liquid is related to the SO_2 interfacial partial pressure by Eqs. (5.17) and (5.20). Because there are three unknowns, $[H^+]$, $[M^+]$, and $p^*_{SO_2}$, another equation is needed relating these quantities. At the top of the tower $[H^+] = [H^+]_0$; however, as soon as the falling liquid encounters SO_2, absorption takes place and the hydrogen ion concentration begins to change. To calculate how $[H^+]$ changes with z perform an overall material balance on SO_2 between the top of the tower ($z = 0$) and any level z. For dilute solutions—which implies constant gas flow rate—that balance takes the form

$$W([SO_2 \cdot H_2O] + [HSO_3^-] + [SO_3^{2-}]) = G(y - y_1) \tag{5.21}$$

Use the equilibrium relationships on the left-hand side of Eq. (5.21)

$$W K_{hs} p^*_{SO_2}\left(1+\frac{K_{s1}}{[H^+]}+\frac{K_{s1}K_{s2}}{[H^+]^2}\right)=G(y-y_1) \tag{5.22}$$

To place this equation in a more compact form, let

$$\eta = [H+] , \quad A=\frac{K_w}{[H^+]_0}-[H^+]_0 , \quad B = K_{hs} Ks_1 , \quad C = K_{hs} K_{s1} K_{s2}$$

$$D = G/W , \quad E = y_1$$

Equation (5.22) can be written as

$$p^*_{SO_2}=\frac{D\eta^2}{K_{hs}\eta^2+B\eta+C}(y-E) \tag{5.23}$$

Also, combining Eqs. (5.17) and (5.20),

$$p^*_{SO_2}=\frac{\eta^3+A\eta^2-K_w\eta}{B\eta+2C} \tag{5.24}$$

Equations (5.23) and (5.24) yield a single nonlinear algebraic equation relating $[H^+]$ (i.e., η)and y:

$$y=f(\eta)=\frac{(\eta^2+A\eta-K_w)(K_{hs}\eta^2+B\eta+C)}{D\eta(B\eta+2C)}+E \tag{5.25}$$

For dilute solutions, Eq. (5.7) becomes

$$\frac{dy}{dz}=\frac{k_y a}{G}(y-y_i) \tag{5.26}$$

Substitute y from Eq. (5.25) in Eq. (5.26), and

$$y_i = \frac{p_i}{P}$$

$$m_2(\eta)\frac{d\eta}{dz} = \frac{k_y a}{G} m_1(\eta) \tag{5.27}$$

where

$$m_1(\eta) = \left[f(\eta) - \frac{\eta^3 + A\eta^2 - K_w\eta}{P(B\eta + 2C)} \right] \tag{5.28}$$

$$m_2(\eta) = \frac{df(\eta)}{d\eta} \tag{5.29}$$

Integrating Eq. (5.27) over the tower gives:

$$\frac{G}{k_y a}\int_{\eta_0}^{\eta_T} \frac{m_2(\eta)}{m_1(\eta)}\,d\eta = \int_0^{Z_T} dz = Z_T \tag{5.30}$$

The upper limit of the integral in Eq. (5.30) , η_T, is the solution of Eq. (5.25) at $y\ (Z_T) = y_0$, that is,

$$y_0 = f(\eta_T) \tag{5.31}$$

Solve Eq. (5.31) for η_T (and thus the pH of the liquid solution leaving the tower), and then evaluate the integral in Eq. (5.30) to calculate the tower height for a given set of operating conditions.

There is a liquid flow rate, W_{min} , below which a specified separation cannot be achieved. It can be shown (see Problem 5.3) that the minimum liquid flow rate is given by

$$\frac{G}{W_{min}} = \frac{y_0 P}{y_0 - y_1}\left(K_{hs} + \frac{B}{\eta_{max}} + \frac{C}{\eta_{max}^2}\right) \qquad (5.32)$$

where η_{max} is the solution to

$$y_0 P = \frac{\eta_{max}^3 + A\eta_{max}^2 - K_w \eta_{max}}{B\eta_{max} + 2C} \qquad (5.33)$$

Example 5.3 The Use of Seawater for FGD

Cogentrix proposes to use seawater to scrub sulfur dioxide from the effluent of the coal-fired power plant of Examples 5.1 and 5.2 . Seawater is naturally slightly alkaline, with a pH of approximately 8. The effluent from the quencher enters a packed tower where it flows countercourrent to seawater, which enters the tower at 300 K, with a pH = 8, and at a flow rate which is 28% above the minimum. Calculate the flow rate of water and the pH of the water leaving the tower for a removal efficiency of 93%.

Solution

From Examples 5.1 and 5.2, $y_0 = 0.001136$, $y_1 = 0.00008$, $G = 783.17$ kmol/min. Calculate the constants in Eq. (5.25); $E = y_1 = 0.00008$; from Table 5.1, at 300 K, $K_w = 1.167 \times 10^{-14}$, $K_{hs} = 1.164$, $K_{s1} = 0.0127$, $K_{s2} = 6.237 \times 10^{-8}$. Therefore, $A = 1.157 \times 10^{-6}$, $B = 0.01478$, $C = 9.353 \times 10^{-10}$. Substitute in Eq. (5.33) and solve for $\eta_{max} = 0.00413$ kmol/m^3. From Eq. (5.32), $W_{min} = 154$ m^3/min. The actual water flow rate is $W = 1.28(154) = 197$ m^3/min. Therefore, $D = 783.17/197 = 4$. Substituting in Eq. (5.31) and solving, $\eta_T = 0.00334$. The pH of the water leaving the tower is pH = – log (0.00334) = 2.48.

Comments

An important drawback of seawater scrubbing is the low pH of the liquid effluent. The water must be aerated in a separate basin to raise its pH to acceptable levels (from 6 to 7) before discharge back to the ocean. Some of the absorbed SO_2 may be released to the atmosphere during the aeration (Abrams et al. 1988). Another disadvantage is the enormous quantities of water required. Figure 5.4 is a schematic diagram of a seawater FGD scrubber.

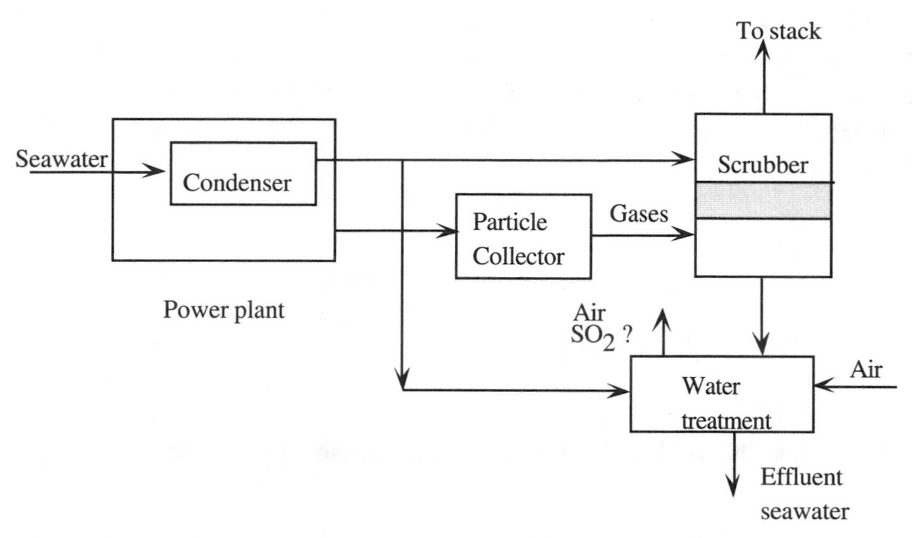

Figure 5.4 Schematic diagram of a seawater FGD system

Example 5.4 Irrigated Packing Height for Seawater Scrubber

The separation of Example 5.3 will occur in a cylindrical tower with a diameter of 20 m. The tower will be packed with 2.54-cm ceramic Raschig rings. Estimate the irrigated packing height required to remove 93% of the SO_2 with the seawater flow rate calculated in Example 5.3. The following correlation is available for absorption of SO_2 at 300 K in towers packed with 2.54-cm Raschig rings (McCabe and Smith 1976):

$$k_y a = 0.09944 L_m^{0.25} G_m^{0.7} \qquad (5.34)$$

where L_m and G_m are the mass flow rates of liquid and gas, respectively, in kg/m^2-h, and $k_y a$ is in kmol/m^3-h-mole fraction.

Solution

Use Eq. (5.30) to estimate Z_T. The average molecular weight of the gases is 28.4 kg/kmol (see Example 5.1). The cross-sectional area of the tower is 314 m^2. Therefore, G_m =

(783.17)(60)(28.4)/(314) = 4,250 kg/m2-h; L_m = (197)(60) (1000)/314 = 37,640 kg/m2-h. From Eq. (5.34), $k_y a$ = 480 kmol/m3-h. Calculate the height of a transfer unit in the gas phase, $H_{tG} = G/k_y a$ = 0.312m. A transformation of variables simplifies calculation of the number of transfer units, N_{tG}, given by the integral on the left-hand side of Eq. (5.30). Because pH = –log η, dη = –η d(pH) and

$$N_{tG} = \int_{\mathrm{pH}_T}^{\mathrm{pH}_0} \eta \frac{m_2(\eta)}{m_1(\eta)} d(\mathrm{pH}) \tag{5.35}$$

After considerable algebraic manipulation, the previous equation becomes:

$$N_{tG} = \int_{\mathrm{pH}_T}^{\mathrm{pH}_0} \frac{\sum_{i=0}^{5} a_i \eta^i}{\left(\eta + \frac{2C}{B}\right) \sum_{i=0}^{4} c_i \eta^i} d(\mathrm{pH}) \tag{5.36}$$

where

$$a_5 = 2K_{hs}$$

$$a_4 = B + \left(A + \frac{6C}{B}\right) K_{hs}$$

$$a_3 = 4C\left(1 + \frac{AK_{hs}}{B}\right)$$

$$a_2 = AC + \frac{2C^2}{B} + BK_w - \frac{2C}{B} K_w K_{hs}$$

$$a_1 = 2CK_w$$

$$a_0 = \frac{2K_w C^2}{B}$$

$$c_4 = PK_{hs} - D \quad , \qquad c_3 = P(B + AK_{hs}) - DA$$

$$c_2 = P(BA - K_w K_{hs} + C) + DK_w + EDB$$

$$c_1 = P(CA - BK_w) + 2EDC\,, \qquad c_0 = -K_w C$$

A computer program was developed to estimate N_{tG} using Gaussian quadrature. Given the gas molar flow rate, initial and final SO_2 molar fractions in the gas phase, initial water pH, operating temperature and pressure, and a multiplier of the minimum water flow rate, it calculates the minimum and actual water volumetric flow rates, the final water pH, and the number of transfer units in the gas phase. Applied to the conditions under consideration, the answer is N_{tG} = 1.619 units. The total irrigated height is Z_T = (1.619)(0.312) = 0.51 m. Figure 5.5 illustrates the effect of the water flow rate on the number of transfer units for the separation specified in this example.

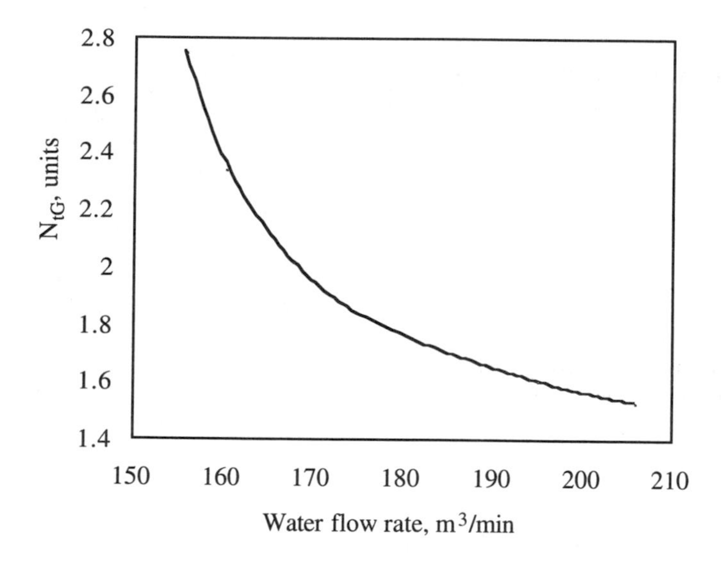

Figure 5.5 Effect of water flow rate on the number of transfer units for a seawater FGD system.

```
      PROGRAM NTG
C     THIS PROGRAM CALCULATES THE NUMBER OF TRANSFER UNITS
C     IN THE GAS PHASE FOR SCRUBBING SO2 FROM FLUE GASES WITH
C     AN AQUEOUS ALKALINE SOLUTION. GAUSS-LEGENDRE QUADRATURE
```

```
C    IS USED FOR NUMERICAL INTEGRATION (N=16).
C
      PARAMETER (N=16)
      COMMON A,B,C,D,KW,KHS,P,Y0,Y1
      EXTERNAL FUNC1,FUNC2
      REAL A,AA(6),B,C,CC(5),D,E,G,KW,KHS,KS1,KS2,P,PHT,PH0,T,
     * W(20),WW,X(20),Y0,Y1,MULT
      PRINT *, 'ENTER ABSOLUTE TEMPERATURE IN K  '
      PRINT *
      READ *, T
      PRINT *, 'ENTER INITIAL PH OF THE WATER   '
      PRINT *
      READ *, PH0
      PRINT *, 'ENTER GAS MOLAR FLOW RATE, G   '
      PRINT *
      READ *, G
      PRINT *, 'ENTER INITIAL GAS SO2 MOL FRACTION, Y0 '
      PRINT *
      READ *, Y0
      PRINT *, 'ENTER FINAL GAS SO2 MOL FRACTION, Y1  '
      PRINT *
      READ *, Y1
      PRINT *, 'ENTER TOTAL PRESSURE IN ATM, P  '
      PRINT *
      READ *, P
      PRINT *
      PRINT *, ' ENTER MULTIPLE OF MINIMUM WATER FLOW RATE '
      PRINT *
      READ *, MULT
      KHS = 10**(1376.1/T-4.521)
      KS1 = 10**(853/T-4.74)
      KS2 = 10**(621.9/T-9.278)
      KW = 10**(-4424.1/T-.01656*T+5.782)
      A = KW/(10**(-PH0))-(10**(-PH0))
      B = KHS*KS1
      C = B*KS2
      E = Y1
      CALL RTFIND(FUNC1,HMAX)
      WWMIN=G*(Y0-Y1)/((Y0*P)*(KHS+B/HMAX+C/HMAX**2))
      WW = MULT*WWMIN
      PRINT *
      PRINT *, 'THE MINIMUM WATER FLOW RATE IS ',WWMIN
      PRINT*
      PRINT *, 'THE ACTUAL WATER FLOW RATE IS  ', WW
      D=G/WW
      CALL RTFIND (FUNC2, HT)
```

```
      PHT =-LOG10(HT)
      PRINT *
      PRINT *, 'THE FINAL WATER PH IS  ', PHT
      AA(6) = 2*KHS
      AA(5) = B+(A+6*C/B)*KHS
      AA(4) = 4*C*(1+A*KHS/B)
      AA(3) = A*C+2*(C**2)/B+B*KW-2*C*KW*KHS/B
      AA(2) = 2*C*KW
      AA(1) = 2*KW*(C**2)/B
      CC(5) = P*KHS-D
      CC(4) = P*(B+A*KHS)-D*A
      CC(3) = P*(B*A-KW*KHS+C)+D*KW+E*D*B
      CC(2) = P*(C*A-B*KW)+2*E*D*C
      CC(1) = -KW*C
      CALL GAULEG (PHT,PH0,X,W,N)
      SUM = 0.0
      DO 30 I=1,N
        XX=10**(-X(I))
        SUM1=0.0
        DO 10 J=1,6
          SUM1=SUM1+AA(J)*XX**(J-1)
 10     CONTINUE
        SUM2 = 0.0
        DO 20 J=1,5
          SUM2 = SUM2+CC(J)*XX**(J-1)
 20     CONTINUE
        SUM = SUM+W(I)*SUM1/((XX+2*C/B)*SUM2)
 30   CONTINUE
      PRINT *
      PRINT *,'NTG =  ', SUM     END
      END

      SUBROUTINE RTFIND(FUNCD,ROOT)
C
C     USING A COMBINATION OF NEWTON-RAPHSON AND BISECTION, FIND
C     THE ROOT OF A FUNCTION. X1 AND X2 ARE INITIAL ESTIMATES OF
C     THE ROOT, WHICH IS REFINED TO AN ACCURACY XACC SPECIFIED
C     BY THE USER. FUNCD IS A USER- SUPPLIED SUBROUTINE WHICH
C     RETURNS BOTH THE FUNCTION VALUE (F) AND THE FIRST DERIVATIVE (DF)
C
      PARAMETER (FACTOR=1.6,NTRY=50)
      COMMON A,B,C,D,KW,KHS,P,Y0,Y1
      REAL A,B,C,D,KW,KHS,P,Y0,Y1
      EXTERNAL FUNCD
C
```

```
C    ROOT BRACKETING ALGORITHM
C
      X1=0.0
      X2=0.001
      XACC=0.00001
      CALL FUNCD (X1, F1, DF1)
      CALL FUNCD (X2, F2, DF2)
      DO 11 J=1, NTRY
        IF (F1*F2 .LT. 0.) THEN
            GO TO 12
        ELSE
            X2=X2+FACTOR*(X2-X1)
            CALL FUNCD(X2, F2, DF2)
        END IF
 11   CONTINUE
C
C    END OF BRACKETING
C
 12   ROOT=RTSAFE (FUNCD,X1,X2,XACC)
      END
C
C
      SUBROUTINE FUNC1(X,F,DF)
      REAL A,B,C,D,KW,KHS,P,Y0,Y1
      COMMON A, B, C,D,KW,KHS,P,Y0,Y1
      F = X**3+A*X**2-(KW+Y0*P*B)*X-2*C*Y0*P
      DF = 3.*X**2+2.*A*X-(KW+Y0*P*B)
      RETURN
      END

C
C
      SUBROUTINE FUNC2(X,F,DF)
      COMMON A,B,C,D,KW,KHS,P,Y0,Y1
      REAL A,B,C,D,KW,KHS,P,Y0,Y1
      F=(X**2+A*X-KW)*(KHS*X**2+B*X+C)-(Y0-Y1)*D*X*(B*X+2*C)
      DF=(X**2+A*X-KW)*(2*KHS*X+B)+(KHS*X**2+B*X+C)*(2*X+A)
     * -2*(Y0-Y1)*D*(B*X+C)
      RETURN
      END

      FUNCTION RTSAFE(FUNCD,X1,X2,XACC)
      COMMON A,B,C,D,KW,KHS,P,Y0,Y1
      REAL A,B,C,D,KW,KHS,P,Y0,Y1
      PARAMETER (MAXIT=100)
      EXTERNAL FUNCD
```

```
      CALL FUNCD(X1,FL,DF)
      CALL FUNCD(X2,FH,DF)
      IF(FL*FH.GE.0.) PAUSE 'root must be bracketed'
      IF(FL.LT.0.)THEN
       XL=X1
       XH=X2
      ELSE
       XH=X1
       XL=X2
       SWAP=FL
       FL=FH
       FH=SWAP
      ENDIF
      RTSAFE=.5*(X1+X2)
      DXOLD=ABS(X2-X1)
      DX=DXOLD
      CALL FUNCD(RTSAFE,F,DF)
      DO 11 J=1,MAXIT
       IF(((RTSAFE-XH)*DF-F)*((RTSAFE-XL)*DF-F).GE.0.
     *     .OR. ABS(2.*F).GT.ABS(DXOLD*DF) ) THEN
        DXOLD=DX
        DX=0.5*(XH-XL)
        RTSAFE=XL+DX
        IF(XL.EQ.RTSAFE)RETURN
       ELSE
        DXOLD=DX
        DX=F/DF
        TEMP=RTSAFE
        RTSAFE=RTSAFE-DX
        IF(TEMP.EQ.RTSAFE)RETURN
       ENDIF
       IF(ABS(DX).LT.XACC) RETURN
       CALL FUNCD(RTSAFE,F,DF)
       IF(F.LT.0.) THEN
        XL=RTSAFE
        FL=F
       ELSE
        XH=RTSAFE
        FH=F
       ENDIF
 11   CONTINUE
      PAUSE 'RTSAFE exceeding maximum iterations'
      RETURN
      END
```

5.3.3 Pressure Drop through Packed-Bed Absorbers

The diameter of a packed-bed absorber is determined by pressure drop considerations. The cross-sectional area should result in gas and liquid mass velocities high enough to stimulate good interfacial contact, but not in excessive pressure drop. There is a trade-off between high power costs at high gas velocities and high capital costs at low gas velocities.

Typically, absorbers are designed for gas-pressure drops of 200 to 400 Pa per meter of packed depth (Treybal 1980). For simultaneous countercourrent flow of liquid and gas, the pressure-drop data of various investigators show wide discrepancies owing to differences in packing density and manufacture, such as changes in wall thickness. Estimates, therefore, can not be expected to be very accurate. For most purposes, the generalized correlation of Figure 5.6 will serve. The values of C_f for some packings commonly used in air pollution control applications are found in Table 5.2.

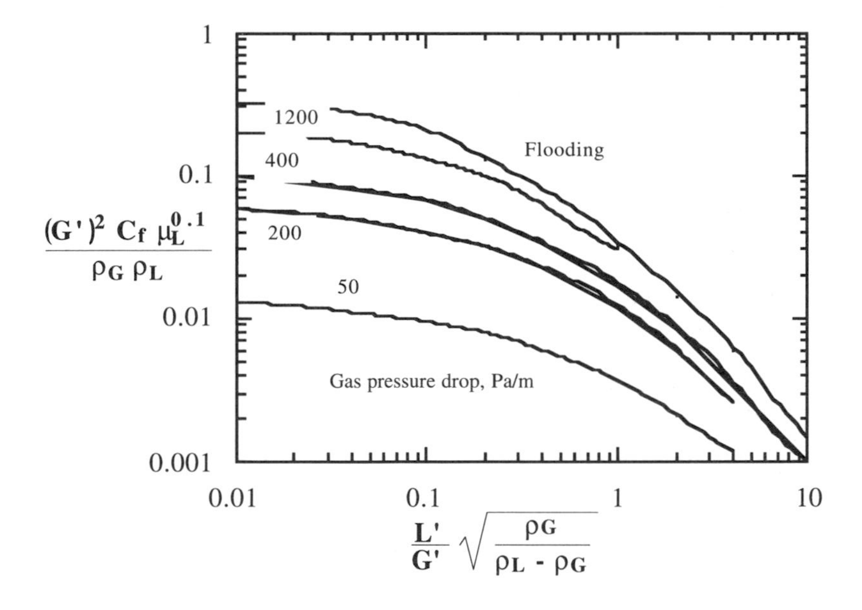

Figure 5.6 Flooding and pressure drop in random-packed towers; L' and G' are in kg/m2-s. (Treybal, R. E. *Mass Transfer Operations*, 3rd. ed., p. 195, McGraw-Hill, New York, ©1980; reprinted with permission from McGraw-Hill, Inc.)

Table 5.2 Characteristics of Random Ceramic Packings

Packing	13 mm	25 mm	50 mm	76 mm
Raschig rings				
C_f	580	155	65	37
ε	0.63	0.73	0.74	0.78
a, m2/m3	364	190	92	62
Intalox saddles				
C_f	200	98	40	22
ε	0.75	0.775	0.79	
a, m2/m3	623	256	118	

Source: From Treybal, R.E. Mass Transfer Operations, 3rd. ed., p. 196, McGraw-Hill, New York, ©1980; reprinted with permission from McGraw-Hill, Inc

.Example 5.5 Pressure Drop through Packed-Bed FGD Scrubber

Estimate the gas pressure drop through the seawater scrubber of Example 5.4, and the power to operate the system fan, located at the scrubber outlet. Assume that 1 m of dry packing is used above the liquid inlet as an entrainment separator. Add 50 Pa to the calculated total pressure drop through the packing to account for inlet expansion and outlet contraction energy losses. The mechanical efficiency of the fan-motor, E_m, is 65%.

Solution

Estimate the pressure drop through the irrigated packing with Figure 5.6. From Example 5.4, $G' = 4{,}250/3{,}600 = 1.18$ kg/m2-s; $L' = 37{,}640/3{,}600 = 10.46$ kg/m2-s. The ideal gas law gives the density of the gas: $\rho_G = PM_{av}/RT = (28.4)(101.3)/(8.314)(300) = 1.15$ kg/m3. The density of seawater at 300 K is approximately $\rho_L = 1{,}000$ kg/m3. Therefore,

$$\frac{L'}{G'}\sqrt{\frac{\rho_G}{\rho_L-\rho_G}} = 0.301$$

From Table 5.2, for 25-mm Raschig rings, $C_f = 155$. The viscosity of the liquid is $\mu_L = 0.001$ kg/m-s.

$$\frac{(G')^2 C_f \mu_L^{0.1}}{\rho_G \rho_L} = 0.094$$

From Figure 5.6, $\Delta P/Z_T = 1{,}200$ Pa/m, which is dangerously close to the flooding line. The pressure drop for the irrigated packing is (1,200)(0.51) = 612 Pa. Equation (4.42) of Chapter 4 gives the pressure drop through the dry packing used as entrainment separator. The particle size in that equation, d_p, is not the packing nominal size, but is defined as :

$$d_p = \frac{6(1 - \varepsilon)}{a} = \frac{6(1 - 0.73)}{190} = 0.00853 \text{ m}$$

$$\text{Re}_p = \frac{G' d_p}{\mu} = \frac{(1.18)(8.56 \times 10^{-3})}{1.85 \times 10^{-5}} = 544$$

$$\frac{\Delta P}{Z} = \left[\frac{150\,(1 - \varepsilon)}{544} + 1.75\right] \frac{(1 - \varepsilon)\, G'^2}{\varepsilon^3\, d_p\, \rho_G}$$

Because $Z = 1$ m for the dry packing, $\Delta P = 180$ Pa. The total pressure drop is $\Delta P_{TOT} = 612 + 180 + 50 = 842$ Pa. Assuming that the gases leave the scrubber in thermal equilibrium with the incoming seawater, the power to operate the fan is

$$\dot{W} = \frac{Q \Delta P_{TOT}}{E_m} = \frac{(375)(300)(0.842)}{(350)(0.65)} = 416.3\ kW$$

5.4 OVERVIEW OF FGD PROCESSES

Your objectives in studying this section are to

1. Distinguish between throwaway and regenerative FGD processes.
2. Describe the design and operating characteristics of the most frequently used FGD systems.

A basic method of classifying FGD systems is throwaway or regenerative. A process is of the throwaway type if the sulfur removed from the flue gas is discarded as a waste. A process is considered regenerative if the sulfur is recovered in a usable form. In general, at the present time, regenerative processes have higher costs than throwaway ones. However, they produce a reusable product, conceptually more satisfying than throwing away a resource such as sulfur. In Japan regenerative processes are used almost exclusively (Cooper and Alley 1986). As a result, Japan produces enough sulfuric acid for its internal industrial consumption and still has some for export.

Following is a brief description of the most frequently used throwaway and regenerative FGD processes.

5.4.1 Limestone Scrubbing (Throwaway)

In this process, a limestone slurry contacts the flue gas in a spray tower. The sulfur dioxide is absorbed and reacts with the alkaline reagent in the liquid phase. The main reaction product is calcium sulfite ($CaSO_3$), although some calcium sulfate ($CaSO_4$) is also formed. The overal stoichiometry can be represented by:

$$CaCO_3(s) + H_2O + 2SO_2 \rightarrow Ca^{+2} + 2HSO_3^- + CO_2(g) \qquad (5.37)$$

$$CaCO_3 + 2HSO_3^- + Ca^{+2} \rightarrow 2CaSO_3 + CO_2 + H_2O \qquad (5.38)$$

$$CaSO_3 + \frac{1}{2}O_2 \rightarrow CaSO_4 \qquad (5.39)$$

A simplified schematic diagram of a limestone scrubbing system is shown in Figure 5.7. The major advantage of this process is that the absorbent is abundant and inexpensive. The disadvantages include scaling inside the tower, equipment plugging and corrosion. It is the most widely used of all FGD systems.

5.4.2 Lime scrubbing (throwaway)

Lime scrubbing is very similar in equipment and process flow to limestone scrubbing, except that lime is much more reactive than limestone. The net reactions for lime scrubbing are:

$$CaO + H_2O \rightarrow Ca(OH)_2 \qquad (5.40)$$

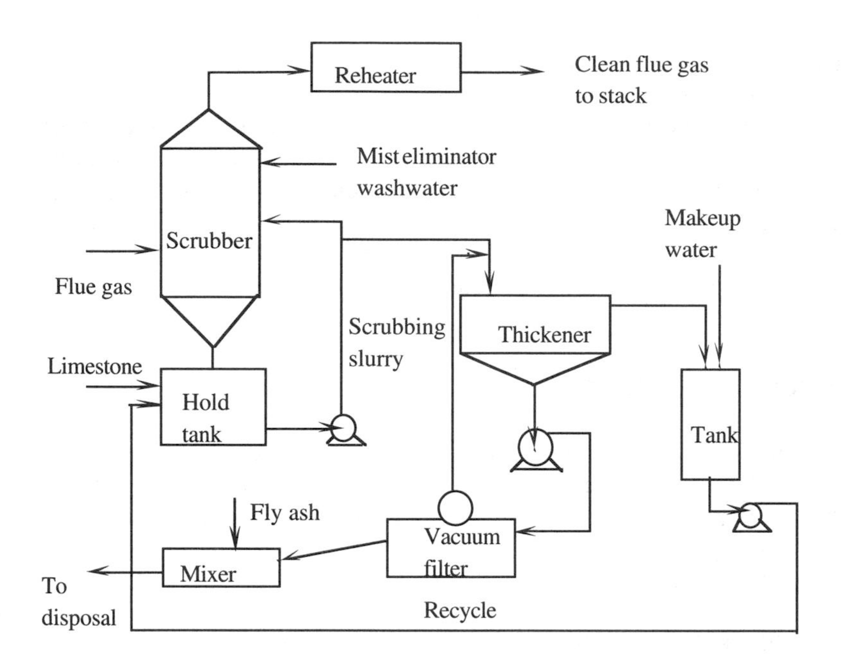

Figure 5.7 Schematic flow diagram of a limestone-based FGD system

$$SO_2 + H_2O \rightarrow H_2SO_3 \tag{5.41}$$

$$H_2SO_3 + Ca(OH)_2 \rightarrow CaSO_3 \cdot 2H_2O \tag{5.42}$$

$$CaSO_3 \cdot 2H_2O + \frac{1}{2}O_2 \rightarrow CaSO_4 \cdot 2H_2O \tag{5.43}$$

Lime scrubbing can achieve higher SO_2 removal efficiencies than limestone scrubbing, but lime is more expensive and hence is not as widely used as limestone. The optimal operating pH for limestone scrubbing is between 5.8 and 6.2, whereas that for lime scrubbing is about 8.0.

The formation of $CaSO_4$ (gypsum) in lime and limestone scrubbing can be a serious problem. Typically, 13% of all the sulfur in the final waste product from lime and limestone scrubbing is in the form of sulfate (Fellman and Cheremisinoff 1977). Gypsum forms a hard, stubborn scale on the surface of the scrubber, and its formation must be avoided. The solubility of $CaSO_3$ increases markedly as the pH decreases (100 ppm at pH 5.8 and 1,000 ppm at pH 4.4). Thus, the rate of oxidation of sulfite to sulfate increases as pH decreases. Limestone scrubbing systems operating at a pH of around 6.0 can successfully avoid gypsum scale formation.

Lime scrubbing systems operate at much higher pHs. Under these conditions, the solubility of the sulfite is very low, leading to a phenomenon known as soft pluggage—the formation of large leafy masses of $CaSO_3$ inside the scrubber. As long as a pH of 8.0 is not exceeded, lime scrubbing can avoid soft pluggage.

Example 5.6 Limestone Scrubbing for FGD

Consider limestone scrubbing of the flue gases of Example 5.1. The limestone is 95% $CaCO_3$ and 5% inerts. The removal efficiencies desired are 93% for SO_2 and 100% for HCl. Enough limestone must be supplied to provide 10% excess alkalinity, based on the desired removal efficiencies. Assume that 13% of the sulfur removed is in the form of $CaSO_4 \cdot 2H_2O$, the rest is in the form $CaSO_3 \cdot 2H_2O$; all the chlorine removed is in the form $CaCl_2 \cdot H_2O$. The sludge formed is dewatered to 60% solids. Calculate

(a) Amount of limestone supplied
b) Dewatered sludge flow rate (to final disposal)
c) Makeup water required

Solution

(a) From Example 5.2, the molar flow rate of SO_2 is 0.89 kmol/min and that of HCl is 0.05 kmol/min. Ninety-three percent of the SO_2, and 100% of the HCl must be neutralized by reaction with the limestone.The overall stoichiometry can be represented by:

$$CaCO_3 + SO_2 + 2H_2O \rightarrow CaSO_3 \cdot 2H_2O + CO_2$$

$$CaCO_3 + 2HCl \rightarrow CaCl_2 \cdot H_2O + CO_2$$

with some of the sulfite oxidized to sulfate according to Eq. (5.43). The $CaCO_3$ required is (0.89)(0.93)(1)(100) + (0.05)(1)(100) = 87.8 kg/min. The limestone required is (87.8)/(0.95) = 92.4 kg/min. The limestone supplied is (1.1)(92.4) = 101.6 kg/min.

(b) Calculate the sludge flow rate from material balances on the solid products.

$CaSO_3 \cdot 2H_2O$ formed = (0.89)(0.93)(0.87)(156.18) =	112.5 kg/min
$CaSO_4 \cdot 2H_2O$ formed = (0.89)(0.93)(0.13)(172.18) =	18.5 kg/min
$CaCl_2 \cdot H_2O$ formed = (0.05)(1)(129) =	6.5 kg/min
Unreacted $CaCO_3$ out	8.7 kg/min
Inert substances out	5.1 kg/min
Total solids in the sludge	151.3 kg/min

Since the sludge is dewatered to 60 percent solids, the sludge flow rate is 151.3/0.6 = 252 kg/min (363 metric ton/d). If the fly ash is added, the total solid residue for final disposition is 252 + 1,905(0.12) = 480.6 kg/min.

c) Even though most of the water in the scrubbing slurry recirculates, some leaves the system in the sludge and with the "clean" flue gases. The water in the sludge is in two forms: free humidity and hydration water. The sum of the two is: 252(0.4) + 112.5 (36)/(156.18) + 18.5 (36)/(172.18) + 6.5 (18)/(129) = 131.6 kg/min. To calculate the water evaporated in the scrubber, the temperature and humidity of the exit gases must be determined. Because of the intimate contact with the slurry, assume that the gases undergo adiabatic saturation in the tower. A combined material and energy balance for this process is (Treybal 1980)

$$T_{as} = T_{G1} - \left(Y_{as}' - Y_1'\right)\frac{\lambda_{as}}{C_{S1}} \tag{5.44}$$

where

T_{as} = adiabatic saturation temperature in K
T_{G1} = initial gas temperature in K
Y_{as}' = saturation humidity, in kg water/kg dry gases
Y_1' = initial humidity, kg water/kg dry gas
λ_{as} = latent heat of vaporization of water at T_{as}, kJ/kg
C_{S1} = humid heat capacity at the initial conditions, kJ/kg dry gas-K
= $1.005 + 1.884\ Y_1'$

Because the final mixture is saturated with water vapor,

$$Y_{as}' = \frac{p_{as}^*}{P - p_{as}^*}\frac{M_w}{M_{dg}} \tag{5.45}$$

where

p_{as}^* = water vapor pressure at T_{as} in Pa
M_w = molecular weight of water
M_{dg} = average molecular weight of the dry gases

The vapor pressure is related to the temperature through Antoine's equation, which is, for water (Himmelblau 1989),

$$\ln p_{as}^* = 23.1961 - \frac{3,816.4}{T_{as} - 46.13} \quad (5.46)$$

For the range between 300 and 350 K—where most FGD adiabatic saturation temperatures lie—the temperature-dependence of the latent heat of vaporization is

$$\lambda_{as} = 3,163.2 - 2.4194\, T_{as} \quad (5.47)$$

From the results of Example 5.1, calculate the average molecular weight of the dry gases, M_{dg} = 30.8 kg/kmol. Calculate, also, the initial humidity, Y_1' = 0.102 kg water/kg dry gas. The initial gas temperature is T_{G1} = 350 K. Calculate C_{S1} = 1.005 + (1.884)(0.102) = 1.197 kJ/kg-K. Calculate the saturation temperature and humidity by trial and error.

1. Assume T_{as}.
2. Calculate p_{as}^* from Eq. (5.46).
3. Calculate Y_{as}' from Eq. (5.45).
4. Calculate λ_{as} from Eq. (5.47).
5. Check the value of T_{as} assumed through Eq. (5.44).

Convergence occurs at T_{as} = 329 K and Y_{as}' = 0.1127 kg water/kg dry gas. At the scrubber outlet, the mass flow rate of dry gases is approximately 20,510 kg/min. Therefore, the mass flow rate of water in that stream is 2,312 kg/min.At the scrubber inlet, the water mass flow rate is (116.8)(18) = 2,100 kg/min. The rate of evaporation of water in the tower is 2,312 – 2,100 = 212 kg/min. The total makeup water requirement is 212 + 131.6 = 343.6 kg/min.

Comments

This example illustrates some important drawbacks of limestone-based scrubbing. Enormous quantities of solid wastes are generated; their final disposition in an environmentally sound manner can be an expensive problem. Significant amounts of water are required. The flue gases leave the scrubber at a relatively low temperature and saturated with water vapor. They must be reheated by, at least, 25 K to provide buoyancy and to prevent acid water condensation in the stack. This process requires a significant amount of energy (see Problem 5.10) reducing the thermodynamic efficiency of the power cycle.

5.4.3 Bechtel's Seawater Scrubbing Process (Throwaway)

The conventional seawater scrubbing process, which relies on the natural alkalinity of seawater, was discussed in Example 5.3. Bechtel's concept obviates some of the drawbacks of that process, which were pointed out then. In the Bechtel process, less than 2% of the cooling seawater from the power plant's condensers flows through the scrubber, the rest being used for dissolving the gypsum produced in the scrubbing system (Abrams et al. 1988). Most of the alkali for SO_2 removal is added to the scrubbing system as lime or a mixture of lime and limestone. The soluble magnesium from the seawater ($MgCl_2$ and $MgSO_4$) reacts with the added alkali to produce magnesium hydroxide, which absorbs SO_2 from the flue gases. Figure 5.8 is a schematic diagram of the process showing the principal chemical reactions involved.

Some of the advantages of this process, as compared to conventional seawater scrubbing and lime-limestone FGD, are

1. Magnesium hydroxide and magnesium sulfite react rapidly with the absorbed SO_2 reducing the recirculation rate of the scrubbing slurry to about 25% of that required by a regular lime-limestone scrubber.
2. The scrubbing liquor is well buffered so the pH changes relatively little as the liquid passes through the absorber and reacts with SO_2.
3. Scale formation is avoided by eliminating calcium hidroxide from the main scrubber circuit.

The liquid effluent of seawater scrubbing systems consists mostly of a low concentration of dissolved gypsum, with small amounts of fly ash and trace metals. According to Bechtel, a typical 250-MW unit burning 1.25% sulfur coal with 90% SO_2 removal discharges 525,000 L/min of seawater with an added gypsum concentration of 148 ppm (Abrams et al. 1988). The proponents of seawater scrubbing argue that, because dissolved sulfates are an abundant natural constituent of seawater, the effluent is completely safe. This way of thinking is similar to the claim that CO_2 emissions to the atmosphere are totally harmless because it is such an abundant constituent of atmospheric air. The truth is that burning of fossil fuels result in enormous additions of mass to the carbon and sulfur elemental cycles. The degree of disruption of those cycles will depend on their capacity to dilute and assimilate those massive additions of foreign material, a capacity that is finite.

5.4.4 The Wellman-Lord Process (Regenerative)

The Wellman-Lord process is the best known regenerative FGD process. It can be divided into five subprocesses which are: (1) flue gas pretreatment, (2) SO_2 absorption by a sodium sulfite solution, (3) purge treatment, (4) sodium sulfite regeneration, and (5) sulfur recovery.

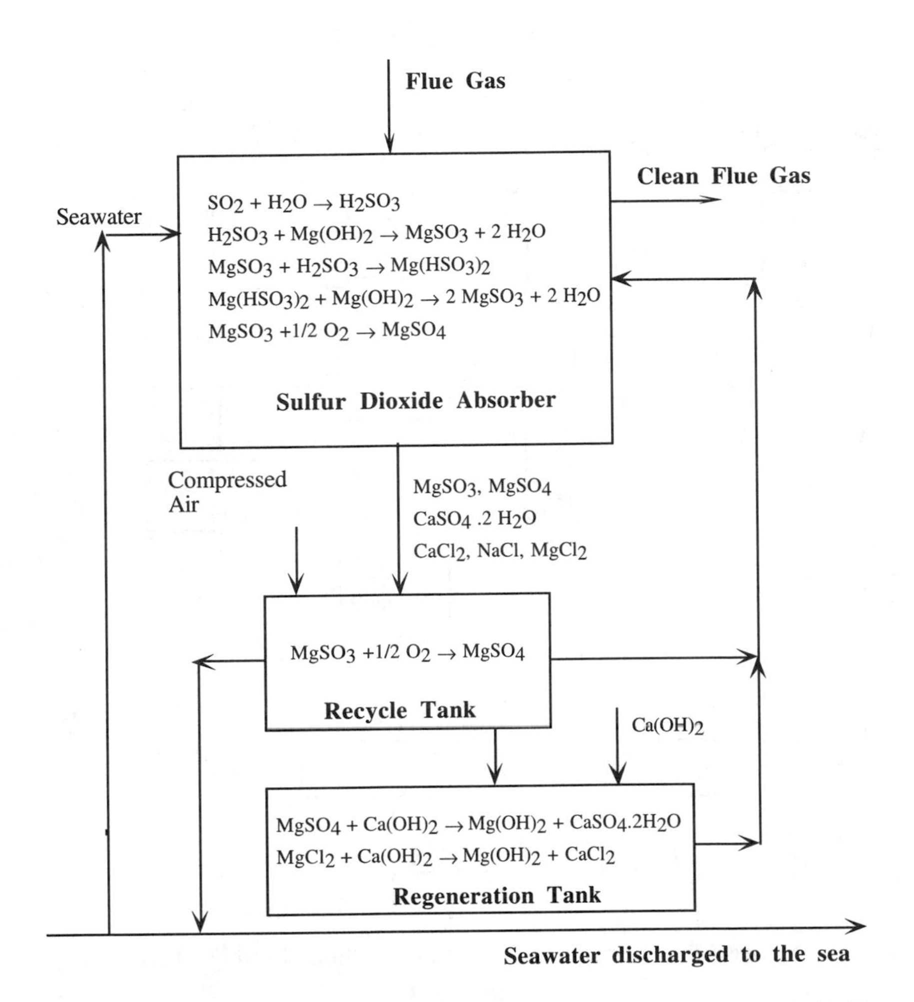

Figure 5.8 Bechtel's seawater FGD process; principal chemical reactions (From Abrams, J.Z. et al. *JAPCA*, 38:969, 1988; reprinted with permission from *JAPCA*).

A simplified schematic diagram of the Wellman-Lord process is shown in Figure 5.9. Flue gas from a particulate removal process is blown through a venturi prescrubber where it contacts a fine spray of water droplets. The prescrubber removes most of the remaining particles as well as any SO_3 and HCl, which would upset the absorption chemistry in the main scrubber. It also cools and humidifies the flue gas. Typical inlet temperatures and relative humidities are 425

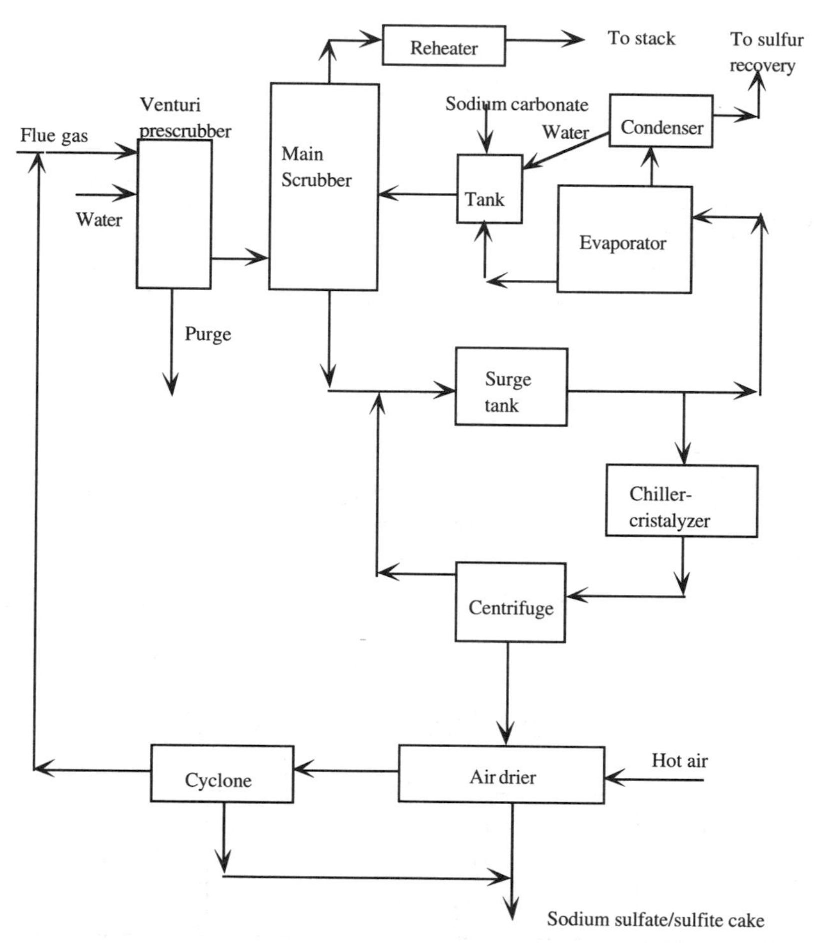

Figure 5.9 Schematic diagram of the Wellman-Lord FGD process.

K and 20%, and outlet values are 325 K and 95%. A liquid purge stream from the prescrubber removes solids and chlorides.The flue gas flows next to the main scrubber. Tray towers are the most common SO_2 absorber units.The flue gas contacts an aqueous sodium sulfite solution and the SO2 is absorbed and reacts in the clear liquid to form sodium bisulfite. The reaction is:

$$Na_2SO_3 + SO_2 + H_2O \rightarrow 2NaHSO_3 \tag{5.48}$$

Some of the sulfite is oxidized to sulfate by oxygen. Also, any sulfur trioxide that passes through the prescrubber results in aqueous sulfate. The reactions are:

$$Na_2SO_3 + 1/2\ O_2 \rightarrow Na_2SO_4 \tag{5.49}$$

$$2Na_2SO_3 + SO_3 + H_2O \rightarrow Na_2SO_4 + 2NaHSO_3 \tag{5.50}$$

The sodium sulfate does not contribute to further SO_2 absorption and must be removed. Excessive sulfate accumulation is prevented by a continuous purge from the bottom of the absorber.

The absorber bottoms stream is rich in bisulfite, so the purge is sent to a chiller-crystallizer where the less soluble sodium sulfate crystals are formed. The slurry is centrifuged and the solids are air-dried and discarded. The centrifugate, which is rich in bisulfite, is returned to the process. The rest of the absorber bottoms stream goes to a heated evaporator-crystallizer where SO_2 is liberated and sodium sulfite crystals are regenerated. The reaction is

$$2NaHSO_3 \rightarrow Na_2SO_3 + SO_2 + H_2O \tag{5.51}$$

The water vapor condenses and returns to the process, leaving a concentrated stream of SO_2 gas (about 85% SO_2 and 15% H_2O) This gas can be reduced to elemental sulfur or oxidized to sulfuric acid either on-site or at a nearby chemical plant.

Because some of the sodium is removed from the process via the sodium sulfate purge, soda ash (Na_2CO_3) is added to provide makeup sodium. The soda ash reacts readily in the absorber as follows:

$$Na_2CO_3 + SO_2 \rightarrow Na_2SO_3 + CO_2 \tag{5.52}$$

A typical makeup rate is 1 mol of sodium per 42 mol of SO_2 removed (Electric Power Research Institute, EPRI 1983).

Several other regenerative processes have been developed. None have been commercialized to the same extent as the Wellman-Lord, although other recently developed processes compare very favorably with it.

Example 5.7 Wellman-Lord FGD Process

The flue gas of Example 5.1 is to be desulfurized by application of the Wellman-Lord process. The expected SO_2 removal efficiency is 93%.

(a) Calculate the soda ash requirement of the process, in kg/min.

(b) Calculate the rate of production of the sodium sulfate/sulfite cake, which constitutes the only solid waste product of the process. Assume the dry solids contain 90% sulfate and 10% sulfite. The cake is dried to a 70% solids content.

Solution

a) Calculate the soda ash requirement from the EPRI suggestion of 1 mol of makeup sodium per 42 mol of removed sulfur dioxide.

Basis: 1 min

From previous examples, the amount the SO_2 to be removed is 0.827 kmol. The soda ash required for make—up sodium is given by:

$$0.827 \text{ kmol } SO_2 \times \frac{1 \text{ kmol Na}}{42 \text{ kmoles } SO_2} \times \frac{1 \text{ kmole } Na_2CO_3}{2 \text{ kmol Na}} \times \frac{106 \text{ kg } Na_2CO_3}{\text{kmol}}$$

$= 1.05$ kg Na_2CO_3 /min

(b) Determine the ratio of kg of waste solids generated to kmoles of Na added. Sodium is a tie-element between the soda ash and the solid waste streams. The desired ratio is obtained through the following tabulated calculations.

Basis: 100 kg of bone-dry solid waste

Component	Mass (kg)	MW	kmol
Na_2SO_4	90.0	142.05	0.634
Na_2SO_3	10.0	126.05	0.079
Total	100.0		0.713

Therefore, 100 kg of waste dry solids are formed per (2)(0.713) = 1.426 kmol of sodium added.

Basis: 1 min

The rate of formation of the bone-dry solid waste is (0.827)(100)/[(42)(1.426)] = 1.23 kg solids/min. Because the cake is 70% solids, the rate of formation of the cake is 1.23/0.70 = 1.76 kg/min = 2.54 metric ton/d.

5.4.5 Magnesium Oxide Process (Regenerative)

The magnesium oxide (MgO) process has an absorption step similar to that of the Bechtel process. Wet scrubbing with a slurry of $Mg(OH)_2$ produces $MgSO_3$ and $MgSO_4$ solids. The absorbate enters a centrifuge where the hydrated crystals of magnesium sulfite and sulfate separate from the mother liquor. The liquor returns to the absorber and the centrifuged wet cake goes to a drier. It is calcined in the presence of coke or other reducing agent, generating SO_2 and regenerating the MgO. The reactions are

$$MgSO_3 \rightarrow MgO + SO_2 \tag{5.53}$$

$$MgSO_4 + \frac{1}{2}C \rightarrow MgO + SO_2 + \frac{1}{2}CO_2 \tag{5.54}$$

$$MgO + H_2O \rightarrow Mg(OH)_2 \tag{5.55}$$

The flue gas from the calciner contains about 15% SO2, which can then be used for sulfuric acid production.

The main advantage of the MgO process over the Wellman-Lord is that there is little, if any, solid waste. The primary disadvantages are that a high-temperature calciner is needed, and that the SO_2 product stream is not as concentrated.

5.4.6 Lime-Spray Drying (Throwaway)

This is a semidry process in which the flue gas and a lime slurry mix in a spray dryer. The flue gas SO_2 and lime react to form a solid that is collected with the fly ash in a fabric filter inmediately downstream. Its capital and operating costs and maintenance and energy requirements are lower than those of conventional limestone scrubbing. Some disadvantages are that the filter bags can blind if the flue gas approaches the saturation temperature, and scaling can occur in the spray drier.

Recent efforts related to this process include the development of sorbents that are even more reactive than lime.Josewicz et al. (1988a, 1988b), and Chu and Rochelle (1989) found that slurrying of fly ash and lime can enhance significantly the reaction of $Ca(OH)_2$ with SO_2. The main reason for this fly ash-enhanced conversion is believed to be the pozzolanic reaction (formation of high surface area calcium silicates) with fly ash being the source of silica (Peterson and Rochelle 1988). Hall et al. (1992) reported bench-and pilot-plant developments in preparation of advanced calcium silicate sorbents for FGD in a process called ADVACATE.

5.4-7 Nahcolite or Trona Injection (Throwaway)

Truly dry scrubbing involves trona or nahcolite sorption of SO_2. Trona is naturally occurring Na_2CO_3. It is mined commercially and sold as a bulk commodity. Nahcolite is naturally occuring $NaHCO_3$. Its mining is not commercially established, but in principle, it would be similar to trona mining. In the dry scrubbing process, pulverized reagent (trona or nahcolite) is injected into the flue gas. Dry sorption occurs and the solid particles are collected in a baghouse. Further SO_2 removal occurs as the flue gas flows through the filter cake on the bags. The advantages are lower capital costs and maintenance requirements. The disadvantages are high reagent costs and possible waste disposal problems because of leaching of soluble sodium salts. A recent modification of the process proposes an ion-exchange resin-based system to regenerate the sorbent and recover the sulfur in the form of $(NH_4)_2SO_4$ which can be sold as a fertilizer (Sheth and Modeste 1990, see Problem 5.18).

5.4.8 Shell Flue Gas Treating System (Regenerative)

This is a commercially demonstrated dry process for simultaneous removal of SO_2 and NO, therefore it is of considerable interest for high-sulfur coal-fired power plants (Jahnig and Shaw 1981). Flue gas is introduced at 673 K into two or more parallel passage reactors containing copper oxide (CuO) supported on alumina (Al_2O_3), where the SO_2 reacts with the copper oxide to form copper sulfate ($CuSO_4$).

$$CuO + \frac{1}{2}O_2 + SO_2 \rightarrow CuSO_4 \tag{5.56}$$

The copper sulfate act as a catalyst in the reduction of NO with added ammonia.

$$4NO + 4NH_3 + O_2 \rightarrow 4N_2 + 6H_2O \tag{5.57}$$

When the reactor is saturated with copper sulfate, the flue gas is switched to a fresh reactor, and the spent reactor is regenerated. The regeneration cycle uses hydrogen to reduce the copper sulfate to copper,

$$CuSO_4 + 2H_2 \rightarrow Cu + SO_2 + 2H_2O \tag{5.58}$$

producing an SO_2 stream of sufficient concentration for conversion to sulfur or sulfuric acid. The copper is then oxidized back to copper oxide.

A variation of this technology is the fluidized bed CuO process, under development at the U. S. Department of Energy's Pittsburgh Energy Technology Center (DOE/PETC). As shown in Figure 5.10, the sulfated sorbent is transported from the fluidized bed adsorber to a solids heater and then to a regenerator (see Problem 5.21). Regeneration occurs by reaction with methane.

5.4.9 Electron Beam Treatment (Throwaway)

In this innovative process, an electron beam penetrates into the effluent gas stream where collisions between the electrons and gas molecules produce ions. These interact with the gas to create free atoms and radicals. The free atoms and radicals react with NO and SO_2 to form the corresponding acids, which are then removed by neutralization with added alkali reagents.

An electron beam is generated by accelerating electrons through a potential field. The depth of penetration of the beam into a gas stream is proportional to the electron energy and inversely proportional to the gas density. Primary reactions induced by the electron beam are the decomposition of water and oxygen molecules into H, O, and OH. These react with pollutant species according to

$$OH + NO \rightarrow HNO_2 \tag{5.59}$$

$$O + NO \rightarrow NO_2 \tag{5.60}$$

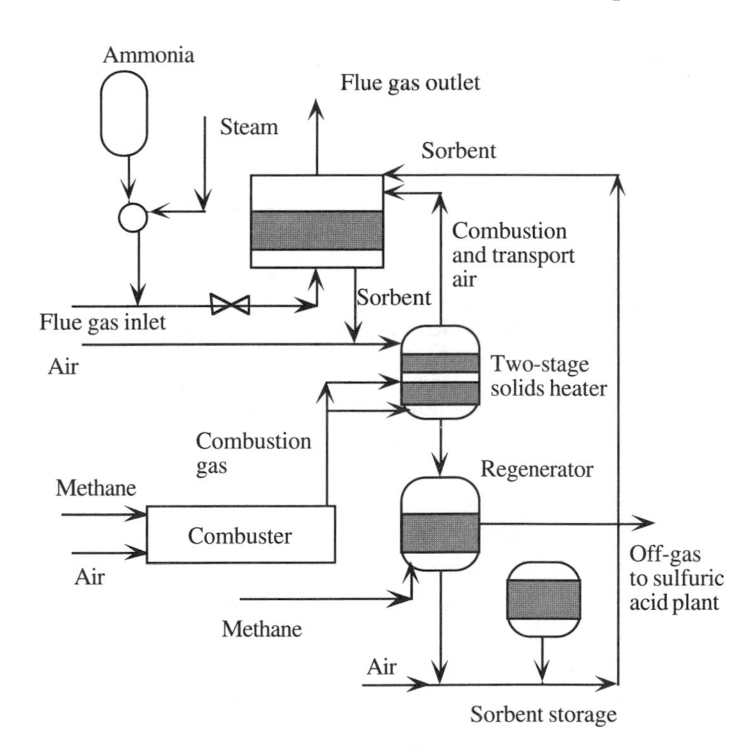

Figure 5.10 Schematic diagram of the PETC fluidized bed CuO process (From Frey, H. C., and Rubin, S. E. *J. Air Waste Manage. Assoc.*, **41**:1585, 1991; reprinted with permission from *J. Air Waste Manage. Assoc.*)

$$OH + NO_2 \rightarrow HNO_3 \tag{5.61}$$

$$SO_2 + O \rightarrow SO_3 \tag{5.62}$$

With the addition of $Ca(OH)_2$, neutralization of the acids occurs, followed by filtering of the solid product:

$$2HNO_3 + Ca(OH)_2 \rightarrow Ca(NO_3)_2 + 2H_2O \tag{5.63}$$

$$SO_3 + H_2O + Ca(OH)_2 \rightarrow CaSO_4 \cdot 2H_2O \tag{5.64}$$

Although the electron beam reactor is a relatively new concept, simultaneous NO and SO_2 removal at efficiencies exceeding 90% have been demonstrated for high-sulfur fuels (Gleason and Helfritch 1985, Dalton et al. 1992).

Example 5.8 Kinetic Model of FGD by Dry Injection of Nahcolite (Gårding and Svedberg 1988)

A first attempt to get a theoretical description of the dry injection kinetics is as follows. The first step is presumed to be the calcination of the $NaHCO_3$.

$$2NaHCO_3 \rightarrow Na_2CO_3 + CO_2 + H_2O \tag{5.65}$$

The next step is the sulfation:

$$Na_2CO_3 + SO_2 + 1/2\ O_2 \rightarrow Na_2SO_4 + CO_2 \tag{5.66}$$

It can be shown (see Problem 5.19) that the calcination step is much faster than the other steps of the proposed kinetic model. Therefore, the time needed for calcination is neglected in the following scheme.

The reaction process is represented by the unreacted-core model. According to it, the reaction occurs first at the outer skin of the particle. The reaction zone then moves into the solid and leaves behind completely converted material. The original particle shape is presumed to be kept throughout the reaction. The rate for the gas-solid reaction is determined by the following resistances: (1) external film diffusion, (2) diffusion through the reacted material, and (3) chemical reaction. The solid particles are presumed to be spherical and of uniform particle size. The sulfation reaction is presumed to be irreversible and of the first order with respect to the partial pressure of SO_2. The reacting solid material is injected into a plug flow of gas. The reaction starts inmediately after injection and continues through the pipe from the injection point into a fabric filter compartment. The gas passes through the filter where the solid material builds up as a cake. At any time during this process the gas reactant concentration is related to the fractional solid conversion through

$$C = 1 - M_R x \tag{5.67}$$

where

$C = C_g/C_{g0}$ = dimensionless gas reactant concentration
C_{g0} = initial gas reactant concentration
M_R = molar ratio, that is, mol of Na_2CO3 per mol of SO_2
x = fractional Na_2CO_3 conversion.

The rate of reaction may be expressed in terms of the above mentioned resistances in series.

$$-\frac{dr_c}{dt} = \frac{\dfrac{C_g M_s}{\rho_p}}{\dfrac{r_c^2}{R_0^2 k_g} + \dfrac{(R_0 - r_c)\, r_c}{R_0 D_e} + \dfrac{1}{k_c}} \tag{5.68}$$

where

r_c = shrinking radius of the unreacted core
M_s = molecular weight of the solid reactant
ρ_p = particle density
R_0 = initial particle radius
k_g = gas film mass transfer coefficient (m/s)
D_e = effective diffusion coeficient for the gas reactant through the porous solid
k_c = reaction rate constant (m/s)

Using the expression

$$x = 1 - \left(\frac{r_c}{R_0}\right)^3 \tag{5.69}$$

and introducing the dimensionless variable τ gives

$$\frac{dx}{d\tau} = \frac{3(1-x)^{1/3}}{1-(1-1/\text{Bi})(1-x)^{1/3} + \text{Da}^{-1}(1-x)^{-1/3}} C \tag{5.70}$$

where

$$\tau = \frac{D_e M_s C_{g0} t}{R_0^2 \rho_p}$$

Bi = Biot number = $k_g R_0 / D_e$

Da = Damköhler number = $k_c R_0 / D_e$

The initial condition is: $\tau = 0$, $x = 0$, $C = 1.0$. The solution to Eqs. (5.67) and (5.70) with the specified initial condition is:

$$\tau = (1+\alpha^3)\{g_1\psi_1 + g_2\psi_2 + g_3(g_4 - \psi_3)\} \tag{5.71}$$

where

$$g_1 = \frac{1}{2\alpha}\left(1 - \frac{\text{Da}^{-1}}{\alpha}\right)$$

$$g_2 = \frac{1}{3}\left(1 - \text{Bi}^{-1}\right) - \frac{1}{6\alpha}\left(1 - \frac{\text{Da}^{-1}}{\alpha}\right)$$

$$g_3 = \frac{1}{\alpha\sqrt{3}}\left(1 + \frac{\text{Da}^{-1}}{\alpha}\right)$$

$$g_4 = \tan^{-1}\left(\frac{2-\alpha}{\alpha\sqrt{3}}\right)$$

$$\psi_1 = \ln\left(\frac{\gamma+\alpha}{1+\alpha}\right), \qquad \psi_2 = \ln\left(\frac{\gamma^3+\alpha^3}{1+\alpha^3}\right)$$

$$\psi_3 = \tan^{-1}\left(\frac{2\gamma-\alpha}{\alpha\sqrt{3}}\right), \qquad \alpha = \left[\frac{1-M_R}{M_R}\right]^{1/3}, \qquad \gamma = (1-x)^{1/3}$$

Consider a flue gas at 473 K and 101.3 kPa, with 1,000 ppm of SO_2. $NaHCO_3$ pulverized to a particle size of 38 μm is injected at a molar ratio of 1.2. The density of the particles is approximately 1,400 kg/m3. The gas film mass transfer coefficient is 1.5 m/s, the sulfation reaction rate constant is 0.1 m/s. The effective SO_2 diffusivity in the porous solid is 3 $\times 10^{-6}$ m2/s. Estimate the reaction time for 93% removal of the SO_2.

Solution

Calculate Bi = $k_g R_0 / D_e$ = (1.5)(19 × 10–6)/(3 × 10–6) = 9.5. Calculate Da = $k_c R_0 / D_e$ = (0.1)(19 × 10–6)/(3 × 10–6) = 0.63. For 93% SO_2 removal, $C = 0.07$. From Eq. (5.67), $x = (1 - C)/M_R$ = 0.93/1.2 = 0.775. Calculate $\gamma = (1 - x)1/3$ = 0.608. Calculate $\alpha = [(1 - M_R)/M_R]1/3$ = –0.550. Calculate the following: $g_1 = -3.533$, $g_2 = 1.476$, $g_3 = 1.98$, $g_4 = -1.213$, $\psi_1 = -2.044$, $\psi_2 = -2.664$, $\psi_3 = -1.076$. From Eq. (5.71), $\tau = 2.531$. Calculate C_{g0} from the ideal gas law, $C_{g0} = 25.8 \times 10^{-6}$ kmol/m3. From the definition of τ,

$$t = \frac{\tau R_0^2 \rho_p}{D_e M_s C_{g0}} = \frac{(2.531)(19 \times 10^{-6})^2 (1{,}400)}{(3 \times 10^{-6})(106)(25.8 \times 10^{-6})} = 156 \text{ s}$$

5.5 FGD SYSTEMS COST ESTIMATES

Your objective in studying this section is to

Estimate the total capital investment and total annual cost for some FGD systems.

Costs estimates of FGD systems can be highly variable, depending on factors such as type of process, size of plant, percentage of sulfur in the fuel, location of the plant, costs of raw materials, values of by-products, type of ultimate disposal required for waste products, and so forth (Cooper and Alley 1986). In addition, many of the technologies mentioned previously have only been demonstrated at the pilot scale. Thus, projections of their costs to full-sized units are more uncertain than cost estimates for commercially demonstrated processes. The technology of FGD is undergoing considerable evolution. Costs are likely to change significantly with future technological changes.

The economics of several FGD systems were estimated in detail in a study sponsored by the Electric Power Research Institute (EPRI 1983). In that study, costs were developed for a hypothetical 1,000-MW power plant in Kenosha, Wisconsin, burning a 4.0%-sulfur bituminous coal. Each FGD process removed 90% of the SO_2. Costs were developed for each alternative starting with a detailed process sheet, material balances, equipment list, and utility consumption list. Capital and levelized busbar costs were developed in December 1982 dollars. The levelized busbar cost is a PW annual cost, which factors in inflation assuming a 30-yr plant life. Levelized annual costs are always much higher than the TAC calculated on a constant-dollar basis. The EPRI study notes that the TAC for each of the alternatives is approximately 40% of the corresponding levelized busbar cost.

Table 5.3 shows total capital investment and total annual cost for some FGD systems, in dollars of June 1990. Capital costs were updated via the CE Index (composite), annual costs were updated via the Consumer Price Index (CPI-U). As mentioned before, these costs are for a given plant size and percent sulfur in the fuel. Jahnig and Shaw (1981) found that FGD costs could be prorated according to the power law with different exponents for different items. For example, costs increase as (1) plant size to the 0.8 power, and (2) with the percent of sulfur in the fuel to the 0.3 power.

They also noted that the estimated costs depend on the stage of technology development. Actual full-scale cost can turn out to be more than 50% higher than the estimated cost based on pilot-scale work.

Table 5.3 FGD System Cost Estimates (in June 1990 dollars)[a]

Process	TCI($/kW)	TAC[b](mill/kWh)
Limestone	165	6.7
Wet lime	188	10.0
Lime spray dryer	125	3.8
Nahcolite/trona injection	28.2	3.3–4.2[c]
Seawater scrubbing[d]	80	4.5
Wellman-Lord	312	13.0
MgO	302	9.6

[a] Data are for a 1000-MW plant burning 4.0%-sulfur bituminous coal with 90% sulfur removal.

[b] 1 mill = one-tenth of a cent

[c] Lower cost pertains to trona injection; higher cost, to nahcolite injection.

[d] Data are for a 300-MW plant burning 1.5%-sulfur bituminous coal with 93% sulfur removal (Cogentrix 1990).

Source: Based on data from EPRI, 1983

Example 5.9 TCI and TAC for FGD Systems

Consider the flue gases of Example 5.1. Estimate the TCI and TAC for flue gas desulfurization through
(a) Seawater scrubbing
(b) Limestone scrubbing
(c) MgO process
Assume the power plant operates 80% of the time.

Solution

(a) The power plant of Example 5.1 has a capacity of 300 MW and burns coal with a 1.5% sulfur content. These conditions are analogous to those of Table 5.3 for the seawater scrubbing entry. Therefore the cost data can be used without any modifications. The total capital investment is

$$\text{TCI} = (80)(300 \times 10^3) = \$24.0 \times 10^6$$

The total annual cost is

$$\text{TAC} = (4.5 \times 10^{-3})(300 \times 10^3)(24)(365)(0.8) = \$9.46 \times 10^6/\text{yr}$$

(b) The capital investment entry for limestone scrubbing in Table 5.3 is for a 1000-MW plant burning 4.0% sulfur coal. It must be prorated to the conditions of Example 5.1

$$\text{TCI} = 165\,(1.5/4.0)^{0.3}\,(1{,}000 \times 10^3)(300/1{,}000)^{0.8} = \$46.7 \times 10^6$$

$$\text{TAC} = (6.7 \times 10^{-3})(300 \times 10^3)(24)(365)(0.8) = \$14.1 \times 10^6/\text{yr}$$

(c) For the MgO process:

$$\text{TCI} = 302(1.5/4.0)^{0.3}\,(1{,}000 \times 10^3)(300/1{,}000)^{0.8} = \$86.0 \times 10^6$$

$$\text{TAC} = 9.6 \times 10^{-3}\,(300 \times 10^3)(24)(365)(0.8) = \$20.2 \times 10^6/\text{yr}$$

The TCI data of Table 5.3 seems too high, specially for the regenerative options. McGlamery and Torstrick (1976) presented detailed economic evaluations of various FGD systems. Updating their results, the following correlation was obtained for the total capital investment for MgO scrubbers:

$$\text{TCI} = 0.744\,(MW)^{0.63}\,(\%\text{S})^{0.27} \tag{5.72}$$

where

TCI = total capital investment, in millions of dollars of June 1990, for 90% removal, including a 98% H_2SO_4 recovery plant

MW = plant size, in MW

Example 5.10 Total Capital Investment for MgO FGD System

Repeat part c of Example 5.9 using Eq. (5.72) to calculate the TCI. Compare your results with Cogentrix's estimate of $\$39 \times 10^6$ (1990) for 93% SO2 removal.

Solution

Substituting values of 300 MW and 1.5% S in Eq. (5.72), the answer is TCI = $\$30.2 \times 10^6$ of June 1990. According to Example 5.9, the total capital investment from Table 5.3 is $86 million. It overestimates the TCI by 143%, as compared with Cogentrix's estimate of $39 million. Eq. (5.72) underestimates the TCI by 29%, which is within the uncertainty of this type of capital estimate (± 30%).

5.6 CONCLUSION

The utility and industrial sectors continue to come under pressure from both national and local regulations to reduce sulfur dioxide emissions. Flue gas desulfurization has been effective in reducing SO_2 emissions to the atmosphere. However, it is a technology plagued by operational problems. Throwaway processes usually result in water pollution or solid waste disposal problems. Regenerative systems are at an economic disadvantage since the market for elemental sulfur or sulfuric acid is highly site-specific and variable.

REFERENCES

Abrams, J.Z., Zaczek,S.J., Benz, A. D., Awerbuch, L., and Haidinger, J. *JAPCA*, **38**:969 (1988).

Chu, P., and Rochelle, G. T. *JAPCA*, **39**: 75 (1989).

Cogentrix Inc. "PSD Permit Application for a 300-MW Coal-Fired Cogeneration Facility," submitted to USEPA, prepared by ENSR Consulting and Engineering, Boston, MA (1990).

Cooper, C. D., and Alley, F.C. *Air Pollution Control; A Design Approach*, Prindle, Weber & Schmidt, Boston, MA (1986).

Dalton, S. M., Toole-O'Neil, B., Gullett, B. K., and Drummond, C. J., *J. Air Waste Manage. Assoc.*, **42**:1110 (1992).

Electric Power Research Institute, Economic Evaluation of FGD Systems, CS-3342, prepared by Stearns-Roger Engineering Corporation, Palo Alto, CA (1983).

Elliot, R. A., Matyas, A. G., Goodfellow, H. D., and Nenninger, E. H. *Environ. Prog.*, **1**:261 (1982).

Fellman, R. T., and Cheremisinoff, P. N. "A Survey of Lime/Limestone Scrubbing for SO2 Removal," in *Air Pollution Control and Design Handbook: Part 2*, P. N. Cheremisinoff and R. A. Young (eds.), Marcel Dekker, NewYork, 813–834 (1977).

Frey, H. C., and Rubin, S. E. *J. Air Waste Manage. Assoc.*, **41**:1585 (1991).

Gleason, R. J. and Helfritch, D. J. *Chem. Eng. Prog.*, 33–38 (October 1985).

Gårding, M., and Svedberg, G. *JAPCA*, **38**:1275 (1988).

Haar, L., Gallagher, J. S., and Kell, G. S. *NBS/NRC Steam Tables*, Hemisphere, New York (1984).

Hall, B. W., Singer, C., Josewicz, W., Sedman, C. B., and Maxwell, M. A. *J. Air Waste Manage. Assoc.*, **42**:103 (1992).

Harned, H. S. and Owen, B. B. *The Physical Chemistry of Electrolyte Solutions*, Van Nostrand Reinhold, NewYork (1958).

Himmelblau, D. M. *Basic Principles and Calculations in Chemical Engineering*, 5th ed. Prentice Hall, Englewood Cliffs, NJ (1989).

Jahnig, C. E. and Shaw, H. *JAPCA*, Vol. 31 (1981).

Josewicz, W., Chang, J. C. S., Sedman, C. B., and Brna, T. G. *JAPCA*, **38**:1027 (1988a).

Josewicz, W., Jorgensen, C., Chang, J. C. S., Sedman, C. B., and Brna, T. G. *JAPCA*, **38**: 796 (1988b).

Maahs, H. G. "Sulfur Dioxide/Water Equilibrium Between 0° and 50° C. An Examination of Data at Low Concentrations," in *Heterogeneous Atmospheric Chemistry*, D. R. Schryer (Ed.), American Geophysical Union, Washington, DC, 187–195 (1982).

McCabe, W. L., and Smith, J. C. *Unit Operations of Chemical Engineering*, 3rd. ed. McGraw-Hill, NewYork (1976).

McGlamery, G. G., and Torstrick, R.L. "Cost Comparisons of Flue Gas Desulfurization Systems," in *Power Generation: Air Pollution Monitoring and Control*, K. E. Noll and W. T. Davis (Eds.), Ann Arbor Science, Ann Arbor, MI (1976).

Peterson, J. R., and Rochelle, G. T. *Environ. Sci. Technol.*, **22**:1299 (1988).

Sheth, A. C., and Modeste, D. C. *J. Air Waste Manage. Assoc.* **40**:1532 (1990).

Treybal, R.E. *Mass Transfer Operations*, 3rd. ed., McGraw-Hill, New York (1980).

PROBLEMS

The problems at the end of each chapter have been grouped into four classes (designated by a superscript after the problem number)

Class a: Illustrates direct numerical application of the formulas in the text.
Class b: Requires elementary analysis of physical situations, based on the subject material in the chapter.
Class c: Requires somewhat more mature analysis.
Class d: Requires computer solution.

5.1^a. Exhaust gases from a coal-fired power plant

Calculate the composition of the exhaust gases from the coal-fired power plant of Example 5.1. Assume 10% excess air. At this low value of excess air, 5% of the carbon in the coal reacts to form CO.

Answer: 17.14 % CO_2

5.2^a. SO_2 emissions from a coal-fired power plant (Flagan, R. C., and Seinfeld, J. H. *Fundamentals of Air Pollution Engineering,* p.160, Prentice Hall, Englewood Cliffs, NJ, 1988)

A bituminous coal has the following analysis: 74.4% C, 5.1% H, 1.4% N, 6.7% O, 0.7% S, and 11.7% ash. Its heating value is 30,700 kJ/kg. It is burned with 17.6% excess air to generate 500 MW of electric power in a cycle with a thermal efficiency of 37%.

(a) Determine the fuel and air feed rates, in kg/s.

Answer: 44.0 kg/s of coal

(b) Determine the product gas composition

Answer: 517 ppm SO2

c) SO_2 is removed from the flue gases with a mean efficiency of 80% and the yearly average output of the plant is 75% of its rated capacity. What is the SO_2 emission rate in metric ton/yr?

Answer: 2,914 metric ton/year

5.3b. Minimum liquid flow rate for absorption with chemical reaction

Show that the minimum liquid flow rate for absorption of SO_2 in an aqueous alkali solution is given by Eq. (5.32).

5.4a. Liquid flow rate for absorption without chemical reaction

A flue gas flows at the rate of 10 kmol/s with a SO_2 content of 1,500 ppm. Ninety percent of the SO_2 will be removed by absorption with pure water at 298 K. The design water flow rate will be 50% higher than the minimum. Calculate the water flow rate, and the SO_2 concentration in the water leaving the tower. The equilibrium line has a slope of 10.0.

Answer: 2,430 kg/s

5.5b. Absorber diameter based on pressure drop specification

The FGD process described in Problem 5.4 is carried out in a tower packed with 50-mm ceramic Raschig rings. Calculate the tower diameter if the pressure drop through the irrigated packing is not to exceed 400 Pa/m. Assume that the properties of the gas mixture are those of air at 298 K and 101.3 kPa.

Answer: Diameter = 17.5 m

5.6b. Absorber irrigated packed height

Calculate the irrigated packed height for the absorber of Problem 5.5. For very dilute solutions, when Henry's law applies (Treybal 1980):

$$Z = N_{tOG} H_{tOG}$$

where:

$$N_{tOG} = \frac{\ln\left[\frac{y_1 - mx_2}{y_2 - mx_2}\left(1 - \frac{1}{A}\right) + \frac{1}{A}\right]}{1 - \frac{1}{A}}$$

$$A = \frac{L}{mG}$$

m = slope of the equilibrium line

$$H_{tOG} = \frac{G}{K_y a}$$

$$\frac{1}{K_y a} = \frac{1}{k_y a} + \frac{m}{k_x a}$$

The following correlations apply for absorption of SO_2 from air with water at 298 K when the packing is 50-mm ceramic Raschig rings:

$$k_y a = 0.0379\,(G')^{0.64}\,(L')^{0.362}$$

$$k_x a = 0.177\,(L')^{0.812}$$

where the volumetric mass transfer coefficients are in kmol/m^3-s, and the mass velocities are in kg/m^2-s.

Answer: Z = 3.61 m

5.7[a]. Gasification of high-sulfur bituminous oil/water emulsions(orimulsions)

A 350-MW power plant to be located on the Baltic coast of Sweeden will gasify a high-sulfur orimulsion from Venezuela (Berglund, B. "The 350 MW NEX Project", presented at the Ninth Annual EPRI Conference on Gasification Power Plants, Oct. 16–19 1990, Palo Alto, CA). The primary fuel, with a heating value of 27.2 MJ/kg, will be gasified with steam and oxygen into an energy-rich gas consisting mainly of hydrogen, CO, and CO_2. The sulfur in the orimulsion (2.7%) will be converted to H_2S in the gasifier and later recovered as elemental sulfur. The desulfurized gas will be sent to a combined gas-steam turbine (combined cycle) for power generation. The thermal efficiency of the power cycle is 40%. Sulfur emissions will be limited to 0.5 mg S per MJ of heat content of the primary fuel. Calculate

(a) Daily mass feed rate of orimulsion

(b) Sulfur removal efficiency

(c) Daily production of elemental sulfur

Answer: 75 metric ton of S/d

5.8[b] Effect of liquid rate on absorber design

Redesign the absorber of Problem 5.6, but for a liquid flow rate which is 10% higher than the minimum. The pressure drop through the irrigated packing should not exceed 400 Pa/m. Calculate the diameter and height of the packing.

Answer: Z = 8.85 m

5.9[d]. The use of seawater for FGD

Consider the FGD process of Examples 5.3 and 5.4. Determine the packing height if the seawater flow rate is 15% above the minimum, absorption occurs at 310 K, and the tower is packed with 50-mm ceramic Raschig rings. Assume that the correlation for k_ya given in Problem 5.6 applies.

Answer: Z = 0.72 m

5.10[a]. Scrubber effluent reheat

Calculate the power to reheat the gaseous effluent from the limestone scrubber of Example 5.6 by 25 K. Assume that the reheating is done with 85 percent efficiency.

Answer: 12.4 MW

5.11[b] Eliminating reheat from FGD systems

It has been proposed that corrosion downstream of FGD absorbers can be prevented without a reheat system (Froelich, D. A., and Graves, G. M. *JAPCA*, **37**:314, 1987). A number of existing material alternatives (mostly alloys) have shown excellent service in the prevention of corrosion in wet ductwork and stacks eliminating the need for reheat.

The final decision on the viability of converting an FGD system with reheat to a “wet stack” should be based on economic considerations.

Consider Mill Creek Station, Unit 3, a power plant in Louisville, Kentucky. The effluent from its FGD system was reheated by 28 K with steam. In 1984 the reheat system required extensive repairs to remain operational. Two alternatives were considered. Alternative A was to repair the reheat system and operate it for 10 yr. Alternative B was to eliminate reheat and convert the system to a”wet stack.” Alternative A required a total capital investment of $1,720,000, mostly for repairs. Annual costs were $442,000 for maintenance and $1,702,000 for operating expenses. Alternative B required a total capital investment of $5,930,000 to line the inside of the stack with a special alloy and to replace the scrubber outlet ductwork with corrosion-resistant materials. No additional annual expenses were required for this option. The minimum attractive return on additional investment for the decision makers was 21%. Choose the best alternative based on the PW measure of merit.

Answer: Choose B

5.12[d] Absorption with chemical reaction

The flue gas from a small power plant flows at the rate of 2,000 m^3/min at 300 K and 1 atm, and it contains 0.3% SO_2. The average molecular weight of the gas is 29.4. It is to be desulfurized to 100 ppm SO_2 by absorption in a packed tower. The absorbing liquid is an aqueous alkali solution with a pH of 10.0. A tower, packed with nonstaggered wood grids, is available. The diameter of the tower is 2.14 m, and the height of the packed section is 4.62 m. The dimensions of the wood grids are

Clearance = 3.81 cm

Height = 10 cm

Thickness = 0.64 cm

The volumetric mass-transfer coefficient for this system is given by (Peters, M. S., and Timmerhaus, K. D. *Plant Design and Economics for Chemical Engineers*, 4th. ed., McGraw-Hill, New York, 1991)

$$k_y a = 0.0266\,(G')^{0.8}$$

where k_ya is in kmol/m^3-s, and G' is in kg/m^2-s. Calculate the water flow rate for the desired separation, and the pH of the liquid effluent.

Answer: pH = 2.24

5.13b. Limestone scrubbing FGD

Abrams et al. (1988) give the following design conditions for a typical 250-MW power plant burning 1.25 percent sulfur coal:

Flue gas flow: 765 kmol/min at 413 K and 1 atm

Flue gas characteristics: 957 ppm SO_2, 8.14% H_2O, average MW 29.33

The gas will be processed in a limestone scrubber to remove 90% of the SO_2. The limestone is 95% $CaCO_3$ and 5% inert material. A limestone excess of 15% will be used, based on the amount required for 90% SO_2 removal. Calculate

(a) Limestone feed rate

Answer: 115.2 metric ton/dy

(b) Solid waste rate of production. Assume that all the SO_2 removed is converted to $CaSO_3 \cdot 2H_2O$. The sludge is dewatered to 60% solids.

Answer: 280.3 metric ton/dy

c) Water makeup requirements. Assume that the gases undergo adiabatic saturation in the scrubber.

Answer: 1,407 metric ton/dy

5.14b. Energy requirements for limestone scrubbing

Estimate the power requirements of the FGD system described in Problem 5.13. The total pressure drop for the gas is 2.8 kPa. The fan, located at the absorber inlet, has a mechanical efficiency of 65%. A reheat of 25 K is needed and the heat is added using a 90%-efficient heat exchanger. The slurry recirculation pump, with a mechanical efficiency of 65%, delivers 28 m of head and the specific gravity of the slurry is 1.09. The ratio of slurry to gas flow is 0.009 m^3 of liquid per m^3 of gas. Assume that all the other pumps and mechanical equipment in the system consume about 20% of the power consumed by the fan and the recirculation pump. Assuming that the thermal efficiency of the power cycle is 35%, estimate the FGD system energy requirements as a fraction of the total thermal energy load of the plant.

Answer: 3.4%

5.15a. Costs of limestone scrubbing

Estimate the total capital investment and total annual costs for the FGD system of Problem 5.13. Assume that the plant operates 85% of the time at its rated capacity.

Answer: TCI = $38,400,000 in June 1990

5.16a. Effect of particle size on the kinetics of FGD by dry injection

Repeat Example 5.8, but for a particle size of 50 μm. Assume that all the other conditions remain equal and calculate the time for 93% removal of the SO_2.

Answer: 202 s

5.17b. FGD by citrate scrubbing

A regenerative FGD process involves SO_2 absorption with a buffered solution of citric acid and sodium citrate (Cooper and Alley 1986). The citrate ions in solution increase the effective solubility of SO^2 by binding some of the hydronium ions created when the SO_2 absorbs into water. The spent citrate solution is regenerated by reduction of the SO_2 with H_2S to form elemental sulfur in a liquid-phase reaction. The H_2S can be generated on site, or it can be obtained from an oil refinery if one is nearby. The elemental sulfur is separated from the regenerated citrate solution by air flotation.

Estimate annual revenues from sulfur sales for a citrate FGD process that removes 90% of the SO_2 from a 40%-efficient 400-MW power plant. The coal is 5% sulfur and has a heating value of 20,000 kJ/kg. Sulfur sells for $150/metric ton, and H_2S costs $160/metric ton at a nearby refinery. Assume that the plant operates 8,760 h/yr.

Answer: $7,800,000/yr

5.18c. Anion-exchange resin-based FGD process

Sheth and Modeste (1990) developed and tested at the bench scale a low-temperature sorbent regeneration process in which commercially available anion-exchange resins remove soluble sulfate ions from aqueous solutions containing alkali metal sulfates. Coupled to dry nahcolite or trona injection, the resulting scheme is a regenerative FGD process in which the sorbent is regenerated and the sulfur is recovered in the form of ammonium sulfate, a valuable by-product.

Consider a flue gas flowing at the rate of 10 kmol/s and which contains 1,000 ppm of SO_2. Pulverized nahcolite is injected to remove 93% of the SO_2. The spent sorbent, mostly in the form of Na_2SO_4, is recovered in a fabric filter and dissolved in water to a concentration of 0.055 kg SO_4^{-2} per liter of solution at 293 K. The solution flows through a bed packed with the anion-exchange resin, kept under a partial pressure of CO_2 of 138 kPa, where the sulfate ions are adsorbed quantitatively. They are substituted in the solution by bicarbonate ions that are then converted to carbonate during the subsequent evaporation-concentration steps, which complete the sorbent regeneration. When the bed reaches breakthrough, it is eluted with NH_4OH, regenerating the resin and forming the ammonium sulfate.

Three similar beds are available, each with a diameter of 3.9 m and a packed depth of 2.0 m. Two beds adsorb in parallel while the third bed regenerates. The bulk density of the resin is 700 kg/m3 and the porosity is 40%. The equilibrium isotherm for this system was found to be of the Langmuir type

$$q = \frac{a_1 C}{1 + a_2 C}$$

where

q = concentration of sorbate in the resin phase, kg SO_4^{-2}/L resin

C = concentration of sorbate in the solution phase, kg/L solution

a_1 = constant, 25.9 L of solution/L of resin

a_2 = constant, 323.6 L of solution/kg SO_4^{-2}

For this system, the rate-controlling step is intraparticle diffusion. Sheth and Modeste (1990) found that, for these operating conditions, $D_p/d_p^2 = 0.15 \times 10^{-3}$ s-1, where D_p and d_p are the effective intraparticle diffusivity and particle size, respectively. Estimate the breakthrough time ($C/C_0 = 0.01$) using the Thomas solution.

Answer: 87 min

5.19b. Dry injection FGD: kinetics of the calcination of $NaHCO_3$

Calcination of NaHCO3 according to Eq. (5.65) is the first step in FGD by nahcolite injection.The effective rate constant for this reaction, k_{eff}, is given by Arrhenius equation (Gårding and Svedberg 1988):

$$k_{eff} = k_{0eff} \exp\left(-\frac{E_a}{RT}\right)$$

where

$k_{0eff} = 2.65 \times 10^{12}\ s^{-1}$

$E_a = 111.5$ MJ/kmol

T = absolute temperature in K

Estimate the time for 99% conversion to Na_2CO_3 at 473 K, and compare it to the time for the sulfation step as given by Example 5.8.

Answer: 3.6 s

5.20b. Economics of MgO-FGD process

Consider the MgO system of Example 5.10. According to Cogentrix (1990), the TCI for the system is $39 million including a plant for recovery of 98% sulfuric acid. The selling price for sulfuric acid is $82.50/metric ton as 100% H_2SO_4. Estimate the total annual cost for this system, assuming that the following data apply.

Materials and utilities	Annual quantity[a]	Unit cost ($)
MgO	414 metric ton	443/metric ton
Coke	290 metric ton	55/metric ton
Catalyst for SO_2 oxidation	756 L	10/L
Fuel oil (No.6)	8.515×10^6 L	0.13/L
Steam	8.4×10^7 kg	8.80/metric ton
Process water	3.52×10^9 L	0.50/m^3
Electricity	3.0×10^7 kW-h	0.05/kW-h

[a] McGlamery and Torstrick, 1976

Operating and supervision labor requirements are estimated as 16,500 man-h/yr at a combined rate of $25/man-h. Maintenance labor and materials are estimated as 7% of the TCI. The system has a 15-yr life expectancy with no salvage value. The minimum attractive return on investment for the company is 12%. Assume that the plant operates 8,760 h/yr.

Answer: TAC = $14 million

5.21c. Fluidized bed CuO FGD process

A granular packed bed begins to expand when the pressure drop owing to the upward flow of gas equals the weight of the packing. At this point, the velocity of the gas is known as the *minimum fluidization velocity*. At higher gas velocities, as the bed expands, the porosity increases and the individual particles move under the influence of the passing fluid. The bed has many of the appearances of a boiling liquid and is referred to as being "fluidized." The bed will disintegrate and stream with the fluid when the gas velocity equals the velocity of free fall of the particles. From the point of initial fluidization to bed disintegration the gas pressure drop remains relatively constant.

(a) The fluidized bed described by Frey and Rubin (1991) consists of copper-impregnated, 3 mm-diameter alumina spheres. The bulk density of the fixed bed is 880 kg/m^3, and its external porosity is 40%. The gas phase is mostly air at 101.3 kPa and 650 K. If the pressure drop through the fixed bed is given by Eq. (4.42), estimate the minimum fluidization velocity, and the corresponding pressure drop per unit bed depth.

Answer: 1.28 m/s

(b) Calculate the gas velocity to expand the bed to a bulk density of 400 kg/m^3.

Answer: 6 m/s

(c) The conversion rate of the reaction in the fluidized bed adsorber described by Eq. (5.56) is based on the SO_2 emission control requirement. The copper-to-sulfur (Cu/S) molar ratio required to achieve a specified SO_2 reduction requirement is estimated based on a first-order sulfation reaction kinetics model. The model assumes an ideal plug flow reactor and may be written as

$$\eta_s = \frac{\exp\left[B\left(1 - r^{-1}\right)\right] - 1}{\exp\left[B\left(1 - r^{-1}\right)\right] - r^{-1}}$$

where

η_s = fractional SO_2 conversion

r = bed inlet Cu/S molar ratio,

$B = \dfrac{k\rho_b RTZC_0}{PM_{Cu}u}$

k = reaction rate constant, time^{-1}

ρ_b = fluidized bed bulk density

R = ideal gas constant

T, P = gas temperature and pressure

Z = fluidized bed height

C_0 = weight ratio of copper oxide to alumina in the sorbent

M_{Cu} = atomic weight of copper

u = superficial (empty-tube) gas velocity

Because incomplete regeneration of the sorbent reduces the copper availability at the bed inlet, the model must explicitly include the effect of the regeneration efficiency on the Cu/S requirement. The following equation relates C0 to the regeneration efficiency, η_r:

$$C_0 = \left(\frac{X_{Cu}}{0.8 - X_{Cu}}\right)\frac{1}{\left[1 + \dfrac{\eta_s}{\eta_r}(1-\eta_r)r^{-1}\right]}$$

where X_{Cu} = weight fraction of copper in the fresh sorbent.

Estimate the molar Cu/S ratio required for 90% SO_2 removal in a bed fluidized to a depth of 1.22 m and a bulk density of 400 kg/m^3 (see parts a and b earlier). Assume that the reaction rate constant is 0.5 s^{-1}. The copper weight fraction in the fresh sorbent is 7%. The regeneration efficiency is 95%.

6

Control of Nitrogen Oxides

6.1 INTRODUCTION

Nitrogen oxides (NO_x), like sulfur oxides, are mainly formed during combustion, although a few industries also emit these gases from process operations. It is estimated that of the 19.5 million metric ton of NO_x emitted in 1987, 43% was from mobile sources, the remainder coming mostly from fuel combustors (McInnes and Van Wormer 1990). Consisting primarily of NO and nitrogen dioxide (NO_2), NO_x are formed by the oxidation of fuel-bound nitrogen (called *fuel* NO_x) and by the fixation of nitrogen in the combustion air at the high temperatures associated with combustion (called *thermal* NO_x). The formation of fuel NO_x depends on such factors as the nitrogen content of the fuel, total excess air, and relative distribution of primary and secondary combustion air. Formation of thermal NO_x is influenced by oxygen availability, temperature, pressure, and residence time in the combustion unit.

In the late 1940s, A.J. Haagen-Smit, of the California Institute of Technology, discovered that a certain type of photochemical smog resulted from atmospheric reactions involving NO_x and reactive hydrocarbons. There are many complex reactions occurring and the exact mechanism that leads to smog formation is, as yet, unknown. It is known that NO_2 functions primarily as the light-energy absorber. The major process by which NO_2 is formed in the atmosphere is (Peavy et al. 1985)

$$O_3 + NO \leftrightarrow O_2 + NO_2 \tag{6.1}$$

Hydroperoxyl radicals may also react with NO to generate NO_2 and hydroxyl radicals

$$\cdot HO_2 + NO \leftrightarrow \cdot OH + NO_2 \tag{6.2}$$

Alkylperoxyl radicals can oxidize NO to generate alkyloxyl radicals and NO_2

$$\cdot RO_2 + NO \leftrightarrow \cdot RO + NO_2 \tag{6.3}$$

Nitrogen dioxide absorbs ultraviolet (UV) radiation and decomposes, followed by ozone formation

$$NO_2 + \text{UV radiation} \leftrightarrow NO + O \quad (6.4)$$

$$O_2 + O + M \leftrightarrow O_3 + M \quad (6.5)$$

The effect is rapid cycling of NO_2, and no overall effect would result if it were not for a series of competing reactions involving hydrocarbons. Atomic oxygen (O) reacts with the hydrocarbons to produce a reactive species called alkylperoxyl radicals ($\cdot RO_2$). These free radicals react rapidly with NO to produce NO_2. This removes the NO from the cycle and, thus, the reaction that would remove O_3 from the system is eliminated, causing the ozone concentration to increase in the atmosphere (see Figure 6.1). The $\cdot RO_2$ radical can also react with O_2 and NO_2 to produce peroxyacetyl nitrates (PAN). The end product of these reactions is photochemical smog consisting of air contaminants such as ozone, PAN, aldehydes, ketones, alkyl nitrates, and CO.

Between 1960 and 1980 tropospheric ozone increased at a rate between 1% and 2% per year, meaning that ozone increased 22% to 48% over that 20-yr period (Fishman and Kalish 1990). By comparison, during the same period, CO_2 increased by 6% to 7%. Trends in tropospheric ozone are being carefully observed since O_3 is a "greenhouse gas," produces serious health effects in plants and animals, and damages materials. There is controversy among the scientific community about what levels of atmospheric ozone are safe. In 1971, the EPA established the primary and secondary standards for ozone as an hourly average of 0.08 ppm not to be exceeded more than once a year. They also established the goal of reducing NO_x emissions from mobile sources by 90% by 1976. In 1973, the NO_x standard was postponed, and in 1982 it was permanently relaxed to 75% reduction. In 1979, the ozone standard was relaxed to an hourly average of 0.12 ppm.

Nonetheless, there are problems with the current standards, according to some scientists studying the effects of ozone. Larsen et al. (1991) and Folinsbee et al. (1988) showed conclusively that even at levels below the 0.12 ppm threshold set by the EPA, healthy exercising adults show signs of lung damage. An even more relevant series of studies was completed over a period of years during the 1980s (Spektor et al. 1988). It assessed the effect of ambient ozone levels on children at summer camp. The researchers found that lung function decreased among the children studied during times of increased ozone levels, although never during the study did the ozone in the air exceed the NAAQS of 0.12 ppm. They concluded that levels of 0.08 to 0.10 ppm are still too high for the safety of active children.

For several reasons, NO_x removal from flue gases is more difficult and expensive than SO_2 removal, and, as a result, technology for NO_x cleaning of flue gases is not as advanced. The key problem is that NO is relatively insoluble and unreactive. In addition, flue gases containing NO often contain H_2O, CO_2, and SO_2 in greater concentrations than NO. These species are more reactive than NO and interfere with its removal.

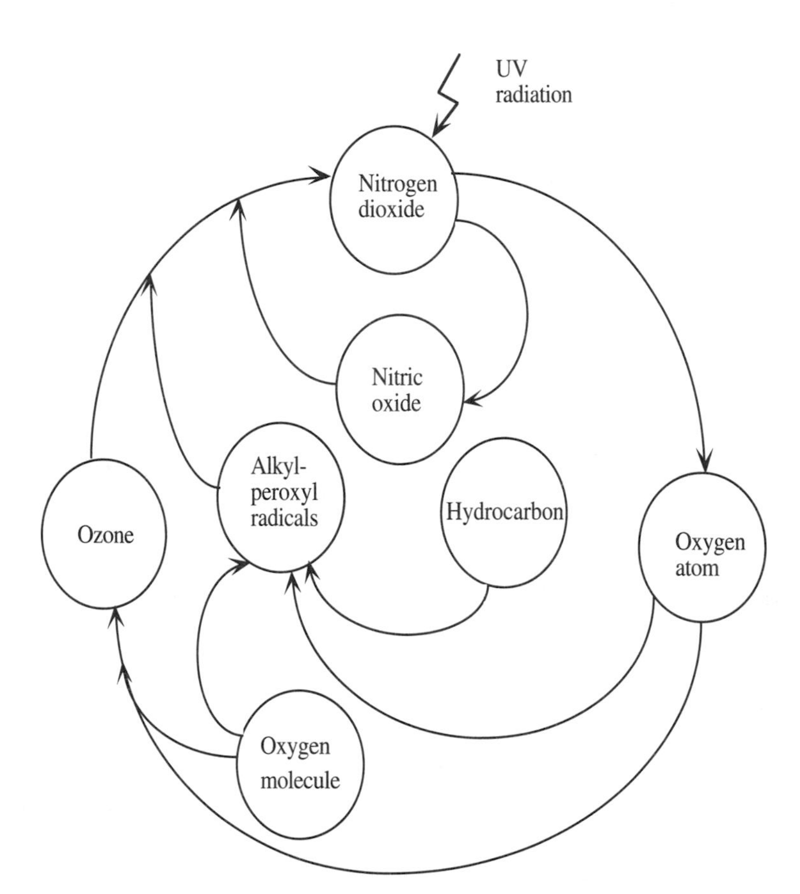

Figure 6.1 Interaction of the atmospheric NO_2 photolytic cycle with hydrocarbons

NO_x can be reduced by the use of low-nitrogen fuels, modifying design and operating features of the combustion unit, and add-on controls. Combustion modifications include flue gas recirculation (FGR), low-NO_x burners, and staged-air combustion. Two add-on control technologies that are gaining widespread acceptance are selective catalytic reduction (SCR) and selective non catalytic reduction (SNCR).

Successful control of NO_x depends on an understanding of the fundamental principles of its formation. This chapter presents an introduction to the chemistry of NO_x formation, followed by an overview of the technologies for NO_x control. Cost information available in the literature is scant and tentative, therefore, no detailed economic analysis will be attempted.

6.2 THERMAL NO_X FORMATION

Your objectives in studying this section are to

1. Understand the Zeldovich mechanism for thermal NO_x formation.
2. Calculate equilibrium concentrations of the chemical species in Zeldovich mechanism.
3. Calculate the rate of formation of thermal NO_x, and the effect on it of the flame temperature and oxygen availability.

The formation of NO by oxidation of atmospheric nitrogen can be expressed in terms of the overall reaction

$$O_2 + N_2 \leftrightarrow 2NO \tag{6.6}$$

which is highly endothermic. As a result, the equilibrium concentration of NO is high at the very high temperatures encountered near stoichiometric combustion and decreases rapidly away from that point.

Even though Eq. (6.6) expresses the overall reaction, the direct reaction of N_2 with O_2 is too slow to account for significant NO formation. Free oxygen atoms, produced in flames by dissociation of O_2 or by radical attack on it, attack nitrogen molecules and begin a chain mechanism that was first postulated by Zeldovich (Zeldovich et al. 1947; Flagan and Seinfeld 1988).

$$N_2 + O \underset{-1}{\overset{+1}{\leftrightarrow}} NO + N \tag{6.7}$$

$$N + O_2 \underset{-2}{\overset{+2}{\leftrightarrow}} NO + O \tag{6.8}$$

The concentration of O_2 is low in fuel-rich combustion, so the reaction of Eq. (6.8) is less important than in fuel-lean combustion. Reaction with the hydroxyl radical eventually becomes the major sink for N

$$\mathrm{N + OH} \underset{-3}{\overset{+3}{\leftrightarrow}} \mathrm{NO + H} \tag{6.9}$$

The rate constants for the Zeldovich mechanism are (Flagan and Seinfeld 1988)

$$k_{+1} = 1.8 \times 10^8 \, e^{-38{,}730/T} \ \mathrm{m^3/mol\text{-}s} \tag{6.10}$$

$$k_{-1} = 3.8 \times 10^7 \, e^{-425/T} \ \mathrm{m^3/mol\text{-}s} \tag{6.11}$$

$$k_{+2} = 1.8 \times 10^4 \, T \, e^{-4{,}680/T} \ \mathrm{m^3/mol\text{-}s} \tag{6.12}$$

$$k_{-2} = 3.8 \times 10^3 \, T \, e^{-20{,}820/T} \ \mathrm{m^3/mol\text{-}s} \tag{6.13}$$

$$k_{+3} = 7.1 \times 10^7 \, e^{-450/T} \ \mathrm{m^3/mol\text{-}s} \tag{6.14}$$

$$k_{-3} = 1.7 \times 10^8 \, e^{-24{,}560/T} \ \mathrm{m^3/mol\text{-}s} \tag{6.15}$$

The high activation energy of the first reaction, resulting from its essential function of breaking the strong N_2 triple bond, makes this the rate-limiting step of the proposed mechanism. Because of the high activation energy, NO production by this route proceeds at a slower rate than the oxidation of the fuel constituents and is extremely temperature sensitive. The production of atomic oxygen is also very sensitive to the temperature.

To understand the rate of NO formation, the rate equations corresponding to the reactions in Eqs. (6.7) to (6.9) follow. Consider, for example, the net rates of formation of NO and N, R_{NO} and R_N respectively

$$R_{NO} = k_{+1} [N_2] [O] - k_{-1} [N] [NO] + k_{+2} [N] [O2] - k_{-2} [NO] [O] + k_{+3} [N] [OH] - k_{-3} [NO] [H] \tag{6.16}$$

$$R_N = k_{+1} [N2] [O] - k_{-1} [N] [NO] - k_{+2} [N] [O2] + k_{-2} [NO] [O] - k_{+3} [N] [OH] + k_{-3} [NO] [H] \tag{6.17}$$

The concentrations of O, H, and OH must be known for calculation of the N and NO formation rates. The high activation energy of the initial attack on N_2 allows an important simplification. Because the reaction rate is high only at the highest temperatures, most of the reaction takes place after the combustion reactions are complete and before significant heat is transferred from the flame. It is a reasonable assumption, therefore, that the O, H, and OH radicals are present in their equilibrium concentrations. At equilibrium

$$k_{+1}\,[N_2]_e\,[O]_e = k_{-1}\,[N]_e\,[NO]_e \tag{6.19}$$

Define the equilibrium, one-way rate of reaction as

$$R_1 = k_{+1}\,[N_2]_e\,[O]_e = k_{-1}\,[N]_e\,[NO]_e \tag{6.20}$$

Similarly, at equilibrium

$$R_2 = k_{+2}\,[N]_e\,[O_2]_e = k_{-2}\,[NO]_e\,[O]_e \tag{6.21}$$

$$R_3 = k_{+3}\,[N]_e\,[OH]_e = k_{-3}\,[NO]_e\,[H]_e \tag{6.22}$$

Define the quantities

$$\alpha = \frac{[NO]}{[NO]_e} \tag{6.23}$$

$$\beta = \frac{[N]}{[N]_e} \tag{6.24}$$

The rate equations may now be expressed in the abbreviated form

$$R_{NO} = R_1 - R_1\alpha\beta + R_2\beta - R_2\alpha + R_3\beta - R_3\alpha \tag{6.25}$$

$$R_N = R_1(1 - \alpha\beta) + R_2(\alpha - \beta) + R_3(\alpha - \beta) \tag{6.26}$$

To calculate the rate of NO formation, the N atom concentration must be determined. Since the activation energy for the oxidation of the nitrogen atom is small and the reaction involves O_2, a major component of the gas for fuel-lean conditions, the free nitrogen atoms are consumed as rapidly as they are generated, establishing a quasi–steady state. Setting the left-hand side of Eq. (6.26) equal to zero and solving for the steady state nitrogen atom concentration, β_{ss}

$$\beta_{ss} = \frac{R_1 + R_2\alpha + R_3\alpha}{R_1\alpha + R_2 + R_3} = \frac{\kappa + \alpha}{\kappa\alpha + 1} \tag{6.27}$$

where

$$\kappa = \frac{R_1}{R_2 + R_3} \tag{6.28}$$

Substituting Eq. (6.27) into Eq. (6.25) yields a rate equation for NO formation in terms of α and known quantities:

$$R_{NO} = \frac{2R_1(1 - \alpha^2)}{1 + \kappa\alpha} \tag{6.29}$$

For constant temperature and pressure, this may be written as a differential equation for α

$$\frac{d\alpha}{dt} = \frac{1}{[NO]_e} \frac{2R_1(1 - \alpha^2)}{1 + \kappa\alpha} \tag{6.30}$$

Equation (6.30) can be integrated analytically to describe NO formation in a constant-temperature system. Assuming that there is no NO present initially (i.e., $\alpha = 0$ at $t = 0$), the result is

$$(1 - \kappa)\ln(1 + \alpha) - (1 + \kappa)\ln(1 - \alpha) = \frac{t}{\tau_{NO}} \tag{6.31}$$

where the characteristic time for NO formation is

$$\tau_{NO} = \frac{[NO]_e}{4R_1} \tag{6.32}$$

Figure 6.2 illustrates the approach to equilibrium for a given value of κ. The characteristic time, τ_{NO}, corresponds to the time that would be required for NO to reach the equilibrium level if the reaction continued at its initial rate and were not slowed by the reverse reactions.

Two major assumptions were made in the derivation of Eq. (6.32): (1) a quasi-steady state for the nitrogen atom concentration and (2) equilibrium concentrations for the O, H, and OH radicals. The validity of the first assumption can be readily examined. Consider the time required to achieve this steady state initially, which is when the NO concentration is small, and only the

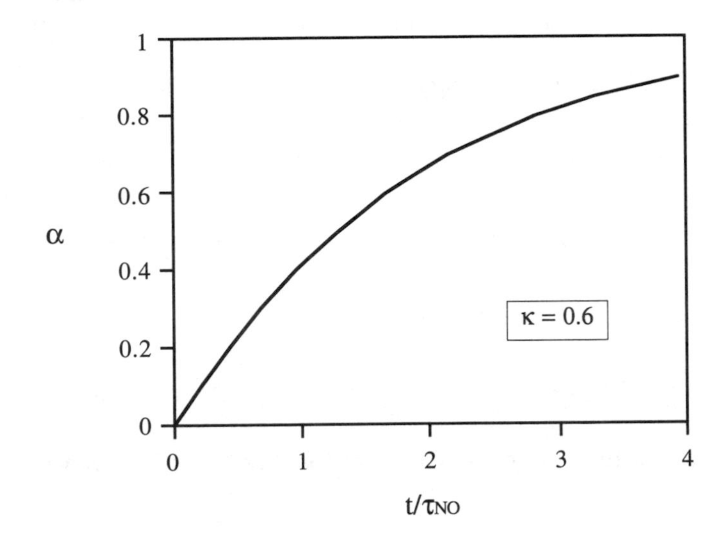

Figure 6.2 Approach of the dimensionless NO concentration to equilibrium

forward reactions need be considered. The rate equation for the nitrogen atom concentration becomes

$$[N]_e \frac{d\beta}{dt} = R_1 - (R_2 + R_3)\,\beta \tag{6.33}$$

with the initial condition of $\beta = 0$ at $t = 0$. Integrating yields

$$\beta = \kappa\left[1 - \exp\left(-\frac{t}{\tau_N}\right)\right] \tag{6.34}$$

where

$$\tau_N = \frac{[N]_e}{R_2 + R_3} \tag{6.35}$$

For the quasi-steady state to be valid, τ_N must be much smaller than τ_{NO}. Comparison of the two time scales for adiabatic combustion shows that is the case throughout the range of air to fuel ratios where formation of NO by the Zeldovich mechanism is significant.

To calculate the NO formation rate, the equilibrium concentrations of NO, O, OH, and H must be calculated. The following reactions occur at the high temperatures associated with combustion:

$$1/2\ N_2 + 1/2\ O_2 \leftrightarrow NO \tag{6.36}$$

$$1/2\ N_2 \leftrightarrow N \tag{6.37}$$

$$1/2\ O_2 \leftrightarrow O \tag{6.38}$$

$$1/2\ H_2O + 1/4\ O_2 \leftrightarrow OH \tag{6.39}$$

$$1/2\ H_2O \leftrightarrow H + 1/4\ O_2 \tag{6.40}$$

The equilibrium constant as a function of temperature for these reactions can be expressed in the general form

$$\ln K_p = -\frac{a}{T} + b \ln T - cT + d \tag{6.41}$$

where

K_p = equilibrium constant in terms of partial pressures (in atm)
T = absolute temperature, in K
a, b, c, and d are constants for a given reaction
Table 6.1 gives values of the constants in Eq. (6.41) for Reactions (6.36) to (6.40).

Table 6.1 Parameters for Estimating Equilibrium Constants Using Eq. (6.41)

Reaction No.	a	b	$c \times 10^5$	d
6.36	10,853	0.0862	3.0	0.9332
6.37	56,720	0.7371	9.0	2.0360
6.38	29,812	0.7205	11.7	2.2838
6.39	19,178	0.5064	12.6	1.2581
6.40	40,403	1.4642	20.7	–1.0702

Source: Calculated from data in Flagan and Seinfeld (1988).

Example 6.1 Equilibrium concentration of the hydroxyl radical at high temperature

An equimolar mixture of water vapor and molecular oxygen is heated at a constant pressure of 1 atm to 2,300 K. Estimate the mole fraction of the hydroxyl radical in the equilibrium mixture.

Solution

Calculate the equilibrium constants for Reactions (6.38) to (6.40) from the data in Table 6.1. They are: $K_{pO} = 4.66 \times 10^{-3}$ atm$^{0.5}$, $K_{pOH} = 0.0317$ atm$^{0.25}$, $K_{pH} = 4.18 \times 10^{-4}$ atm$^{0.75}$. Express the equilibrium constants in terms of the equilibrium molar fractions for all the constituents:

$$K_{pO} = y_O \sqrt{\frac{P}{y_{O_2}}}$$

$$K_{pOH} = \frac{y_{OH}}{\sqrt{y_{H_2O}}} \left(\frac{P}{y_{O_2}}\right)^{1/4}$$

$$K_{pH} = \frac{y_H \, y_{O_2}^{1/4}}{y_{H_2O}^{1/2}} P^{3/4}$$

Express, through material balances, the equilibrium mole fractions in terms of the progress of each of the three reactions.

Basis: 1 mole of the initial equimolar mixture

Let

ε_O = moles of atomic oxygen in equilibrium,

ε_{OH} = moles of hydroxyl radical in equilibrium,

ε_H = moles of atomic hydrogen in equilibrium.

The following table summarizes the material balances in terms of the basis chosen and the three variables defined earlier. For a total pressure of 1 atm, the expressions for the equilibrium constants yield three simultaneous algebraic non-linear equations, which are solved for ε_O, ε_{OH}, and ε_H. The equilibrium concentrations are, then, calculated.

Component	Moles in	Consumed	Generated	In equilibrium
H_2O	0.5	$0.5(\varepsilon_{OH} + \varepsilon_H)$	0	$0.5(1 - \varepsilon_{OH} - \varepsilon_H)$
O_2	0.5	$0.5(\varepsilon_O + \varepsilon_{OH}/2)$	$0.25\ \varepsilon_H$	$0.5(1 - \varepsilon_O - \varepsilon_{OH}/2 + \varepsilon_H/2)$
O	0	0	ε_O	ε_O
OH	0	0	ε_{OH}	ε_{OH}
H	0	0	ε_H	ε_H
Total	1.0			$1 + 0.5\varepsilon_O + 0.25\varepsilon_{OH} + 0.75\varepsilon_H$

The equilibrium relationships are:

$$4.66\ x\ 10^{-3} = \frac{\varepsilon_O}{\sqrt{0.5 n_T(1 - \varepsilon_O - 0.5\ \varepsilon_{OH} + 0.5\varepsilon_H)}} \tag{6.42}$$

$$0.0317 = \frac{\varepsilon_{OH}[0.5(1 - \varepsilon_{OH} - \varepsilon_H)]^{-05}}{[0.5 n_T(1 - \varepsilon_O - 0.5\varepsilon_{OH} + 0.5\varepsilon_H)]^{025}} \tag{6.43}$$

$$4.18\ \text{x}\ 10^{-4} = \frac{\varepsilon_H[0.5(1 - \varepsilon_O - 0.5\varepsilon_{OH} + 0.5\varepsilon_H)]^{025}}{n_T^{0.75}[0.5(1 - \varepsilon_{OH} - \varepsilon_H)]^{05}} \tag{6.44}$$

$$n_T = 1 + 0.5\varepsilon_O + 0.25\varepsilon_{OH} + 0.75\varepsilon_H \tag{6.45}$$

As a first approximation, this system of equations can be simplified assuming that, because the equilibrium constants are relatively small, $n_T \approx 1.0$, $y_{O_2} \approx 0.5$, and $y_{H_2O} \approx 0.5$. From Eqs. (6.42), (6.43), (6.44), and (6.45)

$$\varepsilon_O \approx 4.66\ \text{x}\ 10^{-3}\sqrt{0.5} \approx 0.0033$$

$$\varepsilon_H \approx 4.18\ \text{x}\ 10^{-4}(0.5)^{0.25} \approx 3.51\ \text{x}\ 10^{-4}$$

$$\varepsilon_{OH} \approx 0.0317(0.5)^{0.75} \approx 0.0188$$

$$n_T \approx 1.0066$$

The approximate equilibrium mole fractions are then

Component	O_2	H_2O	O	OH	H
Fraction	0.4905	0.487	0.00328	0.0187	0.000349

To check the accuracy of these values, calculate the equilibrium constants corresponding to these concentrations and compare to the known values. The results are: $K_{pO} = 4.68 \times 10^{-3}$, $K_{pOH} = 0.032$, and $K_{pH} = 4.19 \times 10^{-4}$. These are sufficiently close to the true values of the equilibrium constants, therefore the approximate equilibrium composition calculated previously is appropriate.

Example 6.2 Thermal NO Formation

Aviation kerosene, $CH_{1.88}$, is burned at 2,250 K and 10 atm with a residence time of 0.005 s. If the equivalence ratio is 0.8, calculate the mole fraction of NO in the hot gases. Assume that the combustion reaction is instantaneous compared with the mechanism for NO formation.

Solution

Basis: 1 mol of kerosene

Assuming complete combustion:

$$CH_{1.88} + 1.47\ O_2 \rightarrow CO_2 + 0.94\ H_2O$$

For an equivalence ratio of 0.8, the oxygen supplied is 1.47/0.80 = 1.838 mol. The nitrogen supplied with the air is 1.838 (3.78) = 6.95 mol. From material balance calculations, the composition of the hot gases is (before the Zeldovich mechanism is considered)

Component	Mole fraction
O_2	0.0397
N_2	0.7510
CO_2	0.1081
H_2O	0.1016

Calculate the equilibrium constants for Reactions (6.36) to (6.39):

$K_{pNO} = 0.0371$ $\qquad K_{pN} = 2.085 \times 10^{-8}\ atm^{0.5}$

$K_{pO} = 3.46 \times 10^{-3}\ atm^{0.5}$ $\qquad K_{pOH} = 0.0262\ atm^{0.25}$

Assuming that the equilibrium concentrations of O_2, N_2, and H_2O remain unchanged, the equilibrium concentrations of NO, N, O, and OH are

$y_{NOe} = 6.4 \times 10^{-3}$ $y_{Ne} = 0.57 \times 10^{-8}$

$y_{Oe} = 2.18 \times 10^{-4}$ $y_{OHe} = 2.1 \times 10^{-3}$

From Eq. (6.20), $R_1 = k_{+1}\left(\frac{P}{RT}\right)^2 y_{N_2}\, y_{Oe} = 3.37$ mol/m^3–s

From Eq. (6.21), $R_2 = k_{+2}\left(\frac{P}{RT}\right)^2 y_{O_2}\, y_{Ne} = 3.34$ mol/m^3-s

From Eq. (6.22), $R_3 = k_{+3}\left(\frac{P}{RT}\right)^2 y_{OHe}\, y_{Ne} = 2.03$ mol/m^3-s

From Eq. (6.32), the time scale for NO formation is $\tau_{NO} = 0.0256$ s. Equation (6.28) gives $\kappa = 0.628$. Solve Eq. (6.31) by trial and error for the value of a corresponding to a retention time of 0.005 s. The answer is $\alpha = 0.095$, or $y_{NO} = \alpha\, y_{NOe} = 0.608 \times 10^{-3} = 608$ ppm.

Comments

Flagan and Seinfeld (1988) performed similar calculations, but for a flame temperature of 2,304 K. At that temperature, only 54 K higher than in this example, the NO concentration was 1,180 ppm. Maintaining the temperature at 2,250 K, but increasing the equivalence ratio to 0.95—equivalent to lowering the percent excess air—results in a further reduction of the NO concentration to 317 ppm (see Problem 6.3). These examples illustrate the dramatic effect of the temperature and the oxygen availability on the thermal-NO rate of formation. It must be remembered, however, that those two operating variables are interrelated during adiabatic combustion, with the highest flame temperature occurring near stoichiometric conditions. One method of reducing the temperature without increasing the amount of oxygen available is to inject liquid water into the air-fuel mixture. It has proved to be effective in reducing NO emissions, but usually with a penalty on the thermal efficiency of the combustion process.

6.3 FUEL NO_X FORMATION

Your objectives in studying this section are to

1. Identify the factors which influence the conversion of fuel-nitrogen to NO_x .
2. Estimate the percent conversion of fuel-nitrogen to NO during the combustion of pulverized coal.

Many fuels contain organically bound nitrogen that is readily oxidized to NO during combustion. Crude oils contain 0.1 to 0.2% nitrogen on a mass basis, but when it is refined nitrogen concentrates in the residual fractions, which is the portion of the oil most likely to be burned in power plants and industrial boilers. Coal typically contains 1.2 to 1.6% nitrogen. The range of nitrogen contents of coal is much narrower than that of the sulfur contents. Thus, burning a low-nitrogen coal is not a practical solution to the problem of fuel-NO_x emissions from coal-fired boilers. New fuel sources may further aggravate the problems associated with fuel-nitrogen. Some of the shale oil deposits in the United States contain 2% to 4% nitrogen (Flagan and Seinfeld 1988).

Experiments by Pershing and Wendt (1977) showed the importance of fuel NOx formation during combustion of pulverized coal. Figure 6.3 shows that during their experiments fuel-NO_x accounted for more than 80% of the total NO_x emissions. Even though fuel-nitrogen was the major source of nitrogen oxides, only 20% to 30% conversion of the fuel-nitrogen to NO_x was observed. Conversion efficiencies of fuel-nitrogen to NO for coals and residual fuel oils have been observed between 10% and 60% (USEPA 1983).

Possible fates of fuel-nitrogen are shown in Figure 6.4. Regardless of the detailed mechanism, several general statements can be made regarding fuel-nitrogen oxidation to NO. It is highly dependent on the equivalence ratio. The following regression equation of the data by Pohl and Sarofim (1976) relates the percent conversion of fuel-nitrogen to the equivalence ratio, ϕ, during combustion of certain pulverized coals at a flame temperature of 1,600 K:

$$\text{Percent conversion} = 66.33 - 77.5\,\phi + 34.607\,\phi^2 - 5.2\,\phi^3 \tag{6.46}$$

The degree of fuel-air mixing also strongly affects the percent conversion of fuel-nitrogen to NO, with greater mixing resulting in greater percent conversion. Small temperature differences do not seem to affect production of NO_x from fuel nitrogen. This behavior is in direct contrast to thermal NO_x production, which is highly sensitive to temperature.

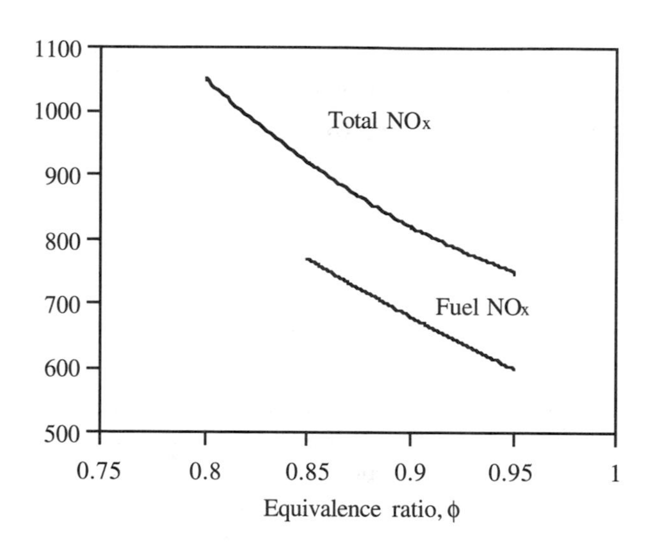

Figure 6.3 Contribution of fuel-NOx to total NOx emissions in the pulverized coal combustion experiments of Pershing and Wendt (1977). Reprinted with permission from The Combustion Institute

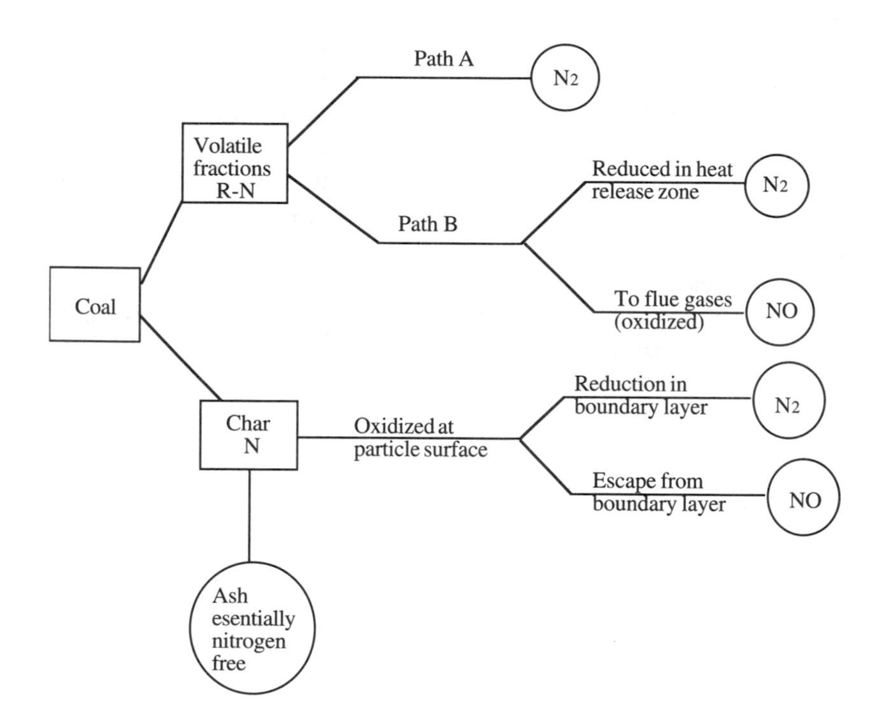

Figure 6.4 Possible fates of nitrogen contained in coal (Heap et al. 1976)

Example 6.3 Total NO_x Formation in Pulverized Coal Burning

A boiler burns 1,000 kg/min of pulverized coal with 2% nitrogen at 1600 K with an equivalence ratio of 0.85. Estimate the total mass emission rate of NO assuming that the fuel-NO_x constitutes 80% of the total NOx emissions.

Solution

From Eq. (6.46), for an equivalence ratio of 0.85, approximately 22.3% of the fuel nitrogen reacts to form NO. Thus, the mass emission rate of fuel-NO is (0.223) (0.02) (1,000) (30/14) = 9.56 kg fuel-NO/min. Assuming that the fuel-NO is 80% of the total, the combined mass emission rate of NO is 12.0 kg/min.

6.4 TECHNIQUES FOR CONTROL OF NOX EMISSIONS

Your objectives in studying this section are to

1. Describe the most common combustion modifications and flue gas treatments for the control of NO_x emissions.
2. Estimate the percentage reduction in NO_x emissions to meet the NSPS.
3. Design SCR reactors based on chemical reaction and mass-transfer rates.

The two broad categories for the control of NO_x emissions are combustion modifications and flue gas treatment techniques. Combustion modifications are used to limit the formation of NO_x during the actual combustion. Flue gas treatment techniques are used to remove nitrogen oxides from the flue gases after these compounds have been formed. Although flue gas treatment systems are in wide commercial use in Japan and Europe, they are only beginning to gain acceptance in the United States (Chen et al. 1990, Cogentrix 1990). The NSPS for NO_x established by the USEPA are so high, especially those for coal-fired power plants, that they are easily accomplished with combustion modifications. That will probably change as evidence of widespread damage to vegetation and human health owing to ozone pollution continues to accumulate.

Example 6.4 NSPS for NOx Emissions from a Coal-Fired Power Plant

Consider the boiler of Example 6.3. Assuming that the heating value of the coal is 26 MJ/kg, calculate the percentage reduction in NO_x emissions required to meet the NSPS for nitrogen oxides (260 g of $NO_2/10^6$ kJ).

Solution

From Example 6.3, the uncontrolled NO emission rate is 12.0 kg/min. Because the NSPS is given in terms of NO_2, the corrected uncontrolled emission rate is 12(42/30) = 16.8 kg/min. To meet the NSPS, the emission rate should not exceed $(0.26/10^6)(1000)(26 \times 10^3) = 6.76$ kg/min. Therefore, to meet the NSPS the percentage reduction is [(16.8 – 6.76)/16.8] (100) = 60%.

Comments

The percentage reduction to meet the NSPS for NOx is modest, even though the nitrogen content of the coal selected is extremely high (2%). Conversely, the allowable emission rate is extremely high: 9,734 kg of NO_2 /d.

6.4.1 Combustion modifications

Combustion controls reduce NOx formation by one or more of the following strategies (Cooper and Alley 1986):

1. **Reduce peak temperatures in the flame zone**
2. **Reduce gas residence time in the flame zone.**
3. **Reduce oxygen concentrations in the flame zone.**

The preceding changes to the combustion process can be achieved by either modification of operating conditions on existing furnaces or by purchase and installation of newly designed low-NO_x burners or furnaces. The following paragraphs summarize some of these modifications.

Low excess-air firing (LEA) is a very simple yet effective technique. Thirty years ago it was common to see furnaces operating with 50 to 100% excess air. As a result of fuel price increases, excess air was reduced to 15 to 30% to save money. Further reductions below the 15% level are now possible with the development of advanced instrumentation which allows continuous automatic furnace monitoring and control. Lim et al. (1980) reported an average 19% reduction of NOx emissions by reducing the percent excess air from 20% to 14%.

Off-stoichiometric combustion (OSC), often called staged combustion, means combustion of the fuel in two or more steps. The initial or primary flame zone is fuel-rich, and the secondary is fuel-lean. New low-NO_x burners incorporate in their designs LEA and OSC (see Figure 6.5). Tests indicate that low-NO_x burners reduce emissions of nitrogen oxides by 40 to 60% compared with older, conventional burners. They are now a standard part of any new design in the electric power industry.

Flue gas recirculation (FGR) is simply the rerouting of some of the flue gas back to the furnace. The flame temperature and oxygen concentration are reduced simultaneously reducing thermal-NOx formation. Flame instability and the decrease in net thermal output limit the recirculation rate. FGR does not affect fuel-NO_x; therefore, it is not used on boilers that burn coal or residual oil, fuels that contain high percentages of nitrogen. Capital and annualized costs for FGR can be high, especially for combustion sources retrofitted with FGR.

Water injection (or steam injection) can be an effective means of reducing flame temperatures, thus reducing thermal-NO_x . Water injection has been shown to be very effective for gas turbines, with NO_x reductions of about 80% for a water injection rate of 2% of the combustion air (Crawford, Manny, and Bartok 1977). The energy penalty for a gas turbine is about 1% of its rated output; however, for a utility boiler, it can be as high as 10% (Cooper and Alley).

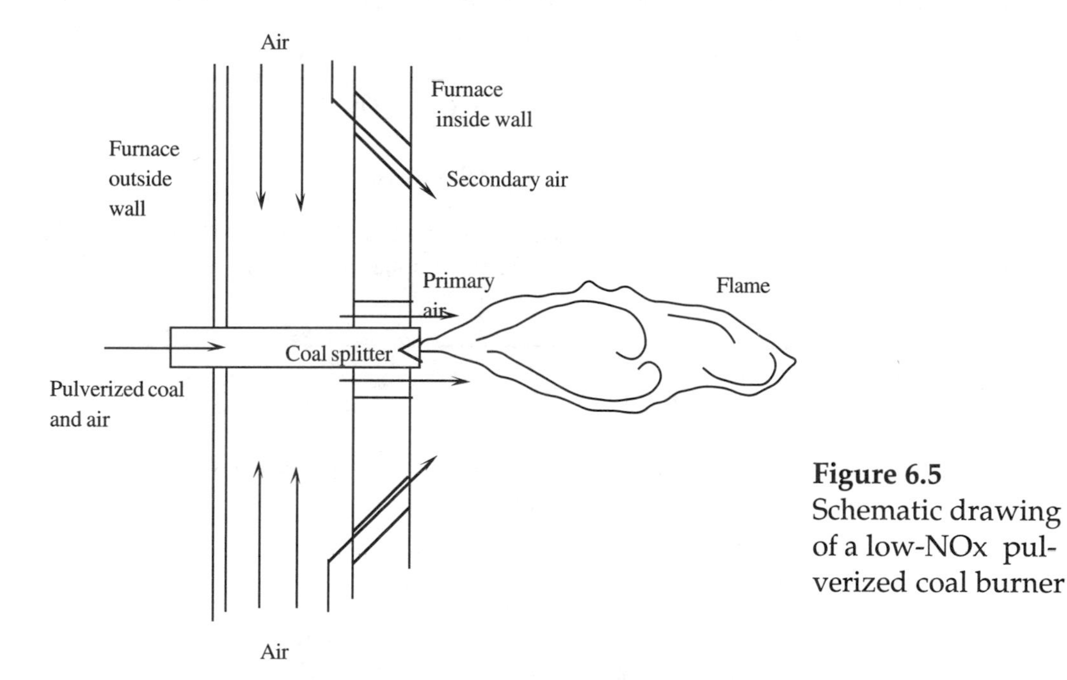

Figure 6.5 Schematic drawing of a low-NOx pulverized coal burner

6.4.2 Flue gas treatment techniques

Flue gas treatment to remove NO_x is useful in cases in which higher removal efficiencies are required than can be achieved with combustion controls. Most of the current processes were developed in Japan. They include selective catalytic reduction, noncatalytic reduction, adsorption, and irradiation.

Selective catalytic reduction (SCR) is the most advanced and effective method for reducing NO_x emissions, achicving removal efficiencies higher than 90%. The concept originated in the United States many years ago (see for example Benítez 1976), but was developed commercially in Japan where awareness of the real threat posed by NO_x emissions is much higher. By the beginning of 1985 about 160 SCR plants were in operation in Japan. About 60% of these are in oil-fired utility burners, 21% in coal-fired boilers, and 19% in gas-fired boilers (Ando 1985). By the beginning of 1990 numerous utilities in Europe had successfully adopted the Japanese SCR technology (Cogentrix 1990).

In selective catalytic reduction, the NO_x species are reduced by NH_3, ultimately to N_2 gas, over a heterogeneous catalyst in the presence of O_2. The process is termed selective because the ammonia preferentially reacts with NO_x rather than with O_2. The oxygen, however, enhances the reaction and is a necessary component of the process. The catalyst is usually a mixture of titanium and vanadium oxides and is formulated in pellets (for gas-fired units) or honeycomb shapes (for coal- or oil-fired units). The predominant reactions are

$$4NO + 4NH_3 + O_2 \rightarrow 4N_2 + 6H_2O \tag{6.47}$$

$$2NO_2 + 4NH_3 + O_2 \rightarrow 3N_2 + 6H_2O \tag{6.48}$$

The best temperature range for SCR catalyst activity and selectivity is from 570 to 720 K. Ammonia is vaporized and injected downstream from the boiler feed water preheater as shown in Figure 6.6.

Example 6.5 NH_3 requirements for SCR

SCR will be used to reduce NO emissions from the boiler of Example 6.3 by 90%. Estimate the stoichiometric amount of ammonia required.

Solution

From Example 6.3, the uncontrolled NO emission rate is 12.0 kg/min. Therefore, the stoichiometric ammonia required for 90% NO removal is (12/30)(1)(17) (0.9)(60)(24) = 8,813 kg/d.

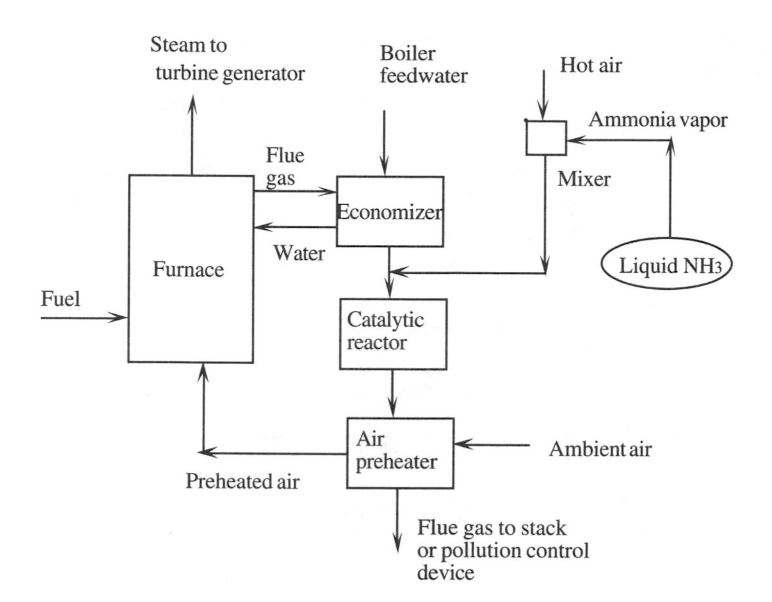

Figure 6.6 Schematic diagram for the SCR method

A problem with SCR processes is the formation of solid ammonium sulfate, $(NH_4)_2SO_4$, and liquid ammonium bisulfate, NH_4HSO_4 , both of which are highly corrosive and interfere with heat transfer. The problem is most severe with high-sulfur oils. With low-sulfur oils, the SO_3 which gives rise to the problem is not present in sufficient quantity. Tests with coal indicate that the ammonium by-products may deposit on the fly ash or be removed from the heat exchanger surfaces by the erosive action of the fly ash (Flagan and Seinfeld 1988).

Another problem with SCR is deactivation of the catalyst during coal combustion caused mainly by various "poisons" in the hot combustion gas, which depend on the type of coal used (Chen et al. 1990; Chen and Yang 1990). According to various researchers, Brönsted acid sites on the catalyst are responsible for the SCR reaction. The Brönsted acid sites are the protons on the surface hydroxyl groups. The NH_3 first chemisorbs on the Brönsted acid site, followed by the bonding of NO on the chemisorbed species and subsequent decomposition of the complex to N_2 and H_2O. Substances that enhance the Brönsted acidity are promoters of the SCR reaction, whereas those that weaken it are poisons.

The effects of alkali and alkaline earth metal oxides on the SCR activity are important because the alkalinity of the U.S. coals are generally high, especially for the eastern bituminous coals. The results of Chen et al. (1990) show that the alkali metal oxides are strong poisons, with the strength of the poison following the order of basicity:

$$Cs_2O > Rb_2O > K_2O > Na_2O > Li_2O$$

It must be noted that most of the alkali metal oxides formed during coal combustion are found in the molten coal ash in the form of "clinker" and never reach the SCR reactor.

Lead oxide is a strong poison for noble metal catalysts used in catalytic converters in automobiles. The poisoning effects of PbO on the SCR activity, however, is not a strong one (Chen et al. 1990). Although arsenic oxides have been considered severe poisons, their poisoning effect was found to be weak as compared with that of the alkali metal oxides. However, unlike the alkali metal oxides, As_2O_3 tends to escape into the combustion gases and reaches the SCR reactor. Sulfur dioxide was found to be a promoter of the catalyst activity because the formation of surface sulfates enhanced the Brönsted acidity. The presence of SO_2 in the gases not only promoted the reaction, but it also reduced dramatically the poisoning effect of As_2O_3.

Example 6.6 Kinetics of the SCR reaction (Chen et al. 1990)

The SCR reaction catalyzed by V_2O_5/TiO_2 is first order with respect to NO concentration and independent of NH3

$$r = -\frac{1}{V}\frac{dN_{NO}}{dt} = k[NO] \qquad (6.49)$$

where

V = reactor volume, m3
N_{NO} = moles of NO
k = rate constant, s-1

Consider a monolithic honeycomb reactor for the control of NOx emissions by SCR. A dimensionless form of the mass balance equation for NO in the wall of the honeycomb is given by (see Figure 6.7):

$$\frac{d^2y^*}{dx^{*2}} = \frac{h^2 k}{D_e} y^* \qquad (6.50)$$

where

x^* = x/h, and $y^* = y_{NO}/y_{NO}{}^b$
h = half-thickness of the wall
$y_{NO}{}^b$ = bulk mole fraction in the cell
D_e = effective NO diffusivity in the wall

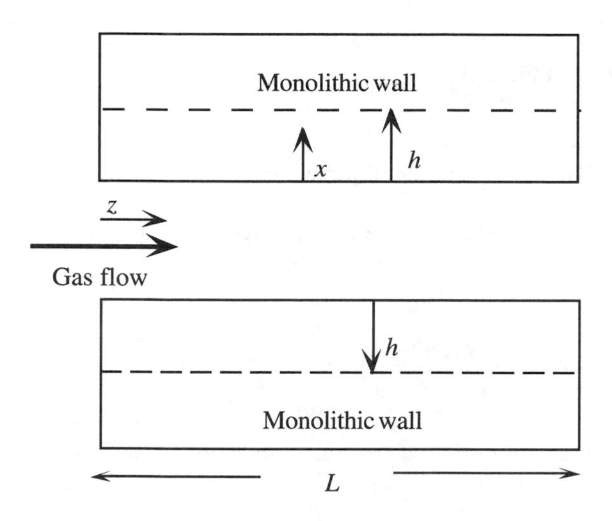

Figure 6.7 Schematic diagram of a cell in a monolithic catalyst

The boundary conditions are

$$x^* = 1,\ dy^*/dx^* = 0 \tag{6.51}$$

$$x^* = 0,\ \ y^* = 1 + \frac{1}{\text{Bi}}\left(\frac{dy^*}{dx^*}\right)_{x^*=0} \tag{6.52}$$

where

$\text{Bi} = k_c h/D_e$

k_c = film mass-transfer coefficient, m/s

The solution gives the concentration profile within the wall (see Problem 6.9).

$$y^* = \frac{e^{\varphi(x^*-2)} + e^{-\varphi x^*}}{1 + e^{-2\varphi} - \dfrac{\varphi}{\text{Bi}}\left(e^{-2\varphi} - 1\right)} \tag{6.53}$$

where φ = Thiele modulus, given by $\varphi^2 = h^2\, k/D_e$. Considering the reactor, the mass balance along its axial direction, z, is

$$\frac{dy^b_{\text{NO}}}{dz} + \frac{k_c\sigma}{uA}\left(y^b_{\text{NO}} - y^s_{\text{NO}}\right) = 0 \tag{6.54}$$

where

u = linear gas velocity in the cell
σ = perimeter length of a cell
A = cross-sectional area of the cell
y_{NO}^s = mole fraction of NO at the surface of the walls.

The boundary condition is $z = 0$, $y_{NO}^b = y_0$. The bulk and surface NO concentrations are related through Eq. (6.52)

$$y_{NO}^b = y_{NO}^s\left(1 - \frac{\varphi}{Bi}\frac{e^{-2\varphi} - 1}{e^{-2\varphi} + 1}\right) \tag{6.55}$$

If y_L is the NO molar fraction at the reactor outlet, the overall conversion, X, is

$$X = \frac{y_0 - y_L}{y_0} \tag{6.56}$$

Combining Eqs. (6.54) to (6.56), the NO conversion is given by

$$X = 1 - \exp\left[-\left(\frac{\sigma L}{uA}\right)\frac{1}{\dfrac{1}{k_c} + \dfrac{1}{\sqrt{D_e k}}\dfrac{1 + e^{-2\varphi}}{1 - e^{-2\varphi}}}\right] \tag{6.57}$$

To calculate X, the values of k_c and D_e must be estimated. The mass-transfer coefficient can be estimated with the correlations presented in Section 3.5.1. The effective diffusivity is estimated with the equations presented in Section 4.3. Chen et al. (1990) obtained experimental values for the intrinsic reaction rate constant, k, at various temperatures from 373 to 573 K. The other parameters can be obtained from the structural information of the honeycomb. The following numerical example illustrates the application of Eq. (6.57) to the design of a SCR reactor.

A flue gas flows at the rate of 100 m^3/s at 573 K and 101.3 kPa. It contains 1,000 ppm of NO, which must be reduced to no more than 100 ppm through SCR. To avoid an excessive pressure drop, the gas velocity through the catalytic bed should not exceed 2 m/s (based on the total cross-sectional area). The catalyst is V_2O_5 supported on a titania monolithic honeycomb with square-shaped cells of the following structural properties (Chen et al. 1990):

BET surface area	**44.4 m^2/g**
Medium pore size	**325 Å**
Internal porosity	**57.8%**
Bulk density	**1,600 kg/m^3**
Wall thickness	**0.038 cm**
Number of cells/cm^2	**31**

The intrinsic reaction rate constant for these conditions is approximately 52.2 s^{-1} (see Problem 6.10). Estimate the dimensions of the catalytic reactor.

Solution

Calculate, first, the reactor diameter from the specified gas velocity through the bed: $A_c = Q/u = 100/2 = 50\ m^2 = \pi D^2/4$, $D = 8.0$ meters. The overall conversion required is X = 0.9. Equation (6.57) gives the reactor length. From the structural information on the honeycomb, the cross-sectional area of a cell is 0.03226 cm^2, the width of a cell is 0.18 cm, and the ratio of perimeter to cell cross-sectional area (σ/A) is 1,753 m^{-1}. The external void fraction is also calculated: 62.2%. From the internal porosity of the wall and the density of solid titania (4,230 kg/m^3), the bulk density of the wall, ρ_p, is 1,800 kg/m^3.

Calculate the effective diffusivity from Eq. (4.24). Calculate the diffusivity of NO in air at 573 K and 101.3 kPa from the Wilke-Lee equation (Treybal 1980): $D_f = 6.92 \times 10^{-5}\ m^2/s$. Calculate the Knudsen diffusion coefficient: $D_K = 6.11 \times 10^{-6}\ m^2/s$. Assuming a tortuosity factor of 2.3 (Beekman 1991), $D_e = D_p = 1.41 \times 10^{-6}\ m^2/s$.

Calculate the Reynolds number for flow inside the channels of the monolith. Since the cross-section is square, the effective diameter is equal to the width of the channel: $d_{eff} = 0.18 - 0.038 = 0.142$ cm. The actual gas velocity through the channels is 2/0.622 = 3.22 m/s. The density and viscosity of the gases are 0.6 kg/m^3 and 2.9×10^{-5} kg/m-s. Therefore, $Re = (0.6)(3.22)(1.42 \times 10^{-3})/(2.9 \times 10^{-5}) = 94$ (laminar)

Calculate k_c from Eqs. (3.56). For mass-transfer from the bulk of the fluid to the walls of the monolith, $a = \sigma/A$, $A = d_{eff}^2$. Hence, $k_c = 14.43\ D_f/\sigma = (14.43)(6.92 \times 10^{-5})/[(4)(0.142 \times 10^{-2})] = 0.176$ m/s.

Calculate the Thiele modulus, $\varphi = [(0.00038/2)^2(52.20)/(1.41 \times 10^{-6})]^{0.5} = 1.160$. Substituting in Eq. (6.57) and solving, $L = 0.374$ m. The reactor volume is (50)(0.374) = 18.7 m^3. This corresponds to an hourly space velocity (HSV) of 19,300 h^{-1}, which is in excellent agreement with the experimental observations of Chen et al. (1990).

Example 6.7 Pressure Drop Through SCR Reactor

Estimate the pressure drop through the reactor of Example 6.6

Solution

For laminar flow the equation for pressure drop is:

$$\Delta P = \frac{32 \mu L u}{d_{eff}^2}$$

Substituting values from Example 6.6, $\Delta P = 0.554$ kPa.

Selective non catalytic reduction (SNCR) is a postcombustion control method that reduces NO_x via injection of ammonia or a urea-based reagent into the combustion chamber or into a thermally favorable location further downstream. At temperatures between 1,200 and 1,300 K, NH_3 reduces NO_x to N_2 without a catalyst. At higher temperatures, the ammonia reacts preferentially with oxygen to form more NO. At lower temperatures, the rate of ammonia reaction declines causing increased NH_3 emissions (ammonia slip) and reducing NO_x control. Maintaining the desired temperature window is therefore one of the most important operating and design considerations. NO_x reductions of 60% to 80% have been achieved through SNCR (McInnes et al. 1990). Capital and operating costs for SNCR are lower than for SCR, mainly because a catalyst is not involved. SCR uses less ammonia, however, and can experience less ammonia slip.

Several dry sorption techniques have been proposed and demonstrated for simultaneous control of NO_x and SO_2. One type of system uses activated carbon with NH_3 injection to simultaneously reduce the NO_x to N_2 and oxidize the SO_2 to H_2SO_4. The system must be operated in the temperature range of 490 to 500 K, and must be regenerated to remove the sulfuric acid (Cooper and Alley 1986). Other dry techniques for simultaneous control of nitrogen oxides and SO_2, the Shell flue gas-treating system and electron beam treatment (see Problem 6.14), were described in Chapter 5.

Wet absorption processes usually remove SO_2 as well as NO_x . The main disadvantage is the low solubility of NO. Nitric oxide must be oxidized to NO_2 in the gas phase, using ozone or chlorine dioxide, before a reasonable degree of absorption can occur in water. Part of the absorbed NO_2 reduces to N_2, and the rest is removed in the waste water as nitrate salts. This process has the potential to remove over 90% of both SO_2 and NO_x from flue gases, however gas-phase oxidants are very expensive. Chlorine dioxide, although cheaper than ozone, adds to the waste water problems created by the nitrate salts.

Absorption-reduction processes circumvent the need for a gas-phase oxidant through the addition of a chelating compound, such as ferrous-EDTA, which has an affinity for the relatively insoluble NO. The NO absorbs into a complex with the ferrous ion, and the SO_2 absorbs as the sulfite ion. Then, the NO complex reduces to N_2 by reaction with the sulfite ion. A series of regeneration steps recovers the ferrous chelating compound and oxidizes the sulfite to sulfate, which is removed as gypsum.

Although this process has the potential to remove 90% of both SO_2 and NO_x from flue gases, a very large absorber is required (see Problem 6.13). Uchida et al. (1983) conducted an experimental study in which they investigated NO absorption into both aqueous $Na_2SO_3/FeSO_4$ and $KMnO_4/NaOH$ solutions. Their experimental trials were at NO concentrations of 400, 900, and 1,790 ppm, which are typical flue gas NO concentrations . Counce and Perona (1983) published a detailed theoretical model as well as experimental data to explain the NO_x-HNO_x-H_2O system. However, their experimental work was at NO concentrations much higher than are found in flue gases from combustion sources.

6.5 COSTS OF NO_x CONTROL TECHNIQUES

Your objectives in studying this section are to

1. Estimate the total capital investment for some of the most common techniques to control NO_x emissions from combustion sources.
2. Estimate the corresponding total annual costs.

Because advanced NOx control techniques have gained widespread acceptance in Japan, are becoming popular in Europe, and are resisted in the United States, capital and operating cost estimates available in the literature are very site-specific and exhibit wide fluctuations.

There is agreement that combustion modifications are inexpensive. The total capital investment for a combination LEA/OSC burner to reduce potential NO_x emissions by 47% on a new, pulverized coal facility was recently estimated as $10.7/kW (Cogentrix 1990). No additional operating costs are required for this option. For retrofit installations, capital costs are highly site-specific and can be as high as $35/kW (Offen et al. 1987).

SCR is one of the most expensive NO_x-control techniques available, requiring high initial capital expenditures for the catalyst and for sophisticated monitors and instrumentation. Operating costs associated with catalyst replacement, ammonia consumption, and electricity use are substantial. Capital costs for SCR on coal-fired power plants reported by European utilities ranged from $65/kW to $125/kW ($ of September 1986). Operational costs varied from 5 to 10 mil/kWh (Offen et al. 1987). The same source reported a Japanese utility claim that new catalyst formulations had reduced the capital cost of SCR systems to $30/kW, and a recent U.S. study estimated capital costs of new plants approaching $80/ kW(Kokkinos et al. 1991).

Capital costs for retrofitting SCR to a 500-MW oil-fired boiler were estimated by the Electric Power Research Institute at $100/kW (McInnes et al. 1990).

Example 6.8 Costs of an SCR system for a coal-fired power plant

Estimate the total capital investment and total annual cost for an SCR system on the new coal-fired, 300-MW power plant proposed for Mayagüez, Puerto Rico. Assume that capital costs are $80/kW (June 1990) and operating costs are 6.0 mil/kWh. The plant will operate 8,760 h / yr.

Solution

TCI = (80) (300,000) = $24,000,000 (June 1990)
TAC = (6×10^{-3}) (300,000) (8,760) = $15,768,000.

6.6 CONCLUSION

NO_x are important air pollutants formed during combustion processes. Their role in the formation of photochemical smog is well established. They also contribute to the formation of acid rain. Removal of NO_x from flue gases is difficult and expensive, and, so far, source performance standards in the United States have been more a reflection of the control costs involved than of the real threat posed by these compounds. Combustion modifications to reduce potential emissions by about 50% are standard practice in the U.S. In Japan and Europe flue gas treatment techniques such as SCR, which can achieve emissions reductions of 90%, are widely used. New processes for simultaneous, high-efficiency removal of nitrogen and sulfur oxides are now operating successfully in the pilot plant stage. They will probably gain widespread full-scale acceptance as source performance standards become more restrictive.

REFERENCES

Ando, J. "Recent Developments in SO_2 and NO_x Abatement Technology for Stationary Sources in Japan," USEPA Project Summary No. EPA-600/57-85-040 (1985).

Beekman, J. W. *Ind. Eng. Chem. Res.*, **30**:428 (1991).

Benítez, J. "Periodic Operation of a Catalytic Nitric Oxide Converter," Ph.D. Thesis, RPI, Troy, New York (1976).

Chen, J. P., Buzanowski, M. A., Yang, R. T., and Chichanowicz, J. E. *J. Air Waste Manage. Assoc.*, **40**:1403 (1990).

Chen, J. P., and Yang, R. T. *J. Catal.* **125**:411 (1990).

Cogentrix Inc. "PSD Permit Application for a 300-MW Coal-Fired Cogeneration Facility," submitted to USEPA, prepared by ENSR Consulting and Engineering, Boston, MA (1990).

Cooper, C. D. and Alley, F. C. *Air Pollution Control: A Design Approach*, PWS Engineering, Boston, MA (1986).

Counce, R. M., and Perona, J. J. *AIChE J*, **29**:26 (1983).

Crawford, A. R., Manny, E. H., and Bartok, W. "Field Testing: Application of Combustion Modifications to Power Generating Combustion Sources", in *Proceedings of the Second Stationary Source Combustion Symposium—Vol. II: Utility and Large Industrial Boilers*, EPA-600/7-77-073b. Washington, D C, USEPA (1977).

Fishman, J. and Kalish, R. *Global Alert, The Ozone Pollution Crisis*, pp. 160–165, Plenum Press, New York (1990).

Flagan R. C., and Seinfeld, J. H. *Fundamentals of Air Pollution Engineering*, pp. 167–200, Prentice Hall, Englewood Cliffs, NJ (1988).

Folinsbee, L. J., McDonnell, W. F., and Hortsman, D. H. *JAPCA*, **38**:28 (1988).

Heap, M. P. et al., "The Optimization of Burner Design Parameters to Control NOx Formation in Pulverized Coal and Heavy Oils Flames," *Proceedings of the Stationary Source Combustion Symposium—Vol. II: Fuels and Process Research and Development,* EPA-600/2-76-152b. Washington, D C, USEPA (1976).

Kokkinos, A., Chichanowicz, J. E., Hall, R. E., and Sedman, C. B. *J. Air Waste Manage. Assoc.*, **41**:1252 (1991).

Larsen, R. I., McDonnell, W. F., Hortsman, D. H., and Folinsbee, L. J. *J. Air Waste Manage. Assoc.*, **41**:455 (1991).

Lim, K. J. et al. *Environmental Assessment of Utility Boiler Combustion Modification NOx Controls—Vol. I: Technical Results*, EPA-600/7-80-075a, Washington, DC, USEPA (1980).

McInnes, R., and Van Wormer, M. B. *Chemical Engineering*, **97**:130 (1990).

Offen, G. R., Eskinazi, D., McElroy, M. W., and Maulbetsch, J. S., *JAPCA*, **37**:864 (1987).

Peavy, H. S, Rowe, D. R., and Tchobanoglous, G. *Environmental Engineering*, pp 457-461, McGraw-Hill, New York (1985).

Pershing, D. W., and Wendt, J. O. L. "Pulverized Coal Combustion: The Influence of Flame Temperature and Coal Composition on Thermal and Fuel NOx," in *Sixteenth Symposium on Combustion*, pp. 491–501,The Combustion Institute, Pittsburgh, PA (1977).

Pohl, J. H. and Sarofim, A. F. "Fate of Coal Nitrogen During Pyrolysis and Oxidation," *Proceedings of the Stationary Source Combustion Symposium—Vol.I: Fundamental Research*, EPA-600/2-76-152a. Washington, D C, USEPA (1976).

Spektor, D. M., Lippmann, M., Lioy, P. J., Thurston, G. D., Citak, K., James, D. J., Bock, N., Speizer, F. E., and Hayes, C. *American Review of Respiratory Disease,* **137**:313 (1988).

Treybal, R.E. *Mass-Transfer Operations*, 3rd ed., McGraw-Hill, New York (1980).

Uchida, S., Kobayashi, T., and Kageyama, S. *Ind. Eng. Chem, Process Design Develop.*, 22:323 (1983).

USEPA, *Control Techniques for Nitrogen Oxide Emissions from Stationary Sources* rev.2nd ed., EPA-450/3-83-002. U.S. Environmental Protection Agency, Research Triangle Park, NC (1983).

Zeldovich, Y. B., Sadovnikov, P. Y., and Frank-Kamenetskii, D. A. *Oxidation of Nitrogen in Combustion,* M. Shelef (trans.), Academy of Sciences of USSR, Institute of Chemical Physics, Moscow-Leningrad (1947).

PROBLEMS

The problems at the end of each chapter have been grouped into four classes (designated by a superscript after the problem number):

Class a: Illustrates direct numerical application of the formulas in the text.
Class b: Requires elementary analysis of physical situations, based on the subject material in the chapter.
Class c: Requires somewhat more mature analysis.
Class d: Requires computer solution.

6.1^{b}. Nitric oxide, tropospheric ozone, and health effects

A very important deleterious effect of atmospheric NO is that it is a precursor of tropospheric ozone. The primary NAAQS for O_3 is 0.12 ppm, not to be exceeded in more than 1-h/yr. A recent work by Larsen et al. (1991) studied the effect of ozone concentration and exposure duration on human lung function. They found that the fractional lung function decrease of active, healthy adults (LFD) is related to the impact parameter (I) through a log-normal model, a two-parameter distribution:

$$\text{LFD} = \frac{1}{2}\left\{1 + \text{erf}\left[\frac{\ln(I/I_{50})}{\sqrt{2}\,\ln\sigma_g}\right]\right\}$$

where

$I = tc^{1.32}$

t = exposure time, hr

c = ozone concentration, ppm

I_{50} = value of I which would result in 50% LFD

σ_g = geometric standard deviation.

$$\text{erf}\, x = \frac{2}{\sqrt{\pi}} \int_0^x e^{-z^2} dz$$

From regression analysis of their experimental data, the best values of the distribution parameters are: $I_{50} = 1.2118$ h-ppm$^{1.32}$ and $\sigma_g = 3.029$. Estimate the fractional lung decrease which results from an 8-h exposure to an ozone concentration of 0.12 ppm.

Answer: 21%

6.2a. Equilibrium NO concentration

Combustion gases in equilibrium at a total pressure of 10 atm contain 71.4% N_2, 3.78% O_2, and 4,640 ppm NO. Estimate the equilibrium temperature.

6.3b. The effect of equivalence ratio on thermal NO formation

Repeat Example 6.2, but for an equivalence ratio of 0.95. Calculate the resulting NO concentration.

Answer: 317 ppm

6.4c. Thermal NO formation during adiabatic combustion (Flagan and Seinfeld 1988)

A furnace is fired with methane. The inlet air and fuel temperatures are 290 K and the pressure is atmospheric. The residence time in the combustor is 0.1 s. Assuming the combustion to be adiabatic, calculate the NO mole fraction at the combustor outlet for combustion at an equivalence ratio of 0.85.

Answer: 420 ppm

6.5b. Total NO emissions from pulverized coal combustion

Estimate the concentration in ppm and the mass emission rate of NO if a pulverized coal containing 1.5% nitrogen is burned at 1600 K using an air-fuel ratio that is 50% higher than the stoichiometric ratio. Assume that the fuel-NO is 80% of the total NO emission rate. The rate of coal combustion is 600 kg/h, and the flue gas (average molecular weight of 27.5) is produced at the rate of 13,000 kg/h.

Answer: 484 ppm

6.6a. NO_x emissions from a source that meets the corresponding NSPS

Consider the coal-fired power plant of Example 5.1. Assuming that it meets the NSPS for NO_x , estimate the mass emission rate of NO_x coming out of its stacks. Express your answer in terms of NO_2 .

Answer: 19,255 kg/d

6.7a. Percentage reduction in NO_x emissions to meet NSPS

Estimate the percentage reduction in NO_x emissions to meet the corresponding NSPS for the power plant of Problem 6.6. Assume that the flame temperature is 1,600 K and that the fuel-NOx is 80% of the total NO_x .

Answer: 47.5%

6.8a. NH_3 requirements for SCR

Calculate the amount of ammonia for 90% reduction of the NOx potential emissions of the power plant of Problem 6.6 through SCR. Assume that all the NO_x is in the form of NO.

Answer: 13,370 kg/d

6.9c. Diffusion and reaction in a monolithic SCR catalyst

Derive Eq. (6.53) starting with a shell mass balance of NO inside the monolithic catalyst wall.

6.10d. Shift in activation energy when intraparticle diffusion effects are significant

Intraparticle diffusion may disguise the true intrinsic chemical kinetics during heterogeneous catalysis. At low temperatures diffusion is rapid compared with chemical reaction and diffusional limitations on the reaction rate will not be observed. In this temperature regime, an Arrhenius plot ($\ln k$ versus T^{-1}) yields the intrinsic activation energy of the reaction. However, because chemical reaction rates increase much more rapidly with increasing temperature than diffusional processes, at higher temperatures mass-transfer limitations are much more likely. In this temperature regime, the apparent activation energy tends to approach one-half of the intrinsic value (Hill, C. G., Jr. *An Introduction to Chemical Engineering Kinetics and Reactor Design*, pp. 453–455, Wiley, New York, 1978).

The following data were published by Chen et al. (1990) for SCR over a V_2O_5/TiO_2 monolithic catalyst:

Temperature, K	First-Order Rate Constant, (cm^3/g-s)
423	3.575
473	8.330
523	10.200
573	10.390

Show that intraparticle diffusion effects are significant at the higher temperatures, and estimate the true intrinsic reaction rate constant at 573 K. The bulk density of the monolithic wall is 1.8 g/cm^3.

Answer: k = *52.2* s^{-1}

6.11b. Effect of SO_2 on SCR activity

Chen et al (1990) found that SO_2 in the flue gases promotes the activity of the V_2O_5/TiO_2 catalyst during NO_x reduction through SCR. For conditions similar to those in Example 6.6, but in the presence of SO_2, the intrinsic reaction rate constant increased from 52.2 s^{-1} to 64.4 s^{-1}. Estimate the overall NO conversion achieved in the reactor of Example 6.6 in the presence of SO_2.

Answer: 92.3%

6.12[b]. Design of a SCR reactor based on space velocity

Consider the flue gases of Example 5.1. Determine the dimensions of a SCR reactor to remove 90% of the NO potential emissions, operating at 573 K. The catalyst is identical to the one described in Example 6.6. Assume that an hourly space velocity of 20,000 h^{-1} results in the desired conversion. The power to overcome the pressure drop through the catalytic bed should not exceed 1.0 MW at a mechanical efficiency of 65%.

Answer: Depth of bed = 0.54 m

6.13[c]. Absorption of NO into aqueous $Na_2SO_3/FeSO_4$ solutions

Simultaneous SO2 and NOx removal through wet scrubbing with $Na_2SO_3/FeSO_4$ aqueous solutions was demonstrated by Uchida et al. (1983). The liquid-phase reactions are:

$$FeSO_4 + NO \rightarrow Fe(NO)SO_4$$

$$Fe(NO)SO_4 + 2Na_2SO_3 + 2H_2O \rightarrow Fe(OH)_3 + Na_2SO_4 + NH(SO_3Na)_2$$

The process was carried out in a bubble chamber with a gas-liquid interfacial area of 1.23 cm^2/cm^3 The rate of NO absorption per unit of interfacial area, for negligible gas-phase resistance, is represented by:

$$R_{NO} = (k\, C_{FeSO4}{}^{1.3}\, C_{Na2SO3}{}^{1.3}\, D_{NO})^{0.5}\, C_{NOi}$$

where

R_{NO} = rate of NO absorption, moles/cm^2 s

C_j = concentration of component j in the liquid phase, mol/cm^3

C_{NOi} = concentration of NO at the liquid interface, mol/cm^3

D_{NO} = diffusivity of NO in the liquid phase, cm2/s

k = rate constant given by

$$k = 1.37 \times 10^{22} \exp(-231/T)$$

Because of the vigorous mixing generated by bubbling the gas through the liquid, the bubble chamber behaves as a continuous flow stirred tank reactor (CSTR).

A flue gas flows at the rate of 250 m3/s at 303 K and 101.3 kPa, and contains 1,000 ppm of NO. Estimate the volume of a bubble chamber to reduce NO emissions by 90% at constant temperature. The liquid concentrations in the chamber are: $C_{FeSO_4} = 5 \times 10^{-5}$ mol/cm3, $C_{Na_2SO_3} = 4 \times 10^{-5}$ moles/cm3, $C_{NOi} = 0.9 \times 10^{-10}$ mol/cm3.

Answer: 95 m3

6.14b. Economics of high-efficiency simultaneous SO_2-NO_x removal by electron beam treatment.

Gleason and Helfritch (*Chem. Eng. Prog*., pp. 33–38, October 1985) demonstrated at a pilot scale that simultaneous removal of NO_x and SO_2 by a combination of dry FGD and electron beam can be sustained at high efficiency. Removal efficiencies higher than 90% were achieved for both pollutants when high-sulfur coal was the fuel. Figure 6.8 illustrates the process.

Consider a new 500-MW power plant that burns coal with a sulfur content of 4% and a heating value of 28 MJ/kg, operating 7,000 h/yr. Gleason and Helfritch estimated that the total capital investment for a combined dry FGD-electron beam system to remove 90% of both SO_2 and NO_x for such a plant was $134 million (in dollars of July 1984). This included on-site waste disposal facilities. Annual operating requirements are as follows (July 1990 unit prices):

Electricity: 1.5×10^8 kWh @ $0.06 / kWh
Lime: 80,000 metric ton @ $44/metric ton
Water: 4×10^6 m3 @ $0.06 / m3
Operating labor: 67,000 h @ $20/h
Maintenance labor and materials: 2.5% of TCI
Capital recovery: 15% of TCI

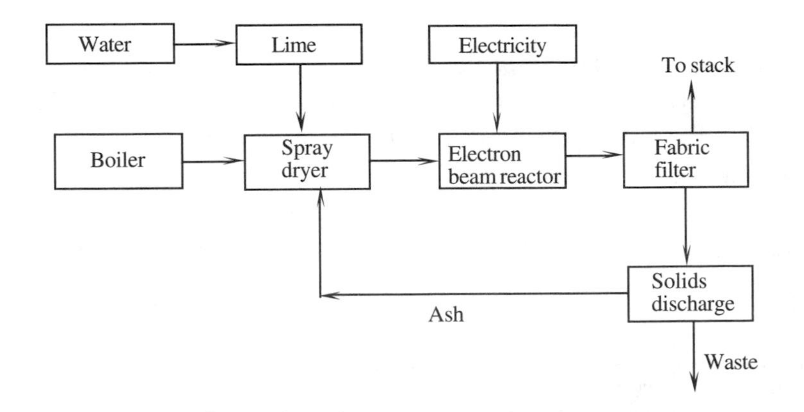

Figure 6.8 Schematic diagram of the electron beam process

Estimate the total annual cost for the system, and express your result in mil / kWh.

Answer: 14 mil/kWh

6.15[b]. Annual costs for a SCR nitric oxide reduction system

Consider the flue gases of Problems (6.8) and (6.12). Estimate the total annual cost for a SCR system to reduce the potential emissions of NO of this 300-MW power plant by 90%. The total capital investment for the system is estimated at \$175/kW. The plant operates 8,760 h/yr. The cost of the catalyst, updated to June 1990, is estimated as \$25,000/m^3 (van der Vaart et al. *J. Air Waste Manage. Assoc.*, **41**:497, 1991). The expected life of the catalyst is 2 yr. The life-expectancy of the rest of the system is 10 yr. High pressure steam is used to vaporize the ammonia at the rate of 2 kg of steam/kg of NH_3. The unit cost of the steam is \$8.80/10^3 kg, that of electricity is \$0.06/kWh. The cost of ammonia is \$121/metric ton. Operating labor requirements are 8,760 h/yr at \$20/h. Maintenance labor and materials are approximately 3% of the TCI. The minimum attractive return on investment is 12%.

Answer: 6.4 mil/kWh

6.16[b]. Formation of NO_2 in automobile exhaust gases

NO_2 can be formed in combustion exhaust gases by the third-order reaction

$$2NO + O_2 \rightarrow 2NO_2$$

Estimate the amount of NO_2 formed by this route under conditions typical of the exhaust of an automobile with no emissions control system. Assume that the exhaust system of a car can be represented as a straight pipe through which the exhaust gases flow in plug flow. Assume that the concentration of NO at the beginning of the exhaust system is 2,000 ppm and that of O_2 is 1,000 ppm. Assume also that the residence time in the exhaust system is 4.0 sec and that the temperature is 573 K. Compute the concentration of NO_2 at the system exit. The following rate constant data are available (Seinfeld, J. H. *Air Pollution, Physical and Chemical Fundamentals*, p. 395, McGraw-Hill, New York, 1975):

Temperature (K)	k (L^2/mol^2-s)
273	17,900
313	13,100
473	6,600
663	5,100

Answer: 0.04 ppm

6.17c. Accidental ammonia releases (modified version of Problems No. 5 and No. 64, *Safety, Health, and Loss Prevention in Chemical Processes: Problems for Undergraduate Engineering Curricula*, © 1990 by AIChE, reproduced by permission of The Center for the Chemical Process Safety of AIChE)

Ammonia is used as the reducing agent in SCR systems for the control of NO_x emissions. However, ammonia itself is an air pollutant and a potentially dangerous chemical. The Occupational Safety and Health Administration (OSHA) limit for exposure to NH_3 is 50 ppm, time weighted average over an 8-h day. The American Conference of Governmental Industrial Hygienists (ACGIH) has set the level that is immediately dangerous to life and health at 500 ppm.

A power plant has a storage tank for the anhydrous ammonia it uses for SCR. The tank is surrounded by a concrete pad with a dike that is high enough so that the contents of the tank could be retained within the diked enclosure. The enclosure is 15 m by 15 m. A tank truck, somewhat carelessly operated, has just backed into the tank and has not only ruptured the tank, but has also damaged the delivery valve on the truck. As a consequence, liquid NH_3 has nearly filled the diked area, and is rapidly boiling, even though it is not an especially hot day. Assume that the air temperature is 283K, and that the temperature of the concrete pad is 288K. The boiling heat-transfer coefficient for this case is 300 W/m^2-K (based on the temperature difference between the wall and the boiling liquid).

Persons exposed to 500 ppm of ammonia will be endangered and anywhere that the concentration might be that high should be evacuated until repairs are made. What recommendation would you make as to how far from the accident site people should be evacuated if the surface wind speed is 2 m/s and there is moderate sunlight intensity?

One of the simpler models to predict atmospheric dispersion is the "Gaussian Plume Model," which expresses the average concentration at a ground-level location downwind of a continuous source as (Cooper and Alley 1986):

$$C = \frac{Q}{\pi\, u \sigma_y \sigma_z} \exp\left[-\frac{y^2}{2\,\sigma_y^2}\right] \exp\left[-\frac{H^2}{2\,\sigma_z^2}\right]$$

where

C = concentration, mg/m^3

Q = source strength, mg/s

u = wind velocity, m/s

H = effective height of the source above ground, m

x = distance downwind from the source to the point of interest, m or km

y = distance crosswind from plume centerline, m

σ_y and σ_z = horizontal and vertical dispersion coefficients, m

The dispersion coefficients are functions of downwind distance, x, and of the prevailing meteorological conditions. For the conditions of the problem (corresponding to a moderately unstable atmosphere):

$$\sigma_y = 156\, x^{0.894}$$

$$\sigma_z = 106.5\, x^{1.149} + 3.3 \qquad x < 1 \text{ km}$$

$$\sigma_z = 108.2\, x^{1.098} + 2.0 \qquad x > 1 \text{ km}$$

where x is in km, σ_y and σ_z are in meters.

Answer: within a radius of 440 m

6.18[b] Ammonia injection in municipal solid waste (MSW) incinerators

MSW incinerators have a potentially important role to play in solving the current solid waste crisis. However, they are prevented from making a significant contribution because of concerns about their emissions, primarily PCDD/PCDF (polychlorinated dibenzo dioxins and dibenzo furans, examples of which are shown in Figures 6.9 and 6.10) and HCl. The most frequently applied air pollution control technology is a spray drier absorber followed by a fabric filter (Frame G. B. *JAPCA*, **38**:1081, 1988). This technology is very capital-intensive and complex. While it considerably reduces emissions to the atmosphere, the PCDD/PCDF formed is not destroyed, but ends up in the solid waste from the filter.

A recent report suggests controlling PCDD/PCDF and HCl emissions from MSW incinerators with ammonia injection followed by SCR for NO_x control (Takacs, L., and Moilanen, G. L. *J. Air Waste Manage. Assoc.*, **41**:716, 1991). These researchers found experimentally that injecting ammonia in the flue gases in the temperature range of 560 to 620 K suppressed almost completely the formation of PCDD/PCDF while reducing HCl emissions by more than 98%. The significance of these findings must be emphasized: *this technology has the potential to suppress the formation of PCDD/PCDF as opposed to others that merely collect and transfer them to solid waste.*

Consider a MSW with a flue gases flow rate of 50,400 DSCM/h (DSCM stands for dry standard cubic meter) and fitted with ammonia injection and SCR. The gases contain 1,520 mg of HCl/DSCM and 800 ppm (on a dry basis) of NO. Calculate the rate of

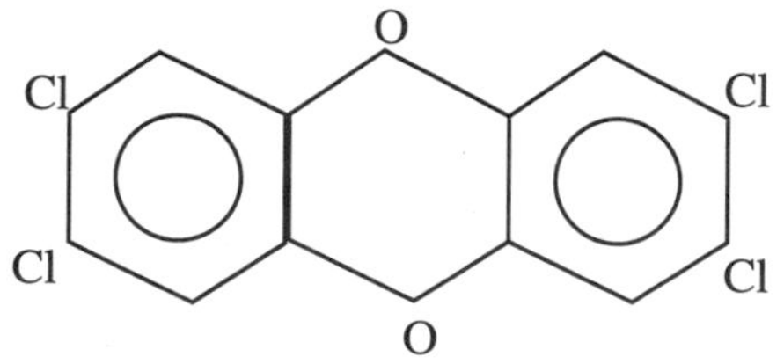

Figure 6.9 Molecular structure of a PCDD (2,3,7,8 TCDD)

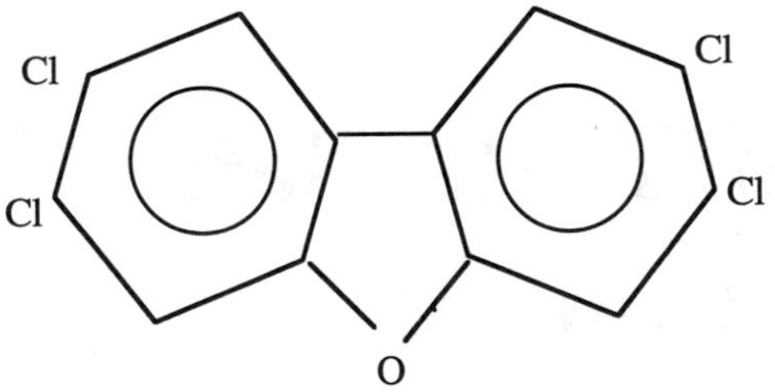

Figure 6.10 Molecular structure of a PCDF (2,3,7,8 TCDF)

ammonia injection to neutralize all the HCl and to achieve a 90% reduction in NO_x emissions in the SCR system.

Answer: 63.2 kg NH_3/h

7

Fundamentals of Particulate Emissions Control

7.1 INTRODUCTION

Particulates constitute a major class of air pollutants. Particles have a diversity of shapes and sizes; they can be either liquid droplets or dry dusts, with a wide variety of physical and chemical properties. They are emitted from many sources including both combustion and non-combustion industrial processes. In addition, primary gaseous emissions may react in the atmosphere to form secondary species that nucleate to form particles or condense on preexisting ones. An important class of industrial gas-cleaning processes remove particles from exhaust gas streams, and such processes are the subject of the following chapters. This chapter presents information about certain characteristics of particles and particulate behavior in fluids, with particular emphasis on those that are relevant to the engineering task of separating and removing particles from a stream of gas.

7.2 CHARACTERISTICS OF PARTICLES

Your objectives in studying this section are to

1. Understand the importance of an aerosol size distribution.
2. Characterize an aerosol size distribution with data from a cascade impactor.
3. Develop and apply a log-normal size distribution function.

An aerosol is a suspension of small particles in air or another gas. Important aerosol characteristics include size, size distribution, shape, density, stickiness, corrosivity, reactivity, and toxicity. From the viewpoint of air pollution, the most important of these is the particle size distri bution. The most common aerosols cover a wide range of sizes—from 0.001 μm to 100 μm. As mentioned in Chapter 1, the effects of aerosols on human health and visibility are strongly size-dependent, with particles in the range of 0.1 to 1.0 μm being the worst.

In addition to average particle concentration per unit atmospheric volume, it is important to note the size distribution by particle count and by mass. Such distributions for a typical atmospheric particulate sample are shown in Table 7.1. From data in the last two entries, particles in the 0 to 1-μm range constitute only 3% by mass. However, the number of particles in that range is overwhelming compared with the rest of the sample. Particles of this size range are capable of entering the lungs. *From a health standpoint, it is not so much a question of lowering the overall atmospheric dust loading in an urban area but of decreasing the heavy particulate count in the smaller size range.*

Table 7.1 Particle Distribution of a Typical Atmospheric Sample

Size range (μm)	Average size (μm)	Particle count[a]	Mass percent
10-30	20	1	27
5-10	7.5	112	53
3-5	4	167	12
1-3	2	555	5
0.5-1	0.75	4,215	2
0-0.5	0.25	56,900	1

[a] Count of other sizes relative to count of 20-mm size.

Source: Wark and Warner (1981).

As can be expected from such a wide range of sizes, one type of particulate collection device might be better suited than others for a specific aerosol. Furthermore, the collection efficiency of these devices depends on particle size, with bigger particles usually removed more efficiently, as shown on Figure 7.1. Thus, to calculate the overall collection efficiency of a device, it is imperative to have good information on the size distribution of the particles.

A good device to obtain this information is a cascade impactor. It separates and sizes suspended particles in a manner similar to the way that sieves separate and size samples of sand. Air with particles is drawn through a series of stages that consist of slots and impaction plates as shown on Figure 7.2. Each successive stage has narrower slots and closer plates so that each successive stage captures increasingly smaller particles. The masses of particles collected on all stages are then used to determine the size distribution of the aerosol. It must be emphasized that a cascade impactor sizes particles according to their aerodynamic diameters rather than their physical size. The aerodynamic diameter is a measure of the reaction of a particle to inertial forces.

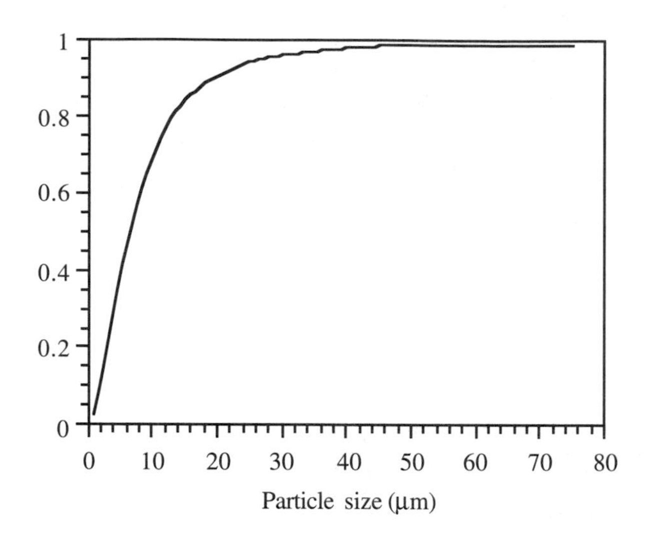

Figure 7.1 Collection efficiency versus particle size for a typical mechanical collector

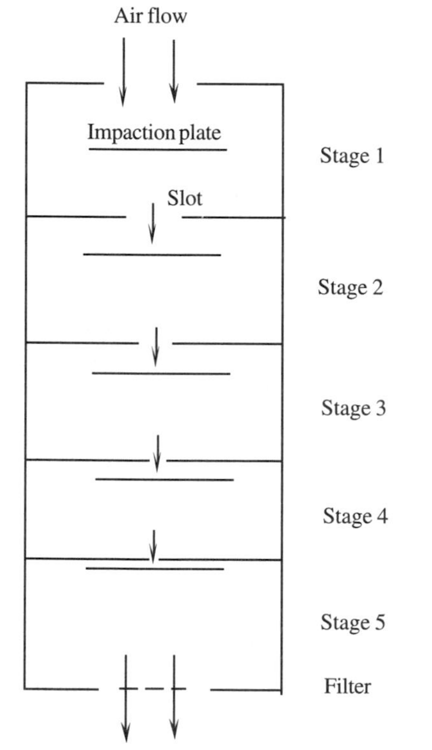

Because the ultimate objective of our analysis is to remove particles from a flowing stream of gas, the aerodynamic diameter is more relevant than the actual dimensions of the particle.

An aerosol population can be treated as if its size variation is essentially continuous. The normalized size distribution function, $n\ (D_p)$, can be defined as follows:

$n(D_p)\ dD_p$ = fraction of the total number of particles per unit volume of air having diameters in the range D_p to $D_p + dD_p$

Figure 7.2 Schematic diagram of a cascade impactor

Integrating over all particle sizes:

$$1.0 = \int_0^\infty n(D_p)\, dD_p \tag{7.1}$$

The units of $n\ (D_p)$ are μm^{-1}. The normalized distribution of particle mass with respect to particle size is defined as follows:

$n_m(D_p)dD_p$ = mass-fraction of particles having diameters in the range D_p to $D_p + dD_p$

The cumulative frequency distribution function, $F(D_p)$, is the fraction of the number of particles with diameters smaller or equal to D_p. Therefore,

$$F(D_p) = \int_0^{D_p} n(D_p')\, dD_p' \tag{7.2}$$

The cumulative mass distribution function, $G(D_p)$, is the mass-fraction of the particles with diameters smaller or equal than D_p. Then,

$$G(D_p) = \int_0^{D_p} n_m(D_p')\, dD_p' \tag{7.3}$$

Because particle sizes in an aerosol population typically vary over several orders of magnitude, it is often convenient to express the size distribution in terms of the natural logarithm of the diameter, ln D_p In a particular incremental particle size range D_p to $D_p + dD_p$ the fraction of particles is a certain quantity, and that quantity is the same regardless of how the size distribution function is expressed. Thus,

$$n(D_p)dD_p = n\ (\ln D_p)\ d(\ln D_p) \tag{7.4}$$

Because $d(\ln D_p) = dD_p/D_p$

$$D_p n(D_p) = n(\ln D_p) \tag{7.5}$$

The next question that arises is: What functions are commonly used to represent particle size distributions? A popular function for this purpose is the log-normal distribution. If a quantity u is normally distributed, the probability function for u obeys the Gaussian distribution:

$$n(u) = \frac{1}{\sqrt{2\pi}\,\sigma_u} \exp\left[-\frac{(u - u_m)^2}{2\sigma_u^2}\right] \tag{7.6}$$

where

u_m is the mean value of the distribution
σ_u is the standard deviation

A quantity that is log-normally distributed has its logarithm governed by a normal distribution. If the quantity of interest is particle diameter D_p, then saying that an aerosol population is log-normally distributed means that $u = \ln D_p$ satisfies Eq. (7.6):

$$n(\ln D_p) = \frac{1}{\sqrt{2\pi}\,\ln \sigma_g} \exp\left[-\frac{(\ln D_p - \ln D_{pm})^2}{2(\ln \sigma_g)^2}\right] \tag{7.7}$$

The physical significance of the parameters D_{pm} and σ_g will be discussed shortly. It is more convenient to express the size distribution function in terms of D_p rather than $\ln D_p$. Combining Eqs. (7.5) and (7.7):

$$n(D_p) = \frac{1}{\sqrt{2\pi}\,D_p \ln \sigma_g} \exp\left[-\frac{(\ln D_p - \ln D_{pm})^2}{2(\ln \sigma_g)^2}\right] \tag{7.8}$$

For a normally distributed quantity, the cumulative frequency distribution function, $F(u)$, is

$$F(u) = \frac{1}{\sqrt{2\pi}\,\sigma_u} \int_{-\infty}^{u} \exp\left[-\frac{(u' - u_m)^2}{2\sigma_u^2}\right] du' \tag{7.9}$$

To evaluate this integral, let $\eta = (u' - u_m)/2^{1/2}\,\sigma_u$, then,

$$F(u) = \frac{1}{\sqrt{\pi}} \int_{-\infty}^{(u-u_m)/\sqrt{2}\sigma_u} \exp\left(-\eta^2\right) d\eta \tag{7.10}$$

Integrating in terms of the error function, erf η (see Problem 6.1),

$$F(u) = \frac{1}{2}\left[1 + \text{erf}\left(\frac{u - u_m}{\sqrt{2}\,\sigma_u}\right)\right] \tag{7.11}$$

For the log-normal distribution, $u = \ln D_p$, so Eq. (7.11) can be expressed as

$$F\left(D_p\right) = \frac{1}{2} + \frac{1}{2}\text{erf}\left[\frac{\ln\left(D_p/D_{pm}\right)}{\sqrt{2}\ln\sigma_g}\right] \tag{7.12}$$

It is evident from Eq. (7.12) that $F(D_{pm}) = 0.5$. Thus D_{pm} is the number median diameter (NMD) defined as the diameter for which exactly one-half the particles are smaller and one-half are larger. To understand the significance of σ_g, consider that diameter $D_{p\sigma}$ for which $\sigma_g = D_{p\sigma}/D_{pm}$. At that diameter

$$F\left(D_{p\sigma}\right) = \frac{1}{2} + \frac{1}{2}\text{erf}\left(\frac{1}{\sqrt{2}}\right) = 0.841 \tag{7.13}$$

Thus, σ_g is the ratio of the diameter below which 84.1% of the particles lie to the number median diameter. $D_{p\sigma}$ is one standard deviation from the median, so σ_g is called the *geometric standard deviation.*

It can be shown (Seinfeld and Flagan 1988) that if the number size distribution function is log-normal, then the mass size distribution function is also log-normal with the same geometric standard deviation and the *mass median diamete*r (MMD) given by

$$\ln \text{MMD} = \ln \text{NMD} + 3\left(\ln \sigma_g\right)^2 \tag{7.14}$$

Therefore,

$$G(D_p) = \frac{1}{2} + \frac{1}{2}\,\text{erf}\left[\frac{\ln(D_p/\text{MMD})}{\sqrt{2}\,\ln\sigma_g}\right] \tag{7.15}$$

The log-normal distribution has the useful property that when the cumulative distribution function, either F or G, is plotted against the logarithm of particle diameter on a special graph paper with one axis scaled according to the error function, so-called *log-probability paper*, a straight line results. Such a plot with actual data from an aerosol population obtained with a cascade impactor serves two purposes: (1) to determine if the log-normal model fits the data, and (2) if so, to estimate the parameters MMD and σ_g.

Example 7.1 Analysis of data from a cascade impactor

The following data were obtained from a cascade impactor run on a sample from an aerosol population (Cooper and Alley 1986):

Size range (μm)	Mass (mg)
0–2	4.5
2–5	179.5
5–9	368
9–15	276
15–25	73.5
> 25	18.5

Show that a log-normal distribution fits the data, and estimate the corresponding values of MMD, NMD, and σ_g.

Solution

Prepare a table of particle size versus cumulative mass fraction, G, less than the stated size, as follows:

Dp (μm)	*G* (%)
2.0	0.5
5.0	20.0
9.0	60.0
15.0	90.0
25.0	98.0

A plot of these data is presented in Figure 7.3, using log-probability scales. The resulting straight line is evidence that a log-normal distribution is an adequate model for the size distribution function.

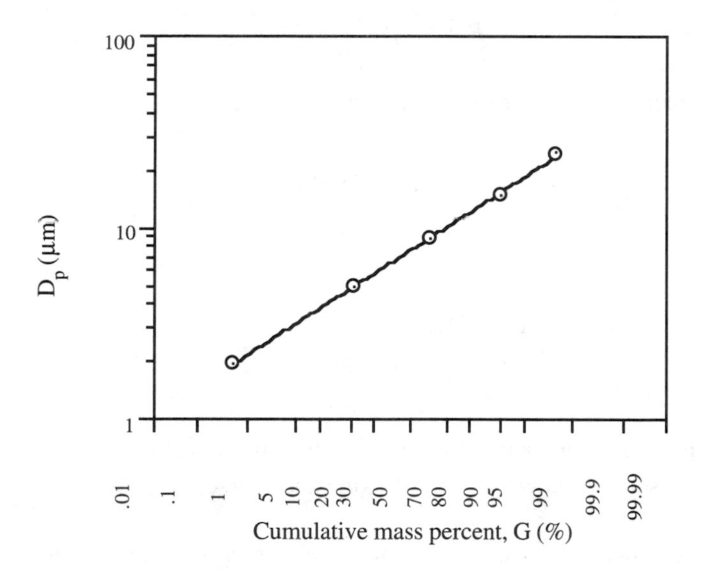

Figure 7.3 Log-normal distribution, data of Example 7.1

From Figure 7.3, MMD = 8.0 μm, $D_{p\sigma}$ = 14 μm. Therefore, σ_g = 14/8 = 1.75. From Eq. (7.14), NMD = 3.13 μm.

Comments

Although there is no completely satisfactory theoretical explanation for it, many investigators have reported that the distribution of several quantities related to environmental pollution, such as particle size, ambient air quality data, indoor radon measurements, stream water quality data, phosphorus in lakes, radio nuclides in soil, trace metals in human tissue, lung function reaction to ozone, and others often appear log-normal (Larsen et al. 1991; Ott 1990).

Example 7.2 The log-bimodal size distribution function

It has been observed that frequently the log-normal size distribution becomes less accurate to describe the end of the distribution representing fine particulate. Kerr (1989) suggested to overcome this difficulty by breaking the size distribution into a fine and a coarse log-normal distributions—a log-bimodal distribution. This is a five-parameter model given by:

$$G\left(D_p\right) = \frac{R\,G_f\left(D_p\right) + G_c\left(D_p\right)}{R + 1} \tag{7.16}$$

where

R = mass ratio of fine particulate source to coarse particulate source

$$G_f\left(D_p\right) = \frac{1}{2} + \frac{1}{2}\,\mathrm{erf}\left[\frac{\ln\left(D_p/\mathrm{MMD}_f\right)}{\sqrt{2}\,\ln \sigma_{gf}}\right]$$

$$G_c\left(D_p\right) = \frac{1}{2} + \frac{1}{2}\,\mathrm{erf}\left[\frac{\ln\left(D_p/\mathrm{MMD}_c\right)}{\sqrt{2}\,\ln \sigma_{gc}}\right]$$

MMD_f and MMD_c = mass median diameter for the fine and coarse fractions

σ_{gf} and σ_{gc} = geometric standard deviations for the fine and coarse fractions

This model can be fitted to actual data from an aerosol population through non-linear regression techniques. Software packages for performing this task are plentiful. Obviously, the validity of this approach should be tested by calculating some criteria, such as the standard error of the estimate, that tests the goodness of fit. Consider the data on Table 7.2 characterizing an aerosol population. Figure 7.4 shows these data plotted on log-probability scales. It is evident from it that the distribution is not log-normal. Using a nonlinear regression program a log-bimodal distribution is fitted to the data. The best estimate of the parameters are: $R = 0.01023$, $\mathrm{MMD}_f = 0.5028$ μm, $\mathrm{MMD}_c = 11.29$ μm, $\sigma_{gf} = 1.202$, and $\sigma_{gc} = 1.353$. The standard error of the estimate is 0.42%. The solid line on Figure 7.4 corresponds to the size distribution predicted by the log-bimodal model whereas the circles illustrate the actual experimental data. The model is remarkably good.

Table 7.2 Log-Bimodal Size Distribution Data

Particle size (μm)	$G \times 100$
0.3	0.00252
0.5	0.4950
0.7	0.9579
1.0	0.9901
2.0	1.012
3.0	1.035
4.0	1.039
5.0	1.343
6.0	2.808
7.0	6.636
8.0	13.59
10.0	34.99
12.0	58.28
15.0	82.67
20.0	97.05
25.0	99.57

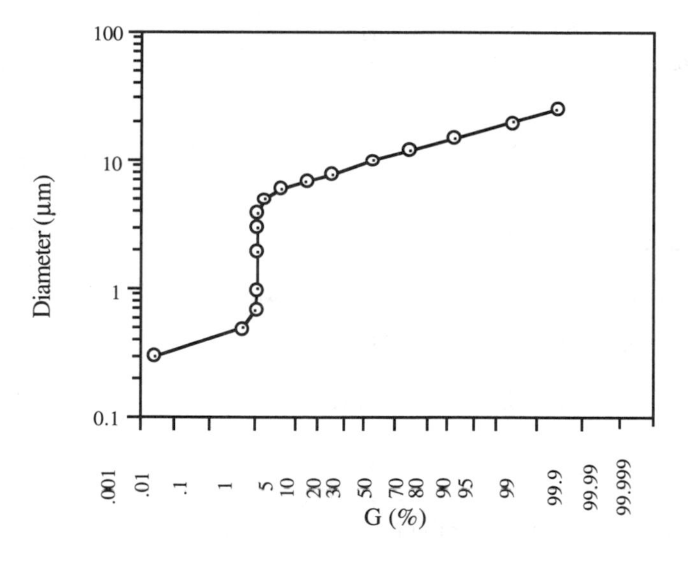

Figure 7.4 Log-bimodal particle size distribution of Example 7.2

Comments

Flagan and Seinfeld (1988) observed that particulate size distributions in the flue gases of pulverized-coal combustion systems exhibit two distinct peaks, one in the submicron-size range, and one in the 3- to 50-μm range. According to them, submicron ash constitutes less than 2% of the total fly ash mass. Ash residue particles that remain when the carbon burns out account for the large-diameter fraction, whereas ash volatilization followed by nucleation and coagulation into very small particles accounts for the fine fraction.

The following computer program, based on a very compact and efficient subroutine presented by Press et al. (1989), estimates the error function with a fractional error everywhere less than 10^{-7}.

```
PROGRAM ERFUNC
PRINT *, ' ENTER VALUE OF X '
READ *, X
IF (X . GT. 0) THEN
ERF = 1.-  ERFCC(X)
ELSE
ERF = ERFCC(X) -1.
END IF
PRINT *, ' THE VALUE OF ERF(X) IS  ', ERF
END
FUNCTION ERFCC(X)
Z=ABS(X)
T=1./(1.+0.5*Z)
ERFCC=T*EXP(-Z*Z-1.26551223+T*(1.00002368+T*(.37409196+
*   T*(.09678418+T*(-.18628806+T*(.27886807+T*(-1.13520398+
*   T*(1.48851587+T*(-.82215223+T*.17087277)))))))))
IF (X.LT.0.) ERFCC=2.-ERFCC
RETURN
```

7.3 DYNAMICS OF PARTICLES IN FLUIDS

Your objectives in studying this section are to

1. Apply Stokes's law to the calculation of the drag force exerted by a fluid on a moving particle.
2. Estimate the drag force when Stokes's law does not apply.
3. Calculate the Cunningham correction factor for small particles.
4. Estimate the settling velocity of particles under the influence of gravity.

Particles are often separated from a fluid as part of a pollution control system. In air pollution control, particles can be removed by gravity settlers, centrifugal collectors, fabric filters, electrostatic precipitators, or wet scrubbers. In all of these devices, particles are separated from the surrounding fluid by the application of one or more external forces. These forces—gravitational, inertial, centrifugal, and electrostatic—cause the particles to accelerate away from the direction of the mean fluid flow. The particles must then be collected and removed from the system to prevent ultimate reentrainment into the fluid. Thus, design and operation of particulate control equipment require a basic understanding of the dynamics of particles in fluids.

7.3.1 Drag Force

A good place to start to study the dynamical behavior of aerosol particles in a fluid is to consider the drag force exerted on a particle as it moves in a fluid. To calculate this force, the equations of fluid motion must be solved to obtain the velocity and pressure fields around the particle, a formidable task. These equations can be solved analytically only at very low velocities, when viscous forces dominate inertial forces. The type of flow that results is called *creeping flow* or low-Reynolds number flow

The solution of the equations of motion for the velocity and pressure distribution around a sphere in creeping flow was first obtained by Stokes. The drag force, which is the net force exerted by the fluid on the particle in the direction of flow, can be calculated once the velocity and pressure fields are known. The result, known as *Stokes's law*, is (Bird et al. 1960):

$$F_D = 3\pi\mu D_p u_r \tag{7.17}$$

where

u_r = relative velocity between the fluid and the particle

μ = fluid viscosity

Stokes's law is valid for $\mathrm{Re} = u_r D_p \rho/\mu < 0.1$. At Re = 1.0, the drag force predicted by Stokes's law is 13% low owing to the neglect of the inertial terms in the equation of motion. To account for the drag force over the entire range of possible Reynolds numbers, express the drag force in terms of an empirical drag coefficient C_D as

$$F_D = C_D A_p \rho \frac{u_r^2}{2} \tag{7.18}$$

where A_p is the projected area of the body normal to the flow. Thus, for a spherical particle of diameter D_p

$$F_D = \frac{\pi}{4} C_D \rho D_p^2 \frac{u_r^2}{2} \tag{7.19}$$

Table 7.3 summarizes some of the correlations available for the drag coefficient as a function of the Reynolds number.

Table 7.3 Correlations for Drag Coefficient of Spherical Particles

Range of Reynolds Number	Correlation for C_D
Re < 0.1 (Stokes's law)	$\frac{24}{\text{Re}}$
0.1 < Re < 2	$\frac{24}{\text{Re}}\left(1+\frac{3}{16}\text{Re}+\frac{9}{160}\text{Re}^2\ln 2\text{Re}\right)$
2 < Re < 500	$\frac{24}{\text{Re}}\left(1+0.15\,\text{Re}^{0.687}\right)$
$500 < \text{Re} < 2 \times 10^5$	0.44

Source: Flagan and Seinfeld 1988.

7.3.2 Noncontinuum Effects

Aerosol particles are small. The particle size is often comparable to the distances that gas molecules travel between collisions with other gas molecules. Consequently, the basic continuum transport equations must be modified to account for non continuum effects. The Knudsen number, $\text{Kn} = 2\lambda_g/D_p$, where λ_g is the mean free path of the gas, is the key dimensionless number in this respect.

The mean free path of a gas can be calculated from the kinetic theory of gases as

$$\lambda_g = \frac{0.1145\,\mu}{P\sqrt{\frac{M}{T}}} \tag{7.20}$$

where P is the gas pressure in kPa, and M is the gas molecular weight. For example, for air at 298 K and 1 atm the mean free path is 6.51×10^{-8} m = 0.0651 μm. Stokes's law derives from the equations of continuum fluid mechanics. When the particle diameter approaches the same order as the mean free path of the suspending gas molecules, the resisting force offered by the fluid is smaller than that predicted by Stokes's law. To account for this effect that becomes important as D_p becomes smaller, a slip factor, C_c, also called Cunningham correction factor, is introduced into Stokes's law:

$$F_D = \frac{3\pi\mu u_r D_p}{C_c} \tag{7.21}$$

where an empirical correlation for C_c was developed, based on experiments performed by Millikan between 1909 and 1923, as (Allen and Raabe 1982)

$$C_c = 1 + \text{Kn}\left[1.257 + 0.40\exp\left(-\frac{1.10}{\text{Kn}}\right)\right] \tag{7.22}$$

Table 7.4 shows the value of the Cunningham correction factor for particles in air at 1 atm and 298 K.

Table 7.4 Cunningham Correction Factor for Air at 1 atm and 298 K

D_p (μm)	Knudsen number (Kn)	Cunningham factor (C_c)
0.01	13.02	22.7
0.05	2.60	5.06
0.10	1.30	2.91
0.50	0.26	1.337
1.00	0.13	1.168
5.00	0.026	1.034
10.00	0.013	1.017

7.3.3 Gravitational Settling

For a relative motion to exist between a fluid and a freely suspended particle, at least one external force must exist. Considering an external force, F_e, which is opposed by the drag force, Newton's second law of motion for a particle of mass mp can be written as

$$F_e - F_D = m_p \frac{du_r}{dt} \tag{7.23}$$

For a spherical particle in the Stokes's region, Eq. (7.21) can be substituted into Eq.(7.23) to yield

$$\frac{du_r}{dt} + \frac{18\mu}{\rho_p C_c D_p^2} u_r = \frac{F_e}{m_p} \tag{7.24}$$

Equation (7.24) can be rewritten as

$$\frac{du_r}{dt} + \frac{u_r}{\tau} = \frac{F_e}{m_p} \tag{7.25}$$

where τ is a characteristic time associated with the motion of the particle given by:

$$\tau = \frac{D_p^2 \rho_p C_c}{18\mu} \tag{7.26}$$

Equation (7.25) is the basic differential equation governing the motion of a particle in a fluid when Stokes's law applies. Consider, for example, the resultant motion when gravity is the only external force (the buoyancy force can be neglected when the fluid is a gas). Equation (7.25) becomes

$$\frac{du_r}{dt} + \frac{u_r}{\tau} = g \tag{7.27}$$

where g is the gravitational constant. If the initial relative velocity is zero, the solution to Eq. (7.27) is

$$u_r = \tau g \left[1 - \exp\left(-\frac{t}{\tau}\right)\right] \tag{7.28}$$

For $t >> \tau$, the particle attains a constant velocity, called its terminal settling velocity, u_t,

$$u_t = \tau\, g = \frac{D_p^2 \rho_p C_c g}{18\mu} \tag{7.29}$$

Figure 7.5 illustrates the transient behavior of a particle settling under the influence of gravity. After four characteristic times, the particle's velocity is virtually equal to its terminal velocity.

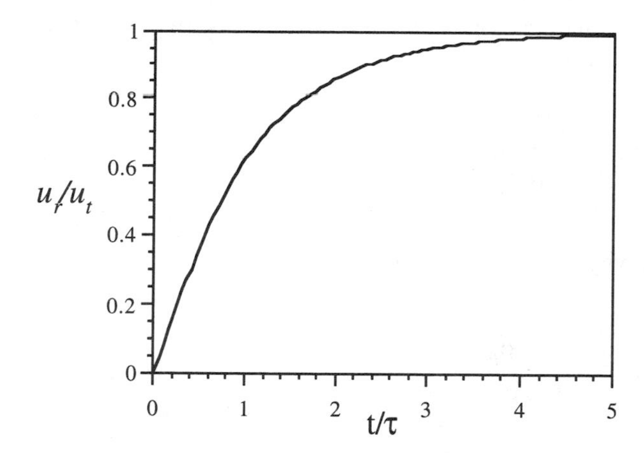

Figure 7.5 Dimensionless particle velocity versus dimensionless time

The transient portion of a particle's settling time is usually ignored. Table 7.5 shows the terminal velocity and the characteristic time for several particles in the size range of interest in air pollution control applications. Because τ is so small, it is justified to assume that the terminal velocity is attained almost instantaneously.

Table 7.5 Gravitational Settling of Unit Density Spheres in Air At 298 K and 1 atm

D_p (μm)	Characteristic time (τ, s)	Terminal velocity (m/s)
0.1	8.8×10^{-8}	8.6×10^{-7}
0.5	1.0×10^{-6}	1.0×10^{-5}
1.0	3.6×10^{-6}	3.5×10^{-5}
5.0	7.9×10^{-5}	7.8×10^{-4}
10.0	3.1×10^{-4}	3.1×10^{-3}

For a particle larger than 10 to 20 μm settling at its terminal velocity, the Reynolds number is too high for the Stokes's regime analysis to be valid. The drag coefficient is a useful way to represent the drag force on a particle over the entire range of Reynolds number. Newton's second law of motion can be rewritten in terms of the drag coefficient:

$$m_p \frac{du_r}{dt} = F_e - \frac{1}{2} C_D \rho A_p u_r^2 \tag{7.30}$$

If the external force is constant, the motion approaches a terminal velocity at which the external force is exactly balanced by the drag force,

$$C_D u_t^2 = \frac{2F_e}{\rho A_p} \tag{7.31}$$

For terminal settling owing to gravity of a spherical particle,

$$C_D u_t^2 = \frac{4 D_p \rho_p g}{3 \rho} \tag{7.32}$$

Because C_D depends on u_t through Re, this equation can not be solved explicitly for u_t. Instead, Eq. (7.32) must be solved for u_t by trial and error or through the following procedure. Define a new dimensionless number, the Galileo number.

$$\mathrm{Ga} = C_D \mathrm{Re}^2 = C_D u_t^2 \left(\frac{D_p \rho}{\mu}\right)^2 \tag{7.33}$$

Substituting Eq. (7.32) in Eq. (7.33),

$$\mathrm{Ga} = \frac{4 D_p^3 \rho \rho_p g}{3 \mu^2} \tag{7.34}$$

Another useful relation between C_D and Re is:

$$\frac{\mathrm{Re}}{C_D} = \frac{\mathrm{Re}^3}{\mathrm{Ga}} = \frac{3 \rho^2 u_t^3}{4 g \rho_p \mu} \tag{7.35}$$

The following correlation due to Koch can be used to relate Re/C_D to Ga (Licht 1980):

$$\ln \sqrt[3]{\frac{\mathrm{Re}}{C_D}} = -3.194 + 2.153 \ln \mathrm{Ga}^{1/3} - 0.238 \left(\ln \mathrm{Ga}^{1/3}\right)^2 + 0.01068 \left(\ln \mathrm{Ga}^{1/3}\right)^3 \tag{7.36}$$

To calculate u_t for a particle of any diameter, first calculate the value of Ga. Then, calculate Re/C_D from Eq. (7.36) and finally obtain u_t from Eq. (7.35).

Example 7.3 Gravitational Settling Velocity

Estimate the terminal settling velocity of a spherical particle with a diameter of 100 μm and a density of 2,600 kg/m^3 falling through air at 373 K and 1 atm under the influence of gravity. Calculate the terminal Reynolds number, Re_t, and the terminal drag force on the particle.

Solution

For air at 373 K and 1 atm, $\rho = 0.947$ kg/m^3, $\mu = 2.1 \times 10^{-5}$ kg/m-s. From Eq. (7.34), $Ga^{1/3} = 4.174$. Equation (7.36) yields $(Re/C_D)^{1/3} = 0.564$. From Eq. (7.35), $u_t = 0.522$ m / s. Calculate $Re_t = u_t D_p \rho/\mu = 2.35$. There are two ways to calculate the terminal drag force. The easy way is to notice that when the particle attains constant velocity, the drag force is perfectly balanced by the external force, gravity in this case. Therefore, the drag force is equal to the weight of the particle: $F_D = m_p g = 1.33 \times 10^{-8}$ N. Remember that this is true only at the terminal velocity. A more general approach to calculate the drag force is through Eq. (7.19). Calculate the drag coefficient with the corresponding correlation from Table 7.2:

$C_D = 24/Re\left(1 + 0.15\, Re^{0.687}\right) = 12.97$. Substituting in Eq. (7.19), $F_D = 1.32 \times 10^{-8}$ N. There is a slight difference between the two results due to the use of Eq. (7.36), an approximate correlation, as part of the procedure to calculate the terminal velocity.

7.3.4 Collection Of Particles By Impaction, Interception, and Diffusion

When a flowing fluid approaches a stationary object such as a fabric filter thread, a large water droplet , or a metal plate, the fluid flow streamlines will diverge around that object. Because of their inertia, particles in the fluid will tend to continue in their original direction. If the particles have enough inertia and are located close enough to the stationary object, they will collide and be collected by it.

Impaction occurs when the center of mass of a particle that is diverging from the fluid streamlines strikes a stationary object. *Interception* occurs when the particle's center of mass closely misses the object, but, because of its finite size, the particle strikes the object. Collection of particles by *diffusion* occurs when small particles (which are subject to random motion about the mean path) diffuse toward the object while passing near it. Once striking the object by any of these means, particles are collected only if there are short-range forces strong enough to hold them to the surface.

A simple means to explain impaction is with the concept of stopping distance. If a sphere in the Stokes's regime is projected with an initial velocity u_0 into a motionless fluid, its velocity as a function of time (ignoring all but the drag force) is

$$u_r = u_0 e^{-t/\tau} \tag{7.37}$$

The total distance traveled by the particle before it comes to rest is

$$x_{stop} = \int_0^{\infty} u_r \, dt = u_0 \tau \tag{7.38}$$

If the particle stops before striking the object, it can be swept around the object by the altered fluid flow. Because τ is very small, x_{stop} is also small. For example, if a 1.0-μm particle with unit density is projected at 10 m/s into air, it will travel only 36 μm.

An impaction parameter, N_t, can be defined as the ratio of the stopping distance of a particle (based on the upstream fluid velocity) to the diameter of the stationary object, d_0, or:

$$N_t = \frac{x_{stop}}{d_0} \tag{7.39}$$

If N_t is large, most of the particles will strike the object, otherwise, most will follow the fluid flow around it.

7.4 EFFECTIVENESS OF COLLECTION

Your objectives in studying this section are to

1. Define fractional efficiency, overall efficiency based on particle number, overall efficiency based on particle mass, and penetration.
2. Develop a relation between overall mass collection efficiency and fractional efficiency for a log-normal aerosol population.
3. Apply Gauss-Hermite quadrature formulas to the evaluation of overall mass collection efficiencies.
4. Estimate the fractional and overall collection efficiencies of settling chambers operating in the turbulent flow regime.
5. Calculate the overall collection efficiency of two or more particulate collection devices operating in series.

The success of a particulate collection system may be expressed either in terms of the amount of aerosol removed from the air stream, or the amount permitted to remain in it. The collection or removal efficiency of a device may be defined in various ways. For instance, the *fractional or grade efficiency* $\eta\,(D_p)$ is defined as:

$$\eta(D_p) = 1 - \frac{\text{number of particles of diameter } D_p \text{ out}}{\text{number of particles of diameter } D_p \text{ in}} \tag{7.40}$$

This efficiency can be expressed in terms of the particle size distribution functions at the inlet and outlet sides of the device:

$$\eta(D_p) = \frac{n_{in}(D_p)\,dD_p - n_{out}(D_p)\,dD_p}{n_{in}(D_p)\,dD_p} \tag{7.41}$$

The overall efficiency based on particle number η_N is defined as

$$\eta_N = 1 - \frac{\text{number of particles out}}{\text{number of particles in}} \tag{7.42}$$

In terms of the particle size distribution functions, the overall efficiency is

$$\eta_N = 1 - \frac{\int_0^\infty n_{out}(D_p)\,dD_p}{\int_0^\infty n_{in}(D_p)\,dD_p} \tag{7.43}$$

Combining Eqs. (7.41) and (7.43):

$$\eta_N = \frac{\int_0^\infty \eta(D_p)\,n_{in}(D_p)\,dD_p}{\int_0^\infty n_{in}(D_p)\,dD_p} \tag{7.44}$$

The overall efficiency based on particle mass η_M is defined as

$$\eta_M = 1 - \frac{\text{mass of particles out}}{\text{mass of particles in}} \tag{7.45}$$

For spherical particles of uniform density:

$$\eta_M = \frac{\int_0^\infty \eta(D_p)\, D_p^3\, n_{in}(D_p)\, dD_p}{\int_0^\infty n_{in}(D_p)\, D_p^3\, dD_p} \tag{7.46}$$

The overall collection efficiency by mass is usually the easiest to measure experimentally. The inlet and outlet streams may be sampled by a collection device, such as a filter, that collects virtually all of the particles.

The collection efficiency is frequently expressed in terms of *penetration*. The penetration is based on the amount emitted rather than on the amount collected; based on particle mass, it is just $Pt_M = 1 - \eta_M$.

The fractional efficiency $\eta(D_p)$ is, for most collectors, a unique single-valued function for a particular set of operating conditions. It depends on such parameters as the nature and design dimensions of the collector, and the rate of flow and particulate loading of the gas stream. The following chapters will develop the fractional efficiency function for the most common devices used for particulate removal from gaseous streams.

For a log-normal particle size distribution, Eq. (7.46) can be written as

$$\eta_M = \frac{\int_{-\infty}^\infty \eta(u)\, e^{3u}\, n_{in}(u)\, du}{\int_{-\infty}^\infty n_{in}(u)\, e^{3u}\, du} \tag{7.47}$$

$$\eta_M = \frac{\displaystyle\int_{-\infty}^{\infty} \eta(u)\, e^{3u}\, e^{-(u-u_m)^2/2\sigma_u^2}\, du}{\displaystyle\int_{-\infty}^{\infty} e^{3u}\, e^{-(u-u_m)^2/2\sigma_u^2}\, du} \tag{7.48}$$

where u = ln D_p, u_m = ln NMD and σ_u = ln σ_g. The denominator of Eq. (7.48) is easily evaluated (see Problem 7.7), leading to:

$$\eta_M = \frac{\displaystyle\int_{-\infty}^{\infty} \eta(u)\, e^{3u}\, e^{-(u-u_m)^2/2\sigma_u^2}\, du}{\sqrt{2\pi}\, \sigma_u\, e^{3u_m}\, e^{9\sigma_u^2/2}} \tag{7.49}$$

Equation (7.49) can be further simplified (see Problem 7.12):

$$\eta_M = \frac{\displaystyle\int_{-\infty}^{\infty} \eta(u)\; e^{-(u-\bar{u})^2/2\sigma_u^2}\, du}{\sqrt{2\pi}\, \sigma_u} \tag{7.50}$$

where $\bar{u}$ = ln MMD. Define a new variable v such that:

$$v = \frac{u - \bar{u}}{\sqrt{2}\, \sigma_u} \tag{7.51}$$

In terms of this variable, Eq. (7.50) becomes:

$$\eta_M = \frac{\displaystyle\int_{-\infty}^{\infty} \eta(v) e^{-v^2} dv}{\sqrt{\pi}} \tag{7.52}$$

For a given fractional efficiency function, the integral of Eq. (7.52) can be approximated numerically using the Gauss-Hermite quadrature method (Carnahan et al. 1969). Kerr (1981,

1989) and Benítez (1988) have used this technique where the log-normal distribution applies. The integral becomes:

$$\int_{-\infty}^{\infty} \eta(v)e^{-v^2}dv \cong \sum_{i=1}^{N} w_i\eta(v_i) \tag{7.53}$$

where w_i and v_i are the weight factors and roots of the Nth degree Hermite polynomial (Abramowitz and Stegun 1972). As a general rule, the accuracy of the numerical integration increases with increasing polynomial degree. Benítez (1988) suggested that an 8-point formula was adequate for preliminary design purposes.

Example 7.4 Overall Mass Collection Efficiency Calculations

The fractional efficiency function of a particulate removal device is given by:

$$\eta(D_p) = 1 - \exp\left(-0.000651D_p^2\right) \tag{7.54}$$

where D_p is in microns. The device processes a log-normally distributed aerosol with a MMD of 50 μm and σ_g of 2.5. Estimate the overall mass collection efficiency. Use an 8-point Gauss-Hermite quadrature formula to estimate the integral.

Solution

The procedure to estimate the overall mass collection efficiency is as follows:

- Choose the number of quadrature points to use, N.
- Obtain the values of the roots, v_i, and weight factors, w_i, of the corresponding Hermite polynomial either from a mathematical table or from a computer program provided subsequently.
- For each of the roots, calculate the corresponding u_i from Eq. (7.51).
- Calculate $D_{pi} = \exp u_i$.
- Calculate $\eta\ (D_{pi})$ from Eq. (7.54).
- Calculate $w_i\ \eta\ (D_{pi})$ for $i = 1, 2, ..., N$.
- Calculate η_M from Eqs. (7.52) and (7.53).

The following table summarizes the calculations for an aerosol population with MMD = 50 μm and $\sigma_g = 2.5$.

vi	D_{pi} (μm)	$\eta\ (D_{pi})$	wi	$wi\ \eta\ (D_{pi})$
–2.93063	1.12	0.0008	0.00020	0
–1.98165	3.83	0.0095	0.01708	0.00016
–1.15719	11.16	0.0779	0.207802	0.01619
–0.38119	30.51	0.4545	0.66115	0.30050
0.38119	81.94	0.9870	0.66115	0.65260
1.15719	223.97	1.0000	0.207802	0.20780
1.98165	651.89	1.0000	0.01708	0.01708
2.93063	2,246.10	1.0000	0.00020	0.00020
				$\Sigma = 1.19453.$

$$\eta_M = 1.19453/\pi^{1/2} = 0.6739\ (67.39\%)$$

Comments

Notice the symmetry of the roots and weight functions of the Hermite polynomials. This characteristic simplifies computer implementation of the method. The following is a computer subroutine to calculate the roots and weight factors of the Nth degree Hermite polynomial.

```
      SUBROUTINE HERMIT(NN,X,A,EPS)
C     CALCULATES THE ZEROES X(I) OF THE NN-TH ORDER
C     HERMITE POLYNOMIAL. ALSO CALCULATES THE CORRESPONDING
C     WEIGHT FACTOR A(I) FOR GAUSS-HERMITE QUADRATURE
C
      DIMENSION X(50), A(50)
      FN = NN
      N1 = NN-1
      N2 = (NN+1)/2
      CC = 1.7724538509*GAMMA(FN)/(2.**N1)
      S = (2.*FN+1.)**.16667
      DO 11 I =1, N2
        IF (I .EQ. 1) THEN
          XT = S**3 - 1.85575/S
        ELSE IF (I .EQ. 2) THEN
            XT = XT- 1.14*FN**.426/XT
        ELSE IF (I .EQ. 3) THEN
              XT = 1.86*XT-0.86*X(1)
```

```
       ELSE IF (I .EQ. 4) THEN
             XT = 1.91*XT-.91*X(2)
       ELSE
             XT = 2.*XT-X(I-2)
       END IF
       CALL HROOT(XT,NN,DPN,PN1,EPS)
       X(I) = XT
       A(I) = CC/DPN/PN1
       NI = NN-I+1
       X(NI) = -XT
       A(NI) = A(I)
11   CONTINUE
     RETURN
     END
     FUNCTION GAM(Y)
C
     GAM = (((((((0.035868343*Y- .193527818)*Y + .482199394)*Y-
    1  .756704078)*Y + .918206857)*Y - .897056937)*Y + .988205891)*Y
    2  -.577191652)*Y + 1.0
     RETURN
     END
     FUNCTION GAMMA(X)
C   COMPUTES THE GAMMA FUNCTION OF X FOR X BETWEEN 0 AND 35.
C
     Z = X
     IF (Z .LE. 0.0 .OR. Z .GT. 35.) THEN
       GAMMA = 0.
     ELSE IF ( Z .EQ.1.) THEN
         GAMMA = 1.
     ELSE IF (Z .LT. 1.0) THEN
         GAMMA = GAM(Z)/Z
     ELSE IF (Z .GT. 1.0) THEN
         ZA = 1.
10        Z = Z -1.
         IF ( Z  .EQ. 1.) THEN
           GAMMA = ZA
         ELSE IF (Z .GT. 1.0) THEN
           ZA = ZA*Z
           GO TO 10
         ELSE IF (Z .LT. 1.0) THEN
           GAMMA = ZA*GAM(Z)
         END IF
     END IF
     RETURN
     END
     SUBROUTINE HROOT(X,NN,DPN,PN1,EPS)
```

```
C   IMPROVES THE APPROXIMATE ROOT X
C   DPN = DERIVATIVE OF H(N) AT X
C   PN1 = VALUE OF H(N-1) AT X
C
    ITER = 0
1   ITER = ITER + 1
    CALL HRECUR(P,DP,PN1,X,NN)
    D = P/DP
    X = X- D
    IF( ABS(D) .GT. EPS .AND. ITER .LT. 10) THEN
      GO TO 1
    ELSE
      DPN = DP
    END IF
    RETURN
    END
    SUBROUTINE HRECUR(PN,DPN,PN1,X,NN)
    P1 = 1.
    P = X
    DP1 = 0.
    DP = 1.
    DO 1 J =2, NN
      FJ = J
      FJ2 = (FJ-1.)/2.
      Q = X*P-FJ2*P1
      DQ = X*DP + P - FJ2*DP1
      P1 = P
      P = Q
      DP1 = DP
      DP = DQ
1   CONTINUE
    PN = P
    DPN = DP
    PN1 = P1
    RETURN
    END
```

Example 7.5 Overall Mass Collection Efficiency of Settling Chamber

The settling chamber is perhaps the simplest of all air pollution control devices. Its main usefulness lies in serving as a preliminary screening device for a more efficient control system. Where the mass of the larger particles is huge, the settling chamber can remove much of the mass of the particulate population which would otherwise choke up the other control device, impairing its operation or requiring too frequent cleaning.

The use of several trays improves the collection efficiency of a settling chamber since the particles have a much shorter distance to travel before reaching the bottom of the passage between trays. Figure 7.6 shows a settling chamber in which trays are provided.
The flow in a rectangular channel, such as the ones in a settling chamber with trays, is turbulent if the Reynolds number, $Re_c > 4{,}000$ (McCabe and Smith, 1976). For the situation depicted in Figure 7.6, the Reynolds number is (Crawford 1976)

$$Re_c = \frac{2Q\rho}{\mu(N_{tr}W + H)} \tag{7.55}$$

where N_{tr} is the number of trays, including the bottom surface of the chamber. For turbulent flow, the fractional efficiency for particulate collection by a settling chamber is given by (see Problem 7.13):

$$\eta(D_p) = 1 - \exp\left[-\frac{N_{tr}LWu_t}{Q}\right] \tag{7.56}$$

where u_t is the terminal velocity of a particle of diameter D_p. Notice that $N_{tr}LW = A_c$ where A_c is the total area available for particle collection. Therefore, Eq. (7.56) can be rewritten as

$$\eta(D_p) = 1 - exp\left[-\frac{A_c u_t}{Q}\right] \tag{7.57}$$

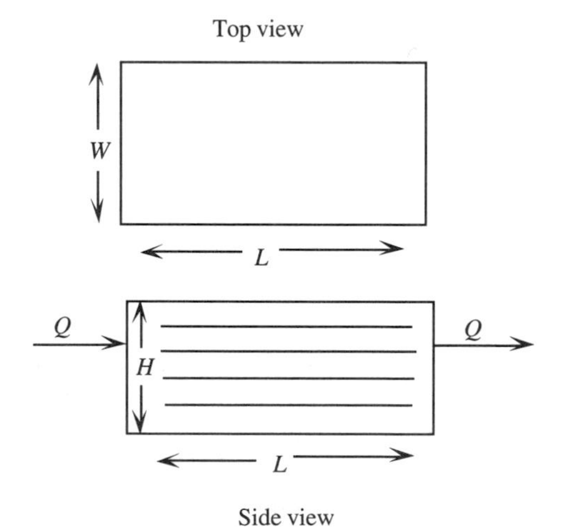

Figure 7.6 Schematic diagram of a settling chamber with trays

Consider a 5-m long settling chamber with 10 square trays. The device processes 10 m^3/s of air at 298 K and 101.3 kPa, carrying a log-normal aerosol with MMD = 25 μm and $\sigma_g = 2.0$. The density of the particles is 2,000 kg/m^3. The total height of the chamber is 3 m. Estimate the overall mass collection efficiency of the chamber.

Solution

The first step in the solution is to determine if the flow through the chamber is turbulent. For air at 298 K and 101.3 kPa, $\rho = 1.185$ kg/m^3 and $\mu = 1.84 \times 10^{-5}$ kg/m-s. The chamber dimensions are: $L = W = 5$ m., $H = 3$ m. For $N_{tr} = 10$, Eq. (7.55) yields $Re_c = 24{,}300$ which means fully developed turbulent flow. Calculate $A_c = (10)(5)2 = 250$ m^2. Equation (7.57) becomes

$$\eta(D_p) = 1 - \exp\left[-25u_t\right] \tag{7.58}$$

To illustrate the concept, choose a 4-point quadrature formula to relate the fractional and overall collection efficiencies. The terminal velocities in Eq. (7.58) must be carefully calculated because Stokes's law is not valid at the higher end of the particle size range covered. At the lower end of the range, the Cunningham correction factor must be included in the calculations.

v_i	D_{pi} (μm)	u_t (m/s)	Re_t	η (D_{pi})	w_i	$w_i\eta(D_{pi})$
−1.65068	5	0.0015	0.0002	0.0368	0.08131	0.0030
−0.52465	15	0.0118	0.0113	0.2548	0.80491	0.2051
0.52465	41.8	0.1020	0.275	0.9220	0.80491	0.7421
1.65068	126	0.6220	5.05	1.0000	0.08131	0.0813
						$\Sigma = 1.0315$

$\eta_M = 1.0315/\sqrt{\pi} = 0.582$ (58.2%)

Frequently, a particulate collection system consists of two or more devices operating in series. The overall collection efficiency is not simply the sum nor the product of the efficiencies of each device. Each device's efficiency is based on the mass loading of particles entering that device, but the overall system efficiency is based on the total mass collected as a fraction of the total mass entering the first device. It can be easily shown that the overall penetration of such a system, Pt_o, is simply the product of the penetrations of all of the individual devices (Cooper and Alley 1986).

$$Pt_o = \prod_{i=1}^{n} Pt_i \tag{7.59}$$

The overall collection efficiency of the system is $\eta_M = 1 - Pt_o$.

Example 7.6 Overall Collection Efficiency with Two Control Devices in Series

A particulate control system consists of a settling chamber with an overall mass collection efficiency of 65%, followed by an electrostatic precipitator with an overall mass collection efficiency of 95%. Calculate the overall collection efficiency for the system.

Solution

Calculate the penetrations of the individual devices: $Pt_1 = 1 - 0.65 = 0.35$, $Pt_2 = 1 - 0.95 = 0.05$. From Eq. (7.59), $Pt_o = (0.35)\,(0.05) = 0.0175$. Therefore, the overall efficiency for the system is $\eta_M = 0.9825$ (98.25%)

7.5 CONCLUSION

Removal of particulate matter from exhaust gases is a very important engineering task because particles constitute a major class of air pollutants. The most important characteristic of an aerosol population is its size distribution function. Not only are the deleterious effects of particulates dependent on their size, but the nature and design of the pollution control device appropriate for a given aerosol are highly dependent on the characteristics of the size distribution function. Most aerosols of interest in air pollution control are log-normally distributed, which is fortunate because such a function is easily characterized in terms of only two parameters.

Most particulate collection devices depend on an external force to impart on the particles a velocity component that is normal to the direction of the gas flow. Settling chambers depend on gravity for that purpose. The following chapters explore the mechanisms through which the most popular particulate control devices operate, and derive the corresponding fractional efficiency equations. Sizing and costing of particulate control equipment are covered in detail.

REFERENCES

Abramowitz, M., and Stegun, I. A. (Eds.) *Handbook of Mathematical Functions with Formulas, Graphs, and Mathematical Tables,* Chap. 25, Wiley, New York, (1972).

Allen, M. D., and Raabe, O. G. *J. Aerosol Sci.*, **13**:537 (1982).

Benítez, J. "The Use of Gaussian Quadrature Formulas for the Optimal Design of Particulate Control Equipment," paper No. 88-72.6, presented at the 81st Annual Meeting of APCA, Dallas, TX (1988).

Bird, R. B., Stewart, W. E., and Lightfoot, E. N. *Transport Phenomena*, Wiley, New York, (1960).

Carnahan, B., Luther, H. A., and Wilkes, J. O. *Applied Numerical Methods*, Chap. 2, Wiley, New York, (1969).

Cooper, C.D., and Alley, F. C. *Air Pollution Control: A Design Approach*, PWS Engineering, Boston, MA (1986).

Crawford, M. *Air Pollution Control Theory,* McGraw-Hill, New York, (1976).

Flagan, R. C., and Seinfeld, J. H. *Fundamentals of Air Pollution Engineering*, Prentice Hall, Englewood Cliffs, NJ (1988)

Hinds, W. C. *Aerosol Technology, Properties, Behavior, and Measurement of Airborne Particles*, Wiley, New York, (1982).

Kerr, C. P. *Environ. Sci. and Technol.*, **15:**119 (1981).

Kerr, C. P. *JAPCA*, **39**:1585 (1989).

Larsen R. I., McDonnell, W. F., Horstman, D. H., and Folinsbee, L. J. *J. Air Waste Manage. Assoc.*, **41**:455 (1991).

Licht, W. *Air Pollution Control Engineering, Basic Calculations for Particulate Collection*, Marcel Dekker, New York (1980).

McCabe, W. L., and Smith J. C. *Unit Operations of Chemical Engineering*, 3rd. ed. McGraw-Hill, New York (1976).

Ott, W. R. *J. Air Waste Manage. Assoc.,* **40**:1378 (1990).

Press, W. H., Flannery, B. P., Teukolsky, S. A., and Vetterling, W. T. *Numerical Recipes: The Art of Scientific Computing,* Cambridge University Press, New York (1989).

Wark, K., and Warner C. F. *Air Pollution Its Origin and Control*, 2nd ed., Harper and Row, New York (1981).

PROBLEMS

The problems at the end of each chapter have been grouped into four classes (designated by a superscript after the problem number).

Class a: Illustrates direct numerical application of the formulas in the text.
Class b: Requires elementary analysis of physical situations, based on the subject material in the chapter.
Class c: Requires somewhat more mature analysis.
Class d: Requires computer solution.

7.1[a]. Analysis of data from a cascade impactor

The following data were obtained when a sample of an aerosol population was analyzed with a cascade impactor:

Size range (μm)	Mass (mg)
0-4	25
4-8	125
8-16	100
16-30	75
30-50	30
> 50	5

Determine whether a log-normal distribution fits this data and, if so, estimate MMD, NMD, and σ_g.

Answer: MMD = 10.8 μm

7.2[b]. Log-normal distribution

For the aerosol population of Problem 7.1, calculate the following:

(a) Mass-fraction of particles with diameters between 5 and 10 μm.

Answer: 32.3%

(b) Fraction of the total number of particles with diameters between 1 and 5 μm.

Answer: 74.6%

7.3[b]. Modal diameter of a log-normal distribution

An important parameter of a size distribution function is the modal diameter, D_{pMO}, defined as the diameter at which the greatest number of particles is clustered. This diameter is located at the maximum point of the curve for $n(D_p)$. Show that, for a log-normal distribution,

$$D_{pMO} = \text{NMD} \exp\left[-(\ln \sigma_g)^2\right]$$

Estimate the modal diameter for the aerosol population of Problem 7.1.

Answer: 1.58 μm

7.4[d]. Log-bimodal size distribution

The following data were obtained with a cascade impactor

D_p (μm)	G (%)
0.5	5.4
1.0	14.4
2.0	23.4
4.0	37.0
8.0	55.1
10.0	61.4
15.0	72.7
20.0	79.6
25.0	84.6
50.0	94.8

Fit a log-bimodal distribution function to these data and comment on the goodness of fit of the model.

7.5[b] Log-bimodal size distribution

An aerosol population results from the combination of particulate matter from two distinct sources. The coarser source is log-normally distributed with MMD = 10 μm and σ_g =3.0. The finer source is also log-normally distributed with σ_g = 2.5, but unknown MMD. Twenty-three percent of all the mass originates at the fine particle source. It was found experimentally that the combined, cumulative mass fraction up to 3 μm was 30.8%. Estimate the value of MMD for the fine source.

7.6[a]. Urban aerosols

The size distribution of urban aerosols containing photochemical smog are usually bimodal. The "fine particle" mode—less than 2μm—contains from one-third to two-thirds of the total mass, with the remainder in the "coarse particle" mode. The fine particles are produced by photochemical atmospheric reactions and the coagulation of combustion products. The coarse particles are mainly of mechanical origin.

The aerosol over Pasadena, California, was sampled on September 3, 1969 under light to moderate smog conditions. The MMD of the fine particle mode was 0.3 μm with a σ_g of 2.05. The corresponding parameters for the coarse particles were 8.0 μm and 2.3, respectively (Hinds 1982). The cumulative mass fraction up to 1.0 μm was 55%.

Estimate what fraction of the total atmospheric aerosol was of photochemical origin.

Answer: 57.3%

7.7[b]. Mass concentration of a log-normal aerosol population

An aerosol with a log-normal size distribution has a NMD of 0.3 μm and a σ_g of 1.5. If the number concentration is 10^6 particles/cm^3, what is the mass concentration? Particles may be assumed spherical with a density of 4,500 kg/m^3. The following identity may be useful:

$$\int_{-\infty}^{\infty} e^{ru} e^{-(u-\bar{u})^2/2\sigma_u^2} du = \sqrt{2\pi}\, \sigma_u e^{r\bar{u}} e^{(r^2\sigma_u^2/2)}$$

Answer: 133 μg/m^3

7.8[a]. Terminal gravitational settling velocity

(a) Estimate the terminal gravitational settling velocity of a unit-density, 200-μm diameter sphere in air at 298 K and 1 atm.

Answer: 0.68 m/s

(b) For a 0.15-μm diameter spherical particle (ρ_p = 2,500 kg/m^3) determine the Cunningham correction factor and the terminal settling velocity in air at 298 K and 1 atm.

Answer: 3.64×10^{-6} m/s

7.9[a]. Dynamic shape factors

A correction factor called the dynamic shape factor, χ, is applied to Stokes's law to account for the effect of shape on particle motion. Stokes's law for irregular particles becomes

$$F_D = \frac{3\pi\mu u_r D_p \chi}{C_c}$$

where D_p is the diameter of a spherical particle with the same volume as the irregularly shaped particle. The following table gives dynamic shape factors for particles of various shapes.

Shape	Dynamic shape factor
Sphere	1.00
Cube (L/D = 4)	1.08
axis horizontal	1.32
axis vertical	1.07
Bituminous coal dust	1.05-1.11
Quartz	1.36
Sand	1.57
Talc	2.04

Source: Davies, C. N. *J. Aerosol Sci.,* **10**:477 (1979).

An old industrial hygiene rule of thumb is that a 10-μm silica particle settles in atmospheric air at a rate of 1 cm/s. What is the true settling velocity of such a particle? The specific gravity of silica is 2.6. Use the dynamic shape factor for quartz.

Answer: 0.566 cm/s

7.10[b]. Gravitational settling velocity

Atmospheric air is dried by bubbling it through concentrated sulfuric acid (ρ_p = 1,840 kg/m^3).The acid container is a 0.1-m diameter, 2-m long tube which holds 1.5 L of acid. The air flow rate is 10 L/min. When the bubbles burst at the liquid surface, they form droplets. What is the largest droplet that can be carried out of this system?

Answer: 19.5 μm

7.11[b].Terminal velocity for electrically charged particles

When a particle possessing an electrical charge q_p enters a region where an electric field of strength E_c is also present, an electrostatic force F will act on the particle. The magnitude of this force is given by $F = q_pE_c$, where F is in newtons, q_p in coulombs, and E_c in volts/m. Estimate the terminal velocity in air at 298 K and 1 atm of a 1.0-μm diameter particle with a charge of 0.3×10^{-15} coulombs under the influence of an electric field of 10^5 V/m.

Answer: 0.2 m/s

7.12[b]. Overall efficiency based on particle mass for log-normal function

Derive Eq. (7.50), which relates overall mass efficiency to fractional efficiency for a log-normal distribution, beginning with Eqs. (7.49) and (7.14).

7.13[c]. Turbulent flow in settling chambers

Derive Eq. (7.56) for the fractional efficiency of particulate collection by a settling chamber operating in the turbulent flow regime. Assume that there is a laminar layer adjacent to the bottom surface of the passage into which turbulent eddies do not penetrate, so that any particle that crosses into this layer is captured shortly. In the remainder of the flow passage, the eddying motion owing to turbulence will cause a uniform distribution of particles of all sizes. The vertical component of the velocity of the particles in the laminar layer is the corresponding terminal settling velocity. Horizontally, the particles move at the average velocity of the gas through the passage.

7.14[d]. Gauss-Hermite quadrature for overall efficiency estimation

Write a computer program to implement the method outlined in Example 7.4 to estimate the overall mass collection efficiency for a device with a given fractional collection efficiency equation when it operates on a log-normally distributed aerosol population. The user should be able to specify the fractional efficiency function, the parameters of the log-normal distribution, and the number of quadrature points up to $N = 50$. Test your program with the information on Example 7.4. Increase the number of points to 16 and observe the effect on the overall efficiency predicted.

7.15[d]. Overall efficiency of a settling chamber

Write a computer program to estimate the overall mass collection efficiency of a settling chamber operating in the turbulent flow regime. Assume that the aerosol is log-normally distributed. Estimate the integral with an N-point Gauss-Hermite quadrature formula. Test your program with the information on Example 7.5. Increase the number of points to 8 and observe the effect on the overall efficiency predicted.

7.16[b]. Overall penetration of a log-bimodal aerosol population

Kerr (1989) showed that the overall mass penetration of a log-bimodal aerosol population through a particulate control device can be estimated by:

$$Pt_M \cong \frac{R\sum_{i=1}^{N} w_i Pt(v_{fi}) + \sum_{i=1}^{N} w_i Pt(v_{ci})}{\sqrt{\pi}\,(R+1)}$$

where v_{ci} and v_{fi} are as defined by Eq. (7.51) for the coarse and fine fractions respectively. Consider a particulate control device with a fractional penetration function given by

$$Pt(D_p) = \exp(-0.066174 - 78{,}371 D_p)$$

where D_p is in meters. The aerosol is log-bimodal with $\text{MMD}_f = 0.5028$ μm, $\text{MMD}_c = 11.29$ μm, $\sigma_{gf} = 1.202$, $\sigma_{gc} = 1.353$, and $R = 0.01023$. Estimate the overall mass collection efficiency of the device based on a 5-point quadrature formula.

Answer: 61%

7.17[b] Particulate matter deposition in the alveolar region

Table 7.6 shows the fraction of inhaled particles deposited in the alveolar region for nose breathing at a rate of 14 L/min as a function of particle size D_p. Estimate how much mass deposits in a person's alveolar region daily owing to breathing local air which contains 150 mg/m^3 of a log-normally distributed aerosol with MMD = 2.72 μm and σ_g = 1.649.

Answer: 0.54 mg/d

Table 7.6 Fractional Deposition of Particles in Alveolar Region

D_p (μm)	Fraction	D_p (μm)	Fraction
0.10	0.20	1.2	0.21
0.20	0.15	1.3	0.22
0.30	0.12	1.4	0.23
0.40	0.12	1.5	0.25
0.50	0.12	2.0	0.28
0.60	0.12	3.0	0.26
0.70	0.15	4.0	0.19
0.80	0.18	5.0	0.11
0.9	0.20	6.0	0.03
1.0	0.20	7.0	0.0
1.1	0.20	8.0	0.0

Source:Hattis et al. *JAPCA* **37**:1060 (1987).

7.18[a]. Particulate control devices in series

Particulate removal efficiency on a certain gas stream must be 98.5% to satisfy emission standards. If a 60%-efficient cyclone precleaner is used with a wet scrubber, what is the required efficiency of the scrubber?

Answer: 96.2%

7.19[c]. Design of a settling chamber

Design a settling chamber to serve as a precleaner for an electrostatic precipitator (ESP). The removal efficiency for the system must be at least 98%. The ESP efficiency is 96.7% for a gas flow rate of 10 m^3/s of air at 298 K and 1 atm. The aerosol entering the settling chamber is log-normally distributed with MMD = 15.0 μm and $\sigma_g = 2.5$, and a particle density of 2,000 kg/m^3. Because of floor space limitations, the dimensions of the trays in the settling chamber cannot exceed 4 m. The tray spacing must be 0.3 m. Calculate the chamber dimensions and the number of trays required.

7.20[b]. Optimal design of a settling chamber

Crawford (1976) showed that, when there are no space limitations, the optimal design of a settling chamber operating in the turbulent flow regime is given by:

$$L = W = \sqrt{\frac{A_c}{N_{tr}}}$$

$$N_{tr} = \frac{A_c^{1/3}}{(2\Delta H)^{2/3}}$$

where A_c is the total collection area and ΔH is the tray spacing. Calculate the optimal dimensions and number of trays of a settling chamber to collect 50-μm particles with 90% efficiency. The gas flow rate is 25 m^3/s of air at 298 K and 1 atm. The particle density is 2,000 kg/m^3. The tray spacing is 0.3 m.

Answer: L = *5.53 m*

8

Cyclonic Devices

8.1 INTRODUCTION

Cyclone separators have been used in the United States for about 100 yr, and are still one of the most widely used of all industrial gas-cleaning devices. The main reasons for the widespread use of cyclones are that they are inexpensive to purchase, have no moving parts, and can be constructed to withstand harsh operating conditions. Cyclonic devices by themselves are generally not adequate to meet stringent particulate emission standards, but they serve an important purpose. Their low capital cost and their maintenance-free operation make them ideal for use as precleaners for more expensive control devices such as fabric filters or electrostatic precipitators.

In the past, cyclones were regarded as low-efficiency collectors. However, efficiency varies greatly with particle size and with cyclone design. Recently, advanced design work has greatly improved cyclone performance. They can now achieve efficiencies greater than 98% for particles larger than 5 μm. In general, as efficiencies increase, operating costs increase, primarily because of the resulting higher pressure drops.

Cyclones use the centrifugal force created by a spinning gas stream to separate particles from a gas. Figure 8.1 shows a tangential inlet, reverse flow cyclone separator. The particulate-laden gas enters tangentially near the top of the device. The cyclone's shape and the tangential entry force the gas flow into a downward spiral. Centrifugal force and inertia cause the particles to move outward, collide with the outer wall, and then slide downward to the bottom of the device. Near the bottom of the cyclone, the gas reverses its downward spiral and moves upward in a smaller inner spiral. The "clean" gas exits from the top through a vortex-finder tube, and the particles exit from the bottom of the cyclone through a pipe sealed by a spring-loaded flapper valve or a rotary valve.

The centrifugal force is proportional to the square of the tangential velocity and inversely proportional to the radius of curvature of the gas trajectory. Therefore, the efficiency of a cyclone increases as the diameter of the device is reduced. To achieve higher efficiencies dictates the use of smaller cyclones. However, the pressure drop through the cyclone increases rapidly as the tangential velocity increases. A way to maintain high efficiencies with a moderate pressure drop is to use a large number of small cyclones placed in parallel.

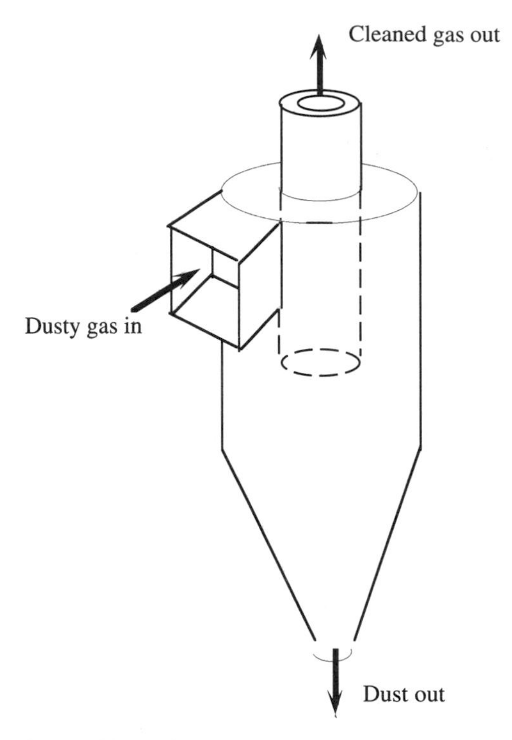

Figure 8.1 Schematic diagram of a tangential inlet reverse flow cyclone

This chapter covers the sizing and costing of single cyclones based on theoretical and empirical considerations, and the optimal design of systems of multiple cyclones to minimize TAC. Because cyclonic devices are usually used as precleaners for more sophisticated particulate control equipment, such as ESP or fabric filters, it is important to be able to characterize the particle size distribution of the aerosol that penetrates the cyclones and enters the next device. This chapter presents computational techniques for that purpose.

8.2 CYCLONIC FLOW

Your objectives in studying this section are to

1. Develop a fractional collection efficiency equation for particulate matter in ideal, laminar cyclonic flow.
2. Develop a fractional collection efficiency equation for particulate matter in ideal, turbulent cyclonic flow.
3. Use a semiempirical practical design equation to predict cyclone performance under real conditions.

Consider a particle entering tangentially onto a horizontal plane of a spinning gas stream at r_3, as shown in Figure 8.2. Because of the centrifugal force, the particle will follow a path outward across the the flow streamlines. Its velocity vector will have a tangential component (u_θ) and a radial component (u_r). The velocity of the spinning gas is assumed to have only a tangential component, v_θ, with $v_r = 0$. Tangential gas flows of this type usually are of the form $v_\theta r^m$ = constant. For an ideal, inviscid fluid in such a vortex flow $m = 1$, although in real flows the value of m may range downward to 0.5. The analysis of cyclone performance that follows begins with ideal, laminar flow. Then, it considers ideal, turbulent flow. Because both of these represent idealized cases that are not attained in real cyclones, it turns finally to a semiempirical theory that has been widely used in practical cyclone design.

8.2.1 Ideal, Laminar Cyclonic Flow

The so-called ideal laminar cyclonic flow refers to a frictionless flow in which the streamlines follow the contours of the cyclone. When the flow enters through a rectangular slot of area $W(r_2 - r_1)$, the gas velocity components are (Crawford 1976):

$$v_r = 0, \qquad v_\theta = v = \frac{Q}{W r \ln\left(\frac{r_2}{r_1}\right)} \tag{8.1}$$

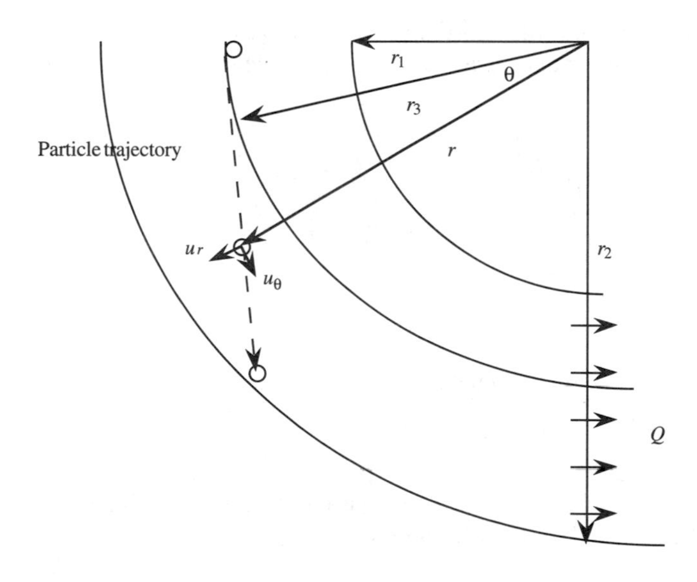

Figure 8.2 Trajectory of a particle in laminar cyclonic flow

To determine the collection efficiency consider a particle entering the cyclone at a radial position r_3 that strikes the wall at an angular position θ_f. The particle's velocity components at any point on its trajectory are u_r and u_θ. The radial velocity component is the terminal velocity of the particle when the centrifugal force, $F_c = m_p u_\theta^2/r$, acts on it. When the drag force can be given by Stokes's law, this velocity is

$$u_r = \frac{F_c}{3\pi\mu D_p} \tag{8.2}$$

Since the θ-component of the particle's velocity is that of the fluid, $u_\theta = v_\theta$, and

$$F_c = \frac{\pi}{6}\rho_p D_p^3 \frac{Q^2}{W^2 r^3 (\ln r_2/r_1)^2} \tag{8.3}$$

Combine Eqs. (8.2) and (8.3) to obtain:

$$u_r = \frac{\rho_p Q^2 D_p^2}{18\mu r^3 W^2 (\ln r_2/r_1)^2} \tag{8.4}$$

The next step is to obtain an equation for the particle's trajectory in the cyclone.The distance traveled in the θ-direction in a time interval dt is $u_\theta\, dt = r d\theta$. Also, the distance the particle moves in the r-direction in time dt is $dr = u_r dt$. Then,

$$\frac{r\, d\theta}{dr} = \frac{u_\theta}{u_r} \tag{8.5}$$

Substituting Eqs. (8.1) and (8.4) in (8.5) gives

$$\frac{d\theta}{dr} = \frac{18\mu W \ln(r_2/r_1) r}{\rho_p Q D_p^2} \tag{8.6}$$

a differential equation describing the particle's trajectory. If the particle enters the cyclone at $r = r_3$ and hits the outer wall at $\theta = \theta_f$, then integrating Eq. (8.6) gives

$$\theta_f = \frac{9\mu W \ln(r_2/r_1)}{\rho_p Q D_p^2}\left(r_2^2 - r_3^2\right) \tag{8.7}$$

An alternative is to solve Eq. (8.7) for r_3 to find the entrance position of a particle that hits the outer wall at $\theta = \theta_f$,

$$r_3 = \left[r_2^2 - \frac{\rho_p Q D_p^2 \theta_f}{9\mu W \ln(r_2/r_1)}\right]^{1/2} \tag{8.8}$$

To obtain an expression for the collection efficiency of a cyclone, assume that the entering particle concentration and gas velocity are uniform across the entrance cross section (see Figure 8.3). If the cyclone has an angle θ_f, all particles that enter the device at $r \geq r_3$ hit the wall over $0 \leq \theta \leq \theta_f$. The collection efficiency is just that fraction of the particles in the entering flow that hit the outer wall before $\theta = \theta_f$. Therefore,

$$\eta(D_p) = \frac{W(r_2 - r_3)}{W(r_2 - r_1)} \tag{8.9}$$

Substituting Eq. (8.8) into (8.9) and rearranging

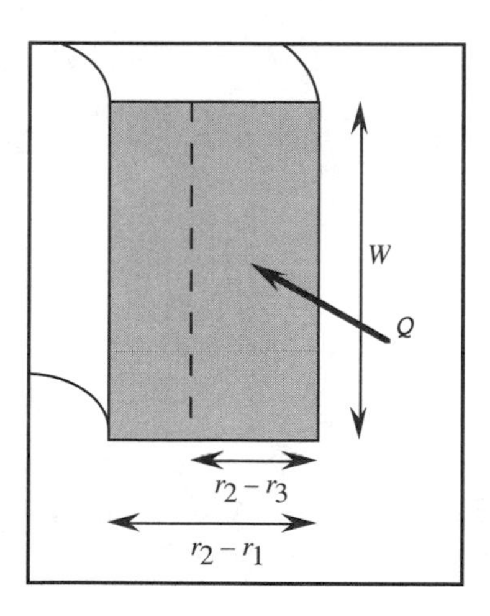

Figure 8.3 Schematic diagram of a cyclone entrance channel

$$\eta(D_p) = \frac{1 - \left[1 - \dfrac{\rho_p Q D_p^2 \theta_f}{9\mu W r_2^2 \ln(r_2/r_1)}\right]^{1/2}}{1 - r_1/r_2} \tag{8.10}$$

The value of θ_f at which $\eta(D_p) = 1$ is given by

$$\theta_f = \left[\frac{\rho_p Q D_p^2}{9\mu W \ln(r_2/r_1)}\right]^{-1} \left(r_2^2 - r_1^2\right) \tag{8.11}$$

Example 8.1 Ideal, Laminar Cyclonic Flow

Air at 298 K and 1 atm flows at the rate of 5.0 m^3/s and carries with it particulate matter with a density of 1,500 kg/m^3. The stream enters a cyclonic region with $r_1 = 0.2$ m and $r_2 =$ 0.4 m in ideal, laminar flow. Through what angle must the flow turn in the cyclone if the efficiency is to be unity for 30-μm particles? The height of the channel, W, is 1 m. Plot the efficiency as a function of particle size for this angle.

Solution

$$\theta_f = \frac{9\left(1.84 \times 10^{-5}\right)(1)\ln 2}{1{,}500(5.0)\left(9 \times 10^{-10}\right)}\left(0.4^2 - 0.2^2\right) = 2.041 \text{ rad } = 117°$$

Equation (8.10) gives the efficiency as a function of particle size. Substituting the value of qf calculated earlier,

$$\eta(D_p) = 2\left(1 - \sqrt{1 - 0.000833 D_p^2}\right)$$

where D_p is in μm. Figure 8.4 is a plot of this equation.

8.2.2 Ideal, Turbulent Cyclonic Flow

Figure 8.5 shows the model of the turbulent flow cyclone separator. Assume that the effect of the turbulent eddies is to distribute the particles uniformly over the cross section, not

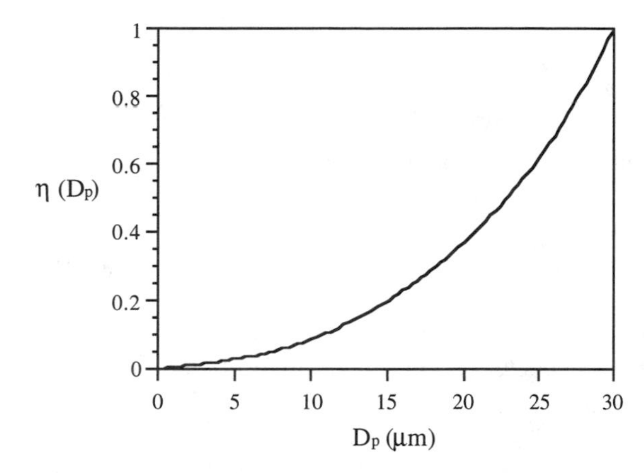

Figure 8.4 Efficiency curve for Example 8.1

only at the entrance, but at any given angle θ. This is a conservative assumption because the centrifugal force effects may serve to damp out the turbulent eddies which naturally occur in turbulent duct flow. Relatively little is known about this area.

Consider the effect of a laminar layer next to the outer edge of the cyclone, as shown in Figure 8.5. Once a particle, vigorously mixed in the core of the flow, enters this layer, it travels to the outer wall and is removed. The distance the particle travels in the θ direction in the laminar sublayer over a time interval dt is $u_{\theta 2}\, dt = r_2\, d\theta$, where $u_{\theta 2}$ is evaluated at $r = r_2$. The thickness of the laminar sublayer is

$$dr = u_{r2}\, dt = u_{r2} \frac{r_2\, d\theta}{u_{\theta 2}} \tag{8.12}$$

where u_{r2} is also evaluated at $r = r_2$. The fractional diminution of particles over the angle $d\theta$ is the fraction of the particles which lies in the boundary layer, or

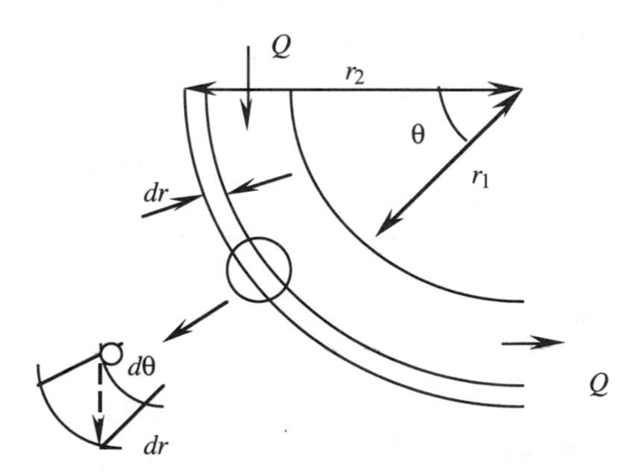

Figure 8.5 Turbulent cyclonic flow

$$-\frac{dn}{n} = \frac{dr}{r_2 - r_1} = \frac{u_{r2}}{u_{\theta 2}} \frac{r_2}{r_2 - r_1} d\theta \tag{8.13}$$

Integrating this equation between the entrance ($\theta = 0$) and any angular position,

$$n = n_0 \exp\left(-\frac{u_{r2}}{u_{\theta 2}} \frac{r_2}{r_2 - r_1} \theta\right) \tag{8.14}$$

where n_0 is the initial number of particles of diameter D_p per unit volume of gas. The collection efficiency of a cyclone that has an angle θ_f is

$$\eta(D_p) = 1 - \frac{n(\theta_f)}{n_0} = 1 - \exp\left[-\frac{u_{r2} r_2 \theta_f}{u_{\theta 2}(r_2 - r_1)}\right] \tag{8.15}$$

For lack of a better approximation, use the inviscid gas velocity components given by Eq. (8.1) to represent the fluid velocity field in the turbulent flow cyclone. Therefore,

$$u_{\theta 2} = \frac{Q}{W r_2 \ln\left(\frac{r_2}{r_1}\right)} \tag{8.16}$$

$$u_{r2} = \frac{\rho_p Q^2 D_p^2}{18\mu r_2{}^3 W^2 (\ln r_2/r_1)^2} \tag{8.17}$$

and the collection efficiency of the cyclone is:

$$\eta(D_p) = 1 - \exp\left[-\frac{\rho_p Q D_p{}^2 \theta_f}{18\mu r_2 W(r_2 - r_1)\ln(r_2/r_1)}\right] \tag{8.18}$$

Example 8.2 Ideal, Turbulent Cyclonic Flow

(a) Consider the data of Example 8.1. Estimate the collection efficiency for 30-μm particles assuming that the flow is turbulent and the cyclone angle is 2.041 rad.

(b) Determine the angle of turn to obtain 99% collection efficiency for the 30-μm particle.

Solution

(a) Through direct substitution in Eq. (8.18):

$$\eta\ (30\text{-}\mu m) = 0.528\ (52.8\%)$$

(b) Solve Eq. (8.18) for the angle of turn, θ_f, for a given efficiency.

$$\theta_f = -\frac{18\mu r_2 W(r_2 - r_1)\ln(r_2/r_1)\ln(1-\eta)}{\rho_p Q D_p^2}$$

Substituting numerical values, the angle for 99% collection efficiency of the 30-μm particles is $\theta_f = 12.53$ rad = 2 full turns.

Comments

A comparison of the results of Examples 8.1 and 8.2 shows that turbulence reduces dramatically the efficiency of a cyclone. A precise criterion for transition from laminar to turbulent flow in a cyclone does not exist (Flagan and Seinfeld 1988). Experimentally determined collection efficiency curves appear to conform more closely to turbulent than to laminar flow conditions.

8.2.3 Practical Cyclone Design Equation

The flow pattern in a cyclonic device is a complex one and the two models presented previously represent extremes in performance. Because operating cyclones do not conform to either of these limiting conditions, semiempirical design equations predict their performance.

Leith and Licht (Licht 1980) developed a theory useful in practical cyclone design. Alexander (1949) found experimentally that the exponent in the fluid tangential velocity profile, $v_\theta r^m$ = constant, is given by:

$$m = 1 - \left(1 - 0.67 D_c^{0.14}\right)\left(\frac{T}{283}\right)^{0.3} \qquad (8.19)$$

where D_c is the cyclone body diameter in meters and T is the gas temperature in K. The collection efficiency, according to the model by Leith and Licht, is given by

$$\eta(D_p) = 1 - \exp\left(-\psi D_p^M\right) \tag{8.20}$$

where

$$M = 1/(m + 1)$$

$$\psi = 2\left[\frac{KQ\rho_p C_c(m+1)}{18\mu D_c^3}\right]^{M/2} \tag{8.21}$$

where K is a dimensionless geometric configuration parameter and C_c is the Cunningham correction factor. *Be consistent with the units in Eqs.* (8.20) *and* (8.21).

Example 8.3 Leith-Licht Model for Efficiency of Cyclone

Consider the gas stream of Examples 8.1 and 8.2. It flows through a cyclone with a body diameter of 2.0 m and a value of K, the geometric configuration parameter, of 551.3. Estimate the removal efficiency of the cyclone for 10-μm particles.

Solution

For a cyclone body diameter of 2.0 m and a gas temperature of 298 K, Eq. (8.19) yields m = 0.734, $M = 0.577$. From Eq. (8.21), $\Psi = 1041$. For 10-μm particles, Eq. (8.20) yields η = 0.742 (74.2%).

8.3 STANDARD CYCLONE CONFIGURATIONS

Your objectives in studying this section are to

1. Understand the concept of a cyclone of standard proportions.
2. Calculate the efficiency of various standard cyclone configurations.
3. Estimate the pressure drop through cyclones of standard proportions.

Extensive work has been done to determine in what manner the relative dimensions of cyclones affect their performance. A number of configurations have been proposed and studied sufficiently to be regarded as "standards." In these *standard cyclone configurations* all dimensions are related to the cyclone body diameter. Table 8.1 presents the dimension ratios of these, along with values of the geometric configuration parameter, K, and a constant, N_H, relating the pressure drop through the cyclone to the inlet velocity head. Figure 8.6 illustrates the various dimensions in Table 8.1.

Table 8.1 Standard Cyclone Configurations

Term	Description	Stairmand[a]	Swift[b]	Lapple[c]
$K_a = a/D_c$	inlet height	0.5	0.44	0.5
$K_b = b/D_c$	inlet width	0.2	0.21	0.25
S/D_c	outlet length	0.5	0.5	0.625
D_e/D_c	outlet diameter	0.5	0.4	0.5
h/D_c	cylinder height	1.5	1.4	2.0
H/D_c	overall height	4.0	3.9	4.0
B/D_c	dust outlet	0.375	0.4	0.25
N_H	Eq. (8.22)	6.4	9.24	8.0
K	Eq. (8.21)	551.3	699.2	402.9

[a]Stairmand (1951) [b]Swift (1969) [c]Shepherd and Lapple (1939).

The other major consideration in cyclone specification, besides collection efficiency, is pressure drop. While forcing the gas through the cyclone at higher velocities results in improved removal efficiencies, to do so increases the pressure drop and the operating costs. There is ultimately an economic trade-off between efficiency and operating cost. Several methods have been proposed to estimate the total pressure drop in the flow of gas through a standard cyclone. Unfortunately, there is no definitive study to determine which is the best to use. Most methods agree to express the pressure drop in terms of a multiple of the inlet velocity head, or

$$\Delta P = N_H \rho_f v_E^2/2 \tag{8.22}$$

where

N_H is a constant which depends on the cyclone configuration (see Table 8.1)

ρ_f is the gas density

v_E is the gas velocity in the cyclone inlet duct = Q/ab.

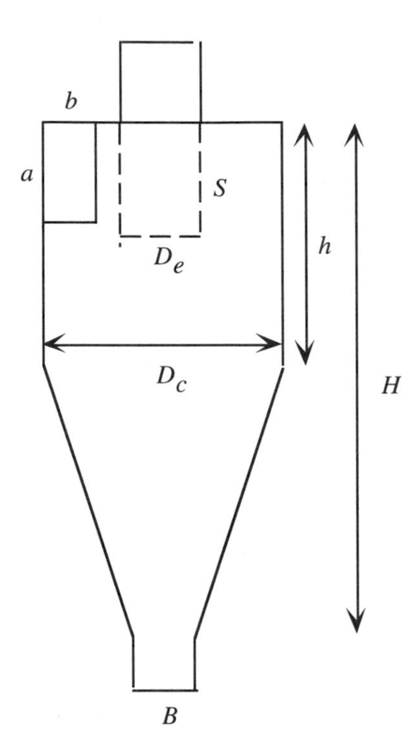

Figure 8.6 Dimensions of a standard cyclone

Equation (8.22) can be rewritten in terms of the cyclone body diameter and the gas volumetric flow rate.

$$\Delta P = \frac{N_H \rho_f Q^2}{2 K_a^2 K_b^2 D_c^4} \tag{8.23}$$

where

$K_a = a/D_c$
$K_b = b/D_c$

It is evident from Eq. (8.23) that the pressure drop is extremely sensitive to the cyclone body diameter, increasing rapidly as the device becomes smaller. Notice in Table 8.1 that, for a given set of operating conditions and body diameter, Swift standard configuration is more efficient (higher value of K), but results in a higher pressure drop (higher value of N_H) . Lapple configuration, with a relatively high pressure drop, is not nearly as efficient as the other two.

Example 8.4 Pressure Drop through Cyclone of Standard Configuration

(a) Estimate the pressure drop for the conditions of Example 8.3. According to the value of K in that example, the cyclone proportions correspond to Stairmand configuration.
(b) If the cyclone body diameter decreases to 1.0 m, estimate the removal efficiency for 30-μm particles, and the corresponding pressure drop.

Solution

a) From Table 8.1, for Stairmand configuration, $N_H = 6.4$, $K_a = 0.5$, $K_b = 0.2$. The ideal gas law gives the fluid density, $\rho_f = 1.186$ kg/m^3. The cyclone body diameter, D_c, is 2.0 m, the gas volumetric flow rate is 5 m^3/s. From Eq. (8.23), $\Delta P = 593$ Pa.
(b) For a body diameter of 1.0 m, and Stairmand standard configuration, $m = 0.665$, $M = 0.6$, $\Psi = 2{,}491$, $\eta(30\ \mu m) = 0.992$. Because the body diameter is twice as small as that of the cyclone of part a, the pressure drop is 16 times higher. Therefore, $\Delta P = 9{,}500$ Pa.

Comments

Usually, the pressure drop is the limiting factor in the design of cyclones. To maintain the pressure drop within acceptable levels, the removal efficiency for small particles must remain relatively low.

8.4 SIZE DISTRIBUTION OF PENETRATING PARTICLES

Your objectives in studying this section are to

1. Develop an equation relating the cumulative mass distribution function of the particles penetrating a cyclone to the inlet conditions and the device grade efficiency function.
2. Estimate the resulting integrals with Gauss-Legendre quadrature formulas.
3. Estimate MMD and σ_g for the aerosol that penetrates the cyclone.

Cyclones are frequently used as precleaners for more sophisticated particulate control devices. The size distribution characteristics of the aerosol population entering the cyclone are known, but they must be calculated for those particles penetrating the device.

The size distribution function of the particles penetrating the cyclone, by cumulative mass fraction less than size, $Go(D_p)$, may be determined by a material balance taken over all particles finer than a given size D_p as shown in Figure 8.7. The mass flow rate of particles finer than $D_p$entering the cyclone is given by $Qc_iG_i(D_p)$, where

Q = volumetric flow rate of the gas
c_i = mass of total particulate matter per unit volume of entering gas
$G_i\,(D_p)$ = cumulative mass fraction of particles finer than D_p at the entrance

The mass flow rate of particles finer than D_p penetrating the cyclone is $Qc_iPt_MG_o(D_p)$, where

Pt_M = overall penetration = 1 - η_M
$G_o(D_p)$ = cumulative mass fraction of particles finer than D_p at the outlet

The mass flow rate of particles finer than D_p in the collected dust is given by

$$Qc_i\int_0^{D_p} \eta(D_p')n_{mi}(D_p')d(D_p')$$

where

$n_{mi}(D_p')d(D_p')$ = the mass fraction of particles with diameters around D_p' at the inlet of the cyclone

For a log-normal size distribution at the cyclone inlet,

$$n_{mi}(D_p') = \frac{1}{\sqrt{2\pi}\,D_p'\ln\sigma_{gi}}\exp\left[-\frac{(\ln D_p' - \ln \text{MMD}_i)^2}{2(\ln\sigma_{gi})^2}\right] \tag{8.24}$$

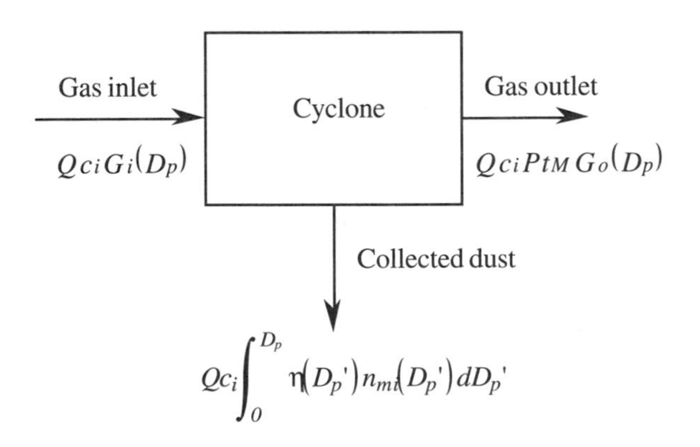

Figure 8.7 Material balance on particles finer than D_p

where MMD_i and σ_{gi} are characteristic of the inlet aerosol population.

Assuming that the size of the individual particles do not change as they flow through the cyclone, a material balance yields

$$Qc_i Pt_M G_o(D_p) + Qc_i \int_0^{D_p} \eta(D_p')n_{mi}(D_p')dD_p' = Qc_i G_i(D_p) \tag{8.25}$$

Simplifying and rearranging Eq. (8.25) becomes

$$G_o(D_p) = \frac{G_i(D_p) - \int_0^{D_p} \eta n_{mi} d(D_p')}{Pt_M} \tag{8.26}$$

Substituting in Eq. (8.26) the definition of $G_i(D_p)$,

$$G_o(D_p) = \frac{\int_0^{D_p} (1-\eta) n_{mi} dD_p'}{Pt_M} \tag{8.27}$$

Define the grade penetration, $Pt(D_p') = 1 - \eta(D_p')$. Then, Eq. (8.27) becomes

$$G_o(D_p) = \frac{\int_0^{D_p} Pt(D_p')n_{mi}(D_p')dD_p'}{Pt_M} \tag{8.28}$$

For a log-normal distribution, the integral in Eq. (8.28) can not be evaluated in closed form. Define a new variable, x, such that

$$x = \frac{2D_p'}{D_p} - 1 \tag{8.29}$$

Then,

$$\int_0^{D_p} Pt(D_p')n_{mi}(D_p')dD_p' = \frac{D_p}{2}\int_{-1}^{+1} Pt(x)n_{mi}(x)\,dx \tag{8.30}$$

The integral on the right hand side of Eq. (8.30) can be approximated numerically with a Gauss-Legendre quadrature formula:

$$\int_{-1}^{+1} Pt(x)n_{mi}(x)\,dx \cong \sum_{i=1}^{N} w_i Pt(x_i)n_{mi}(x_i) \tag{8.31}$$

where w_i and x_i are the weight factors and roots of the Nth degree Legendre polynomial. Equation (8.28) becomes

$$G_o(D_p) \cong \frac{\frac{D_p}{2}\sum_{i=1}^{N} w_i Pt(x_i)n_{mi}(x_i)}{Pt_M} \tag{8.32}$$

The overall penetration in Eq. (8.32) can be estimated with a Gauss-Hermite quadrature formula (as illustrated in Chapter 7) given MMD_i, $\sigma_{gi,}$ and the grade efficiency equation for the cyclone. The following example illustrates the computational scheme to characterize the particle size distribution function at the cyclone gas outlet.

Example 8.5 Size Distribution Function of Particles Penetrating a Cyclone

Consider Example 8.3. The aerosol population entering the cyclone is log-normally distributed with MMD = 8.0 μm and $\sigma_g = 2.5$. Estimate the overall removal efficiency for the device. Calculate the cumulative mass fraction of particles penetrating the cyclone for various particle sizes. Determine if these cumulative results follow a log-normal distribution, and, if so, estimate the corresponding MMD and σ_g values.

Solution

The grade efficiency function for the cyclone at the operating conditions of Example 8.3 is, from Eq. (8.20),

$$\eta\,(D_p') = 1 - \exp\,(-\,1{,}041\ D_p'^{\,0.577}) \qquad (8.20\text{ a})$$

Follow the computational scheme outlined in Example 7.4 to estimate the overall removal efficiency through application of Gauss-Hermite quadrature formula. The result, based on a 16-point formula, is $\eta_M = 0.686$. Therefore, the overall penetration is $Pt_M = 0.314$.

The procedure to characterize the particle size distribution penetrating the cyclone is as follows:

- Choose N, the number of quadrature points for Gauss-Legendre formula.
- Obtain the values of the roots, x_i, and weight factors, w_i, of the corresponding Nth degree Legendre polynomial, either from a mathematical table or from the computer subprogram GAULEG already provided (Section 4.3.1).
- Choose a value of D_p.
- For each of the roots calculate the corresponding D_p' from Eq. (8.29).
- From Eq. (8.20 a) calculate $Pt\,(D_p') = 1 - \eta\,(D_p')$.
- From Eq. (8.24) calculate $n_{mi}\,(D_p')$.
- From Eq. (8.32) calculate $G_o\,(D_p)$.
- Choose other values of D_p and repeat the calculations.
- Plot G_o versus D_p on log-probability paper.
- If a reasonable straight line results, estimate MMD and σ_g at the cyclone outlet.

The following table summarizes the results of the outlet cumulative mass fraction calculations with a 16-point Gauss-Legendre quadrature formula.

D_p (μm)	$G_o\,(D_p)$
1.0	0.0274
2.0	0.1355
4.0	0.3958
5.0	0.5037
7.0	0.6651
10.0	0.809
11.0	0.8404
12.0	0.8659
13.0	0.8868
15.0	0.9184

Figure 8.8 is a log-probability plot of the tabulated results. It shows that the cumulative mass fraction distribution of the particles penetrating the cyclone is log-normal, with MMD = 4.96 μm and σ_g = 2.22.

Comments

The net effects of the cyclone are to remove 68.6% of all the particulate mass, to displace the size distribution toward finer particles, and to reduce the spread in particle sizes exhibited by the original distribution.

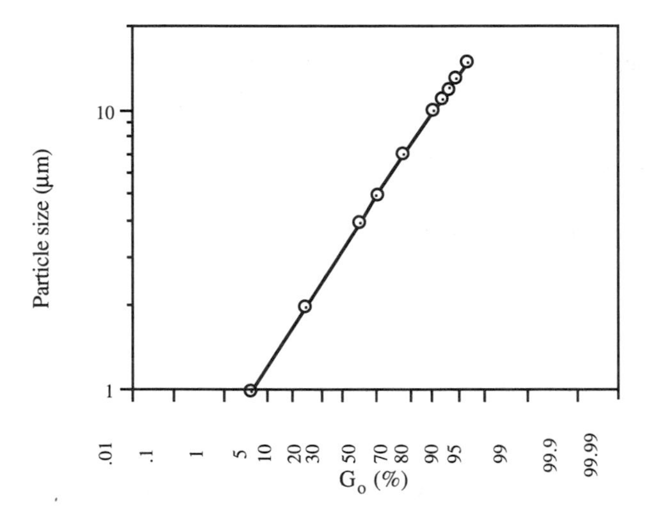

Figure 8.8 Size distribution of penetrating particles in Example 8.5

8.5 MULTIPLE CYCLONES

Your objectives in studying this section are to

1. Understand the advantage of multiple small-size cyclones operating in parallel—as compared to a single cyclone—for high particulate removal efficiency.
2. Calculate the overall removal efficiency and gas pressure drop for a multiple cyclone system.

The centrifugal force—the driving force for particulate removal in cyclones—is proportional to the square of the tangential velocity and inversely proportional to the radius of curvature of the gas trajectory. Therefore, the efficiency of a cyclone increases as the diameter of the device is reduced. To achieve higher efficiencies dictates the use of smaller cyclones. However, the pressure drop through the cyclone increases rapidly as the tangential velocity increases. A way to maintain high efficiencies with a moderate pressure drop is to use a large number of small cyclones placed in parallel. If N_c is the number of cyclones in parallel in a multiple cyclone system, the gas volumetric flow rate through each one is Q/N_c. Equation (8.20) for the grade efficiency of the system can be rewritten as (see Problem 8.4):

$$\eta(D_p) = 1 - \exp\left\{-2\left[\frac{KQ\tau}{MN_c D_c^3}\right]^{\frac{M}{2}}\right\} \tag{8.33}$$

Equation (8.23) for the pressure drop through the multiple cyclone system becomes:

$$\Delta P = \frac{N_H \rho_f Q^2}{2K_a^2 K_b^2 N_c^2 D_c^4} \tag{8.34}$$

Equation (8.34) suggests that the pressure drop decreases rapidly as the number of cyclones increases. The following example shows that a multiple cyclone system can achieve a high removal efficiency for small particles at a much lower pressure drop than a single cyclone of similar efficiency.

Example 8.6 Collection Efficiency and Pressure Drop of Multiple Cyclone System

A paper mill in Louisiana operates a power plant that produces steam for the paper making process, and enough electricity to run a city of 200,000. The plant has four boilers—two recovery boilers that burn "black liquor" from the pulp mill, and two power boilers that burn bark, coal, oil, and gas. The power boilers have a flue gas cleaning system that consists of a multiple cyclone precleaner, followed by an electrostatic precipitator. The flue gas flow rate is 165 m^3/s, measured at the actual conditions of 450 K and 1 atm. The average particle density is 1,600 kg/m^3. The precleaner consists of 2 units in parallel, each with 450 Stairmand cyclones with body diameter of 0.25 m. Assuming that the flue gases behave as air,

(a) Estimate the removal efficiency of the multiple cyclone system for 10-μm particles and the resulting pressure drop.
(b) If a single Stairmand cyclone removes 10-μm particles with the same efficiency as the multiple cyclone system of part a, estimate the corresponding pressure drop.

Solution

(a) The total number of cyclones, N_c, is 900; $D_c = 0.25$ m. From Eq. (8.19), $m = 0.485$, $M = 0.674$. Calculate the characteristic or relaxation time for 10-μm particles. The viscosity of air at 450 K is 2.48×10^{-5} kg/m-s, the Cunningham correction factor is approximately 1.0. Therefore, $\tau = 3.58 \times 10^{-4}$ s. From Eq. (8.33), η (10 μm) = 0.952 (95.2%). The pressure drop is from Eq. (8.34): $\Delta P = 2.16$ kPa.
(b) Calculate, by trial-and-error, the body diameter of a single Stairmand cyclone for 95.2% removal of 10-μm particles. The answer is $D_c = 2.37$ m. The pressure drop is from Eq. (8.23): $\Delta P = 216.8$ kPa.

Comments

This example illustrates the advantages of multiple cyclone systems over single cyclones as high efficiency precleaners. Many combinations of number of cyclones and body diameter result in the desired efficiency. The optimum combination is obtained from a cost analysis.

8.6 COST ANALYSIS FOR CYCLONES

Your objectives in studying this section are to

1. Estimate the initial and annual cost for single and multiple cyclones.
2. Design a multiple cyclone system to minimize its total annual cost.

Cyclones are very inexpensive, having capital costs at least an order of magnitude less than final control devices such as baghouses and electrostatic precipitators. Because of their simplicity, the only significant operating expense is the cost of electricity to overcome the pressure drop through the device.

8.6.1 Costs of Single Cyclones

Vatavuk (1990) gives a correlation to estimate the equipment cost of single cyclones (updated to June 1990) for a range of inlet duct areas. The cost includes a carbon steel cyclone, support stand, a fan and motor, and a hopper for collecting captured dust. The correlation is:

$$EC = 57{,}800\,[ab]^{0.903} \tag{8.35}$$

where

EC = equipment cost, in $ of June 1990
a and b are the inlet height and width, respectively
Equation (8.35) is valid in the range $0.020 \leq ab,\ \text{m}^2 \leq 0.4$

Installation costs and ductwork connections are often more expensive than the cyclone itself. The total capital investment is about twice the purchased equipment cost.

Example 8.7 TCI and TAC for Cyclone

Consider the cyclone of Examples 8.3 and 8.4a. Estimate the total capital investment and annual cost if the device operates 8,000 h/yr. Assume a useful life of 10 yr and a minimum attractive rate of return on investment of 15%. The mechanical efficiency of the motor-fan system is 65%, and the cost of electricity is $0.08/kW-h.

Solution

The cyclone is of Stairmand standard configuration with a body diameter of 2 m. The inlet area is $ab = K_a K_b\, D_c^2 = (0.5)(0.2)(2.0)^2 = 0.4\ \text{m}^2$. The equipment cost is from Eq. (8.35), EC = $25,300. Add 8% to EC to account for freight and taxes, and multiply by 2.0 to estimate the total capital investment: TCI = $54,650.

The only two significant expenses in the total annual cost calculation are the capital recovery cost and the cost of electricity. For $i = 0.15$ and $n = 10$ yr, the capital recovery factor is CRF = 0.20/yr. Therefore, the capital recovery cost is CRC = $10,930/yr. For a volumetric flow rate of the gases of 5 m^3/s and a pressure drop of 0.593 kPa, the power to operate the fan is (5)(0.593)/(0.65) = 4.56 kW. The annual cost of electricity is (4.56)(8,000)(0.08) = $2,920/yr. TAC = 10,930 + 2,920 = $13,850/yr.

8.6.2 Costs of Multiple Cyclones

The following correlation for the equipment cost of multiple cyclone systems is based on suggestions by Crawford (1976) and cost data presented by Cooper and Alley (1986), and Peters and Timmerhaus (1991):

$$\text{EC} = 7{,}000 N_c\, ab + 72 N_c \tag{8.36}$$

where

EC = equipment cost, in \$ of June 1990
N_c = number of cyclones.

Equation (8.36) is valid in the range $1.0 \le N_c ab$, $\text{m}^2 \le 6.0$

Example 8.8 TCI and TAC of a Multiple Cyclone System

Estimate the total capital investment and annual cost for the multiple cyclone system of Example 8.6a. Assume that the conditions are similar to those of Example 8.7.

Solution

For a total of 900 Stairmand cyclones with body diameter of 0.25 m, the total inlet area is $N_c ab = N_c K_a K_b D_c^2 = (900)(0.5)(0.2)(0.25)2 = 5.625\ \text{m}^2$. The equipment cost is from Eq. (8.36): EC = \$104,200. TCI = (2)(1.08)(104,200) = \$225,000.

For a capital recovery factor of 0.2/yr, the capital recovery cost is \$45,000/yr For a pressure drop of 2.16 kPa and a gas flow rate of 165 m^3/s, the power to push the gas through the system is (2.16)(165)/(0.65) = 548 kW. The annual cost of electricity is (548)(8,000)(0.08) = \$351,000/yr. TAC = \$396,000/yr.

8.6.3 Optimization of Multiple Cyclone System Design

It is possible to estimate the most economical design for a multiple cyclone system, given the required collection efficiency. The total annual cost for the system is

$$\text{TAC} = K_1 N_c ab + K_2 \dot{W} t + K_3 N_c \tag{8.37}$$

where

K_1 = the capital recovery factor times the installed cost of a cyclone of unit inlet area, \$/yr-m^2
K_2 = the cost of electric energy, in \$/kW-h
$\dot{W}$ = electric power, in kW
t = the number of hours the unit operates per year
K_3 = the capital recovery factor times the portion of the installed cost of the system that is proportional to the number of cyclones, in \$/yr.

Equation (8.37) can be rewritten as

$$\text{TAC} = K_1 N_c K_a K_b D_c^2 + \frac{K_2 t N_H \rho_f Q^3}{2E K_a^2 K_b^2 N_c^2 D_c^4} + K_3 N_c \tag{8.38}$$

where E = mechanical efficiency of the motor-blower system.

Equation (8.33), which gives the grade efficiency of the multiple cyclone system, may be written as

$$N_c D_c^3 = \frac{K Q \tau}{M\left[-\frac{1}{2}\ln(1-\eta)\right]^{2/M}} \tag{8.39}$$

The only term on the right hand side of Eq. (8.39) that depends on the system design is M. However, it is only a weak function of the cyclone body diameter and may be assumed constant without introducing significant errors in the following analysis. For a given set of operating conditions, then, the combination $N_c D_c^3$ is fixed. Equation (3.38) can be written in terms of ND3 $= N_c D_c^3$.

$$\text{TAC} = \frac{K_1 K_a K_b (\text{ND3})}{D_c} + \frac{K_2 t N_H \rho_f Q^3 D_c^2}{2E K_a^2 K_b^2 (\text{ND3})^2} + \frac{K_3 (\text{ND3})}{D_c^3} \tag{8.40}$$

Define

$K_1' = K_1 K_a K_b\,(\text{ND3})$,

$$K_2' = \frac{K_2 t N_H \rho_f Q^3}{2E K_a^2 K_b^2 (\text{ND3})^2}$$

$$K_3' = K_3(\text{ND3})$$

Then,

$$\text{TAC} = \frac{K_1'}{D_c} + K_2' D_c^2 + \frac{K_3'}{D_c^3} \tag{8.41}$$

To minimize TAC, take the derivative of Eq. (8.41) with respect to D_c and set it equal to zero.

$$\left[\frac{\partial \text{TAC}}{\partial D_c}\right]_{\text{ND3}} = -\frac{K_1'}{D_c^2} + 2K_2' D_c - \frac{3K_3'}{D_c^4} = 0 \tag{8.42}$$

Rearranging Eq. (8.42),

$$D_c^5 - \frac{K_1'}{2K_2'} D_c^2 - \frac{3K_3'}{2K_2'} = 0 \tag{8.43}$$

The optimum cyclone body diameter is the real root of Eq. (8.43).

Example 8.9 Optimal Design of Multiple Cyclone System

Determine the number and body diameter of Stairmand cyclones that minimize the TAC in Example 8.8. Your design should achieve the same efficiency for 10-μm particles, namely, 95.2%.

Solution

Assume $M = 0.7$. From Eq. (8.39), ND3 = 14.1 m^3. The capital recovery factor is 0.20/yr. Based on Eq. (8.36), $K_1 = (7{,}000)(1.08)(2)(0.2) = 3{,}024$ \$/yr-m^2. According to their definitions, $K_1' = (3{,}024)(0.5)(0.2)(14.1) = 4{,}265$ \$-m/yr, $K_2' = (0.08)(8{,}000)(6.4)(0.785)(165)^3/[(2)(0.65)(0.5)^2(0.2)^2(14.1)^2(1{,}000)] = 5.62 \times 10^6$ \$/yr-m^2, $K_3' = (72)(1.08)(2)(0.2)(14.1) = 438.5$ \$-m^3/yr. Equation (8.43) becomes

$$D_c^5 - 0.000378\, D_c^2 - 0.0001167 = 0$$

The real root of this polynomial is $D_c = 0.166$ m. The number of cyclones is $N_c =$

$14.1/(0.166)^3 = 3{,}080$. Check the value of M assumed. From Eq. (8.19), $m = 0.45$, $M = 0.69$, which is sufficiently close to the assumed value. Therefore, the optimal design is 3,080 cyclones with body diameter of 0.166 m. Equation (8.41) gives the total annual cost for this design, TAC = \$276,000/yr. That is \$120,000/yr lower than the TAC for Example 8.8!

Example 8.10 Optimal Design Based on Overall Efficiency

The aerosol population entering the multiple cyclone system of Example 8.9 is log-normally distributed with MMD = 4.0 μm and $\sigma_g = 2.5$. Determine the optimum design that achieves an overall particulate removal efficiency of 70%. Characterize the aerosol population penetrating the system.

Solution

The efficiency specified is not a grade efficiency, but an overall efficiency. The relation between ND3 and η_M is more complex than Eq. (8.39). The solution involves trial and error. The computational scheme is as follows:

- Assume a value of ND3.
- Calculate K_1', K_2', and K_3'.
- Solve Eq. (8.43) for D_c ; calculate M and $N_c = \text{ND3}/D_c^3$.
- Apply Gauss-Hermite quadrature formula to estimate the resulting overall efficiency.
- If the calculated efficiency differs from the specified efficiency by less than a specified tolerance, the design is optimal; characterize the penetrating aerosol according to the scheme presented in Example 8.5.
- Otherwise, assume a new value of ND3 and repeat the procedure.

The computer program MLTCYC implements this scheme. This program calls the subprograms RTSAFE (see Example 2.9), GAULEG (see Section 4.3.1), and HERMIT (see Example 7.4) The input data are defined as follows:

PENR = specified overall penetration, fraction
MMD = mass median diameter of initial aerosol, mm
SIGG = σ_g, dimensionless

Q = volumetric flow rate, m³/s
LAMBDA = mean free path, μm
TEMP = temperature, K

DENP = particle density, kg/m^3
VIS = gas viscosity, kg/m-s
K1P = $K_1 K_a K_b$

K2P = $K_2 t N_H \rho_f Q^3/(2EK_a^2 K_b^2)$
K3P = K_3

The optimal design for an overall efficiency of 70% is N_c = 1,310 cyclones, D_c = 0.30 m. The total annual cost is TAC = $155,700. The penetrating aerosol is log-normally distributed with MMD = 2.39 μm and σ_g = 2.16.

```
BLOCK DATA
  REAL PENR, MMD, SIGG, Q, LAMBDA, TEMP, DENP, VIS, K1P, K2P,K3P,K
  COMMON /BLOCK1/ PENR, MMD, SIGG, Q, LAMBDA, TEMP, DENP, VIS,
 * K1P, K2P, K3P,K1PP, K2PP, K3PP, K
  DATA PENR/0.30/, MMD/4.00/, SIGG/2.50/, Q/165./, LAMBDA/0.1500/,
 * TEMP/450.0/, DENP/1600.0/, VIS/2.48E-5/, K1P/302.40/,
 * K2P/1.11E9/, K3P/31.10/, K/551.3/
  END

PROGRAM MLTCYC
  REAL PENR, MMD, SIGG, K, Q, LAMBDA, TEMP, DENP
  REAL VIS, K1P, K2P, K3P, X1, X2, XACC, K1PP, K2PP, K3PP
  REAL CA, CB, DIAM, ND3, DPI, RTSAFE, RTBIS, SUMA
  REAL W, Y, G, D, TAU, ENE, PEN, Q0
  INTEGER N, NC
  LOGICAL SUCCES
  EXTERNAL FUNCD, FUNCP,FUN
  COMMON /BLOCK1/ PENR, MMD, SIGG, Q, LAMBDA, TEMP, DENP, VIS,
 * K1P, K2P, K3P,K1PP,K2PP,K3PP,K
  COMMON DIAM
  PARAMETER ( N = 10, FACTOR=1.6, NTRY = 50)
  DIMENSION Y(N), W(N)
  TAU1=(MMD*1.0E-6)**2*DENP/(18.*VIS)
  X1=1.43*TAU1*Q*K/(-0.5*LOG(PENR))**2.86
  X2=X1/SIGG
  F1 = FUNCP(X1)
  F2 = FUNCP(X2)
  SUCCES = .TRUE.
  DO 1 J =1, NTRY
    IF(F1*F2 .LT. 0) GO TO 2
    IF(ABS(F1) .LT. ABS(F2)) THEN
       X1 = X1+FACTOR*(X1-X2)
       F1=FUNCP(X1)
```

```
      ELSE
         X2=X2+FACTOR*(X2-X1)
         F2=FUNCP(X2)
      ENDIF
1    CONTINUE
     SUCCES = .FALSE.
2    IF (SUCCES) THEN
      XACC=1.0E-3
      ND3=RTBIS(FUNCP,X1,X2,XACC)
      PRINT *, 'ND3 = ', ND3
         NC=INT(ND3/(DIAM**3))
         DIAM=(ND3/NC)**(1./3.)
         TAC = K1PP/DIAM +K2PP*DIAM**2 + K3PP/DIAM**3
         PRINT *, ' NUMBER OF CYCLONES = ', NC
         PRINT *, ' DIAMETER OF EACH CYCLONE = ', DIAM
         PRINT *, ' TOTAL ANNUAL COST = ' , TAC
         PAUSE 'PRESS THE ENTER KEY TO CONTINUE'
5         PRINT *, 'ENTER PARTICLE SIZE, ENTER 0.0 TO TERMINATE '
         READ *, DPI
         IF (DPI .NE. 0.0) THEN
           Y1=0.0
           CALL GAULEG (Y1,DPI,Y,W,N)
           ENE=1.-(1.-0.67*DIAM**0.14)*(TEMP/283.)**0.3
           SUMA=0.0
           DO 10 I=1,N
            D=Y(I)
            IF (0.55*D/LAMBDA .GT. 80.) THEN
               CUN = 1. + 2.*LAMBDA/D
            ELSE
               CUN=1.+(2.*LAMBDA/D)*(1.257+0.4*EXP(-0.55*D/
    *           LAMBDA))
            ENDIF
            TAU=CUN*(1.0E-6*D)**2*DENP/(18.*VIS)
            ARGUM=2.*((K*Q*TAU*(ENE+1))/ND3)**(1./
    *           (2.*ENE+2.))
            IF (ARGUM .GT. 80.0) THEN
               PEN = 0.0
            ELSE
               PEN = EXP(-ARGUM)
            ENDIF
            Q0=(1./(D*SQRT(2.*3.1416)*LOG(SIGG)))*EXP(-(
    *           LOG(D/MMD)/(SQRT(2.)*LOG(SIGG)))**2)
            SUMA=SUMA+W(I)*PEN*Q0
10          CONTINUE
           G=SUMA/PENR
```

```
          PRINT *, 'MASS CUMULATIVE FRACTION= ', G
          PRINT *, 'FOR PARTICLE SIZE = ', DPI
          GO TO 5
        END IF
        GO TO 15
      END IF
      PRINT *, ' FAILURE IN BRACKETING OPTIMUM DIAMETER'
      GO TO 15
15    STOP
      END

      FUNCTION FUNCP(ND3)
      REAL K1P,K2P,K3P,K1PP,K2PP,K3PP,ND3,K,MMD, LAMBDA, NMD
      LOGICAL SUCCES
      EXTERNAL FUNCD, FUN
      COMMON /BLOCK1/ PENR, MMD, SIGG, Q, LAMBDA, TEMP, DENP, VIS,
     * K1P, K2P, K3P,K1PP, K2PP, K3PP,K
      PARAMETER ( N = 10)
      COMMON DIAM
      DIMENSION X(N), A(N)
        K1PP=K1P*ND3
        K2PP=K2P/ND3**2
        K3PP=K3P*ND3
        CA=K1PP/(2.*K2PP)
        CB=(3.*K3PP)/(2.*K2PP)
      X1=0.0
      X2=1.0
      XACC=1.0E-3
      CALL ZBRAC1(FUN,X1,X2,SUCCES,CA,CB)
      IF (SUCCES) THEN
        DIAM=RTSAFE(FUNCD,X1,X2,XACC,CA,CB)
        EPS=1.0E-6
        CALL HERMIT(N,X,A,EPS)
        ENE=1.-(1.-0.67*DIAM**0.14)*(TEMP/283.)**0.3
        NMD = EXP(LOG(MMD)-3*(LOG(SIGG))**2)
        SUMA=0.0
        DO 10 I=1,N
          D=EXP(SQRT(2.)*LOG(SIGG)*X(I)+LOG(NMD))
          IF (0.55*D/LAMBDA .GT. 80.0) THEN
            CUN = 1. + 2.*LAMBDA/D
          ELSE
            CUN=1.+(2.*LAMBDA/D)*(1.257+0.4*EXP(-0.55*D/
     *         LAMBDA))
          ENDIF
          TAU=CUN*(1.0E-6*D)**2*DENP/(18.*VIS)
          ARGUM=(2.*((K*Q*TAU*(ENE+1))/ND3)**(1./
```

```
     *          (2.*ENE+2.)))
           IF (ARGUM .GT. 80.0) THEN
              PEN = 0.0
           ELSE
              PEN=EXP(-ARGUM)
           ENDIF
           SUMA=SUMA+A(I)*PEN*EXP(3.*SQRT(2.)*LOG(SIGG)*X(I))
 10     CONTINUE
         PENC=SUMA/(SQRT(3.1416)*EXP(4.5*(LOG(SIGG))**2))
         ABC=PENC-PENR
         FUNCP=ABC
       ENDIF
       END

       SUBROUTINE FUNCD(X,FN,DF,CA,CB)
       FN=X**5-CA*X**2-CB
       DF=5.*X**4-2.*CA*X
       RETURN
       END

       FUNCTION FUN(X,CA,CB)
       FUN=X**5-CA*X**2-CB
       RETURN
       END

     FUNCTION RTBIS(FUNCP,X1,X2,XACC)
       © 1986 by Numerical Recipes Software. Reprinted with permission from Numerical
       Recipes: The Art of  Scientific Computing, Cambridge University Press, New York (1986).
       PARAMETER (JMAX=40)
       FMID=FUNCP(X2)
       F=FUNCP(X1)
       IF(F*FMID.GE.0.) PAUSE 'Root must be bracketed for bisection.'
       IF(F.LT.0.)THEN
        RTBIS=X1
        DX=X2-X1
       ELSE
        RTBIS=X2
        DX=X1-X2
       ENDIF
       DO 11 J=1,JMAX
        DX=DX*.5
        XMID=RTBIS+DX
        FMID=FUNCP(XMID)
        IF(FMID.LE.0.)RTBIS=XMID
        IF(ABS(DX).LT.XACC .OR. FMID.EQ.0.) RETURN
```

```
11  CONTINUE
    PAUSE 'too many bisections'
    END
C
    SUBROUTINE ZBRAC1(FUN,X1,X2,SUCCES,CA,CB)
    © 1986 by Numerical Recipes Software. Reprinted with permission from Numerical
    Recipes: The Art of Scientific Computing, Cambridge University Press, New York (1986).
    EXTERNAL FUN
    PARAMETER (FACTOR=1.6,NTRY=50)
    LOGICAL SUCCES
    IF(X1.EQ.X2)PAUSE 'You have to guess an initial range'
    F1=FUN(X1,CA,CB)
    F2=FUN(X2,CA,CB)
    SUCCES=.TRUE.
    DO 11 J=1,NTRY
     IF(F1*F2.LT.0.)RETURN
     IF(ABS(F1).LT.ABS(F2))THEN
      X1=X1+FACTOR*(X1-X2)
      F1=FUN(X1,CA,CB)
     ELSE
      X2=X2+FACTOR*(X2-X1)
      F2=FUN(X2,CA,CB)
     ENDIF
11   CONTINUE
    SUCCES=.FALSE.
    RETURN
    END
```

8.7 CONCLUSION

Cyclones are very simple devices for particulate removal. They are effective to remove relatively big particles from waste gas streams. The pressure drop as the gas flows through the cyclone is the main factor limiting the removal efficiency achieved. Multiple cyclone systems exhibit high removal efficiencies at moderate pressure drops. Careful selection of the number and size of cyclones operating in parallel in a multiple cyclone system can result in significant annual savings. When these devices are used as pre-cleaners for more sophisticated particulate control equipment, such as electrostatic precipitators or fabric filters, the size distribution of the aerosol penetrating the cyclonic device must be characterized. The next chapter shows how the design of an electrostatic precipitator depends on the entering aerosol population characteristics.

REFERENCES

Alexander, R. M. *Proc. Austral. Inst. Min. and Met. (N.S.),* **152**:202 (1949).

Cooper, C. D., and Alley, F. C. *Air Pollution Control: A Design Approac*h, PWS Engineering, Boston, MA (1986).

Crawford, M. *Air Pollution Control Theory*, McGraw-Hill New York (1976).

Flagan, R. C., and Seinfeld, J. H. *Fundamentals of Air Pollutio*n *Engineering*, Prentice Hall, Englewood Cliffs, NJ (1988).

Licht, W. *Air Pollution Control Engineering, Basic Calculations for Particulate Collection*, Marcel Dekker, New York (1980).

Peters, M. S., and Timmerhaus, K. D. *Plant Design and Economics f*or *Chemical Engineer*s, 4th ed., McGraw-Hill, NewYork (1991).

Shepherd, G. B., and Lapple, C. E. *Ind. Eng. Chem.*, **31**:972 (1939).

Stairmand, C. J. *Trans. Inst. Chem. Engrs.*, **29**:356 (1951).

Swift, P. *Steam Heating Eng'g.*, **38**:453 (1969).

Vatavuk, W. M. *Estimating Costs of Air Pollution Control*, Lewis, Chelsea, MI (1990).

PROBLEMS

The problems at the end of each chapter have been grouped into four classes (designated by a superscript after the problem number)

Class a: Illustrates direct numerical application of the formulas in the text.
Class b: Requires elementary analysis of physical situations, based on the subject material in the chapter.
Class c: Requires somewhat more mature analysis.
Class d: Requires computer solution.

8.1[a]. Ideal, laminar cyclonic flow

A stream of 15 m^3/s of air at 298 K and 1 atm flows in laminar cyclonic flow through a duct where the radii are 0.5 m and 1.0 m, and the height is 2.0 m. What angle of turn is necessary to collect 20-μm particles with perfect efficiency? The particle density is 2,000 kg/m^3.

Answer: 14.35 rad

8.2[a]. Ideal, turbulent cyclonic flow

Repeat Problem 8.1 assuming turbulent flow. What angle of turn is necessary to collect 20-μm particles with 99% efficiency?

Answer: 88 rad

8.3[b]. Collection efficiency for a cyclone of standard proportions

A cyclone with a body diameter of 1.0 m and with Stairmand standard proportions processes air at 298 K and 1 atm, which carries particles with a density of 1,000 kg/m^3.

The gas velocity in the entrance duct is 20.0 m/s. The number of turns that the gas makes while flowing through a cyclone of standard proportions, N_t, is given by

$$N_t = \frac{H + h}{2a}$$

and $\theta_f = 2\pi N_t$ rad. Estimate the collection efficiency for 10-μm particles assuming

(a) Ideal, laminar cyclonic flow

Answer: 56%

b) Ideal, turbulent cyclonic flow

Answer: 38.2%

c) Leith-Licht semiempirical model applies

Answer: 81.3%

8.4b. Alternative form of Leith-Licht model

Show that, for a multiple cyclone system, Eq. (8.20) can be rewritten as

$$\eta(D_p) = 1 - \exp\left\{-2\left[\frac{KQ\tau}{MN_cD_c^3}\right]^{\frac{M}{2}}\right\}$$

where

N_c = the number of cyclones

τ = relaxation time for a particle of diameter D_p, which is given by

$$\tau = \frac{D_p^2 \rho_p C_c}{18\mu}$$

8.5b. Design of a cyclone of standard proportions

Design a cyclone to remove 80% of particles of 20 μm diameter and density 1,500 kg/m^3 from a stream of 20 m^3/s of air at 298 K and 1 atm. Determine suitable values for the major dimensions of the cyclone, assuming Swift standard configuration has been chosen.

Answer: H = *18.25 m*

8.6[b]. Design of a cyclone of standard proportions

Repeat Problem 8.5 for Lapple standard configuration.

Answer: h = *7.71 m*

8.7[d]. Overall efficiency of a cyclone of standard proportions

Estimate the overall efficiency for the conditions of Problem 8.5 if the particles are log-normally distributed with MMD = 15 μm and $\sigma_g = 2.0$. Approximate the integral with a 16-point Gauss-Hermite quadrature formula.

Answer: 73.8%

8.8[a]. Pressure drop through a cyclone

Estimate the pressure drop and the power to overcome it if the mechanical efficiency of the motor-blower system is 60%

(a) For Problem 8.5

Answer: 17.8 kW

(b) For Problem 8.6

Answer: 18.3 kW

8.9[d]. Pressure drop and overall efficiency

A stream of air at 298 K and 1 atm flows at the rate of 10 m^3/s and carries with it particles with a density of 2000 kg/m3. The particles are log-normally distributed with MMD = 10 μm and $\sigma_g = 2.5$. You must design a cyclone of Swift standard configuration to serve as a precleaner for this stream. A blower is available with a motor rated at 20 kW and a mechanical efficiency of 65%. Incorporating this blower in your design, specify the cyclone body diameter and the overall efficiency achievable.

Answer: 76.1%

8.10[b]. Saltation effect

Equations (8.20) and (8.21) suggest that as the size of a cyclone decreases, which means higher inlet velocities, the grade efficiency will continue to increase and approach 100% in the limit. However, it is known that this is not so. There is a limit to the inlet velocity above which further increase results in a decrease in collection efficiency. This is due to the reentrainment of particles by the saltation effect described by Kalen and Zenz (Kalen, B. and Zenz, F.A. *A.I.Ch.E. Sympos. Ser.*, **70**:388, 1974). Licht (1980) proposed the following empirical correlation to estimate the inlet velocity, v_M, that results in the maximum cyclone collection efficiency, just before the saltation effect becomes important:

$$v_M = 3{,}025 \frac{\mu \rho_p}{\rho_f^2} \frac{K_b^{1.2}}{1 - K_b} D_c^{0.201}$$

Consider the multiple cyclone system of Example 8.6. Estimate v_M for those conditions, and compare it to the actual inlet velocity.

Answer: v_M *= 26.7 m/s*

8.11[d]. Overall efficiency of a multiple cyclone system.

The aerosol population entering the multiple cyclone system of Example 8.6 is log-normally distributed with MMD = 4.0 μm and $\sigma_g = 2.5$. The particulate loading at the entrance of the system is 0.028 kg/m^3. Estimate the rate at which the collected dust must be removed from the system, in kg/d. Neglect the Cunningham correction factor.

Answer: 308,600 kg/d

8.12[c]. Effects of hopper evacuation on the performance of multiple cyclone systems

Tucker et al. (*JAPCA* **39**:1614, 1989) showed that the collection efficiency of multiple cyclone systems increases when a small amount of the gas flow (typically less than 15%) is withdrawn from the dust hopper. This hopper evacuation flow then goes through a small baghouse before joining the remainder of the system exhaust flow and going out the stack. They demonstrated the concept experimentally in a system containing 10 cyclones of 0.25-m body diameter and 3.61-m overall body height. The air flow rate was 52 m^3/min at 298 K and 101.3 kPa. The particulate matter was spherical glass beads with a density of 2,200 kg/m^3. The size distribution function was approximately log-normal with MMD = 12 μm and $\sigma_g = 2.0$. With no hopper evacuation, the measured overall efficiency was 89.5%. With 14% hopper evacuation, the overall efficiency increased to 95%. The authors suggest that the improved performance with hopper evacuation is due to a higher gas velocity and a more uniform gas distribution through the individual cyclones.

(a) Estimate the value of the geometric configuration parameter, K, for the cyclones in the experimental work described and compare it to the Swift and Stairmand standard configurations.

(b) If the improved performance was due solely to the effect of higher gas velocities, estimate the percent increase in gas velocity to explain the observed efficiency of 95%.

Answer: 190%

8.13[a]. Cost of a single cyclone

Consider the cyclone of Problem 8.9. Calculate the total capital investment and total annual cost. The useful life of the cyclone is 5 yr and the minimum attractive return on investment is 20%/yr. The cyclone operates 8,000 h/yr; the cost of electricity is $0.06/kW-h.

Answer: TAC = $37,800

8.14[b]. Optimization of multiple cyclone system design

Repeat Example 8.9, but for Swift standard configuration. Calculate the number of cyclones and body diameter that minimize the total annual cost.

Answer: TAC = $312,000/yr

8.15[b] Effect of the cost of electricity on multiple cyclone system design

Repeat Example 8.9, but for a cost of electricity of $0.04/kW-h. Calculate the number of cyclones and body diameter that minimize the TAC.

Answer: $D_c = 0.192$ m

8.16[d]. Optimization of multiple cyclone system based on overall efficiency

Repeat Example 8.10 for cyclones of Swift standard configuration. Calculate the number and size of cyclones that minimize TAC. Estimate TAC, and MMD and σ_g for the aerosol penetrating the system.

Answer: TAC = $175,800

8.17[d] Multiple cyclone system for a Portland cement kiln

The cement dust from a Portland cement kiln is log-normally distributed with MMD = 12 μm and σ_g = 3.08, having a density of 1,500 kg/m^3 (Licht 1980). The waste gas from the kiln flows at the rate of 377 m^3/min, at 394.4 K and 101.3 kPa, and its properties are similar to those of air. The particulate loading of the waste gas is 0.023 kg/m^3. To satisfy the corresponding NSPS, 99.8% of the particulate matter in the gas must be removed. A multiple cyclone system, followed by a fabric filter or an ESP is being considered for this purpose.

Design the multiple cyclone system for overall efficiencies of 80%, 85%, and 90%. Specify, for each case, which standard configuration to use, the size and number of cyclones, TCI, and TAC. Characterize the size distribution of the penetrating aerosol, and calculate the particulate loading at the outlet. The following data apply:

- Useful life, 10 yr; no salvage value
- Minimum attractive return on investment, 20%
- 8,400 h/yr of operation
- Cost of electricity, $0.08/kW-h

9

Electrostatic Precipitators

9.1 INTRODUCTION

An electrostatic precipitator (ESP) is a particle control device that uses electrical forces to move the particles out of the flowing gas stream and onto collector electrodes. They range in size from those installed to clean the flue gases from the largest power plants to those used as small household air cleaners. ESPs account for about 95% of all utility particulate controls in the U.S. (Offen and Altman 1991). The particles are given an electrical charge by forcing them to pass through a corona, a region in which gaseous ions flow. The electrical field that forces the charged particles to the walls comes from discharge electrodes maintained at high voltage in the center of the flow lane.

Once the particles are on the collecting electrodes, they must be removed from the surface without reentraining them into the gas stream. This is usually accomplished by knocking them loose from the plates, allowing the collected layer of particles to slide down into a hopper from which they are evacuated. Some ESPs remove the particles by intermittent or continuous washing with water. Precipitators are unique among particulate matter control devices in that the forces of collection act only on the particles and not on the entire gaseous stream. This phenomenon typically results in a high collection efficiency (above 99.5%) with a very low gas pressure drop.

9.2 TYPES OF ESPS

Your objectives in studying this section are to

1. Identify the most common ESP configurations: (1) plate-wire, (2) flat plate, (3) tubular, (4) wet, and (5) two-stage.
2. Describe the operating characteristics and advantages of each configuration.
3. Understand the importance of reentrainment, gas sneakage, and dust resistivity.

ESPs are configured in several ways. Some of these configurations have been developed for special control actions, while others have evolved for economic reasons. A description of the most common configurations follows.

9.2.1 Plate-Wire Precipitators

This configuration is used in a wide variety of industrial applications including coal-fired boilers, cement kilns, solid waste incinerators, paper mill recovery boilers, petroleum refining catalytic cracking units, sinter plants, open hearth furnaces, coke oven batteries, and glass furnaces (Turner et al. 1988 a).

In a plate-wire ESP, gas flows between parallel plates of sheet metal. Weighted high-voltage wire electrodes hang between the plates as shown schematically in Figure 9.1. Within each flow path, the gas must pass each wire sequentially as it flows through the unit.

The plate-wire ESP allows many flow lanes to operate in parallel, and each lane can be quite tall. Therefore, this type of precipitator is well suited for handling large volumes of gas. The need for rapping the plates to dislodge the collected material causes the plates to be divided into sections, often three or four in series, which can be rapped independently. The power supplies are usually sectionalized in the same manner to obtain higher operating voltages.

The power supplies for the ESP convert the industrial AC voltage (220–440 V) to pulsating DC voltage in the range of 20 to 100 kV as needed. The voltage applied to the electrodes causes the gas between the plates to break down electrically, an action known as a "corona." The electrodes usually are given a negative polarity because a negative corona supports a higher voltage than a positive corona before sparking occurs. The ions generated in the corona follow electric field lines from the wires to the collecting plates. Therefore, each wire establishes a charging zone through which the particles must pass.

Particles passing through the charging zone absorb some of the ions. Small aerosol particles (less than 1-μm diameter) can absorb tens of ions before their total charge becomes large enough to repel further ions. Large particles (more than 10-μm diameter) can absorb tens of thousands of ions. The electrical forces are therefore much stronger on the large particles.

As the particles pass each successive wire, they are driven closer and closer to the collecting walls. The turbulence in the gas, however, tends to keep them uniformly mixed with the gas. The collection process is a competition between the electrical and dispersive forces. Eventually the particles approach close enough to the walls so that the turbulence drops to low levels and the particles are collected.

Reentrainment of the particles during rapping reduces the efficiency of the ESP. The rapping that dislodges the accumulated dust also projects some of the particles (typically 12% for coal fly-ash) back into the gas stream. These reentrained particles are then processed again by later sections, but the particles reentrained in the last section of the ESP have no chance to be recaptured and so escape the unit. Part of the gas flows around the charging zone into the space provided to support and align the electrodes. This is called "sneakage" and amounts to 5% to 10% of the total flow. Antisneakage baffles force the sneakage flow to mix with the main gas stream

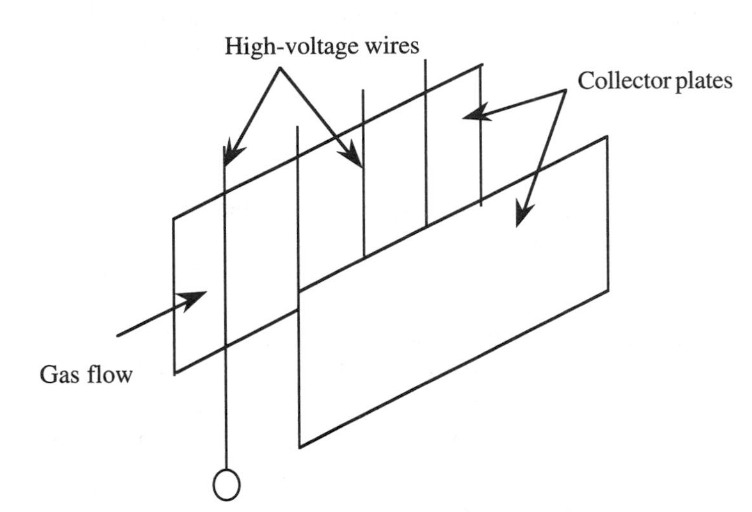

Figure 9.1 Schematic diagram of a plate-wire ESP

for collection in later sections. But, again, the sneakage flow around the last section has no opportunity to be collected.

Another major factor that affects the performance of ESPs is the resistivity of the collected material. Because the particles form a continuous layer on the ESP plates, all the ion current must pass through the layer to reach the ground plates. This creates an electric field in the layer which can become large enough to cause local electrical breakdown called "back corona." Back corona is prevalent when the resistivity of the layer is high, usually above 2×10^{11} ohm-cm. It reduces the collection ability of the unit because severe back corona causes difficulty in charging the particles. Conversely, at resistivities below 10^8 ohm-cm, the particles hold on to the plates so loosely that reentrainment becomes much worse. The resistivity is strongly affected by temperature, moisture, gas composition, particle composition, and surface characteristics.

9.2.2 Flat Plate Precipitators

A significant number of smaller ESPs (50–100 m^3/s) use flat plates instead of wires for the high-voltage electrodes. The flat plates increase the average electric field and the surface area available for the collection of particles. Flat plates can not generate corona by themselves, so corona generating electrodes are placed ahead of and sometimes after the discharge plate zone. These electrodes may be sharp-pointed needles attached to the edges of the plates or independent corona wires. Unlike plate-wire or tubular ESPs, this design operates equally well with either positive or negative polarity. A positive polarity reduces ozone generation.

A flat plate ESP operates with little or no corona current flowing through the collected dust, except directly under the corona needles or wires. This makes the unit less susceptible to back corona than conventional precipitators. However, the lack of current in the collected layer causes an electrical force that tends to dislodge the layer from the collecting electrode leading to high reentrainment losses during rapping. The dislodging electrical force is stronger for large particles.

Flat plate ESPs have wide application for high-resistivity particles with small MMDs (1–2 μm). Fly-ash has been successfully collected with this type of precipitator, but low-flow velocity appears to be critical for avoiding high rapping losses.

9.2.3 Tubular Precipitators

The original ESPs were tubular, with the high-voltage wire running along the axis of the tube. Today they comprise only a small portion of the precipitator population and are most commonly applied where the particulate is either wet or sticky—for example, in sulfuric acid plants. There are no sneakage paths and, because they are usually cleaned with water, reentrainment losses are much lower.

9.2.4 Wet Precipitators

Any of the precipitator configurations described earlier may be operated with wet walls. The water flow may be applied intermittently or continuously to wash the collected particles into a sump for disposal. This type of ESPs has no problems with rapping reentrainment or with back corona. The wash increases the complexity of the device, and the collected slurry must be handled more carefully than a dry product adding to the expense of final disposal.

9.2.5 Two-Stage Precipitators

This configuration separates particle charging and collecting functions to optimize the electrical conditions for each. Charging requires a high current density and electric field, whereas collection requires high electrical fields but much less current. The two-stage ESP is a series device with the discharge electrode, or ionizer, preceding the collector electrodes. Advantages for this configuration include more time for particle charging, less propensity for back corona with high-resistivity ash, and economical construction for small sizes.

This type of precipitator is usually applied for gas flow rates of 25 m^3/s or less. The smaller devices are often sold as preengineered package systems consisting of a mechanical prefilter, ionizer, collecting-plate cell, after-filter, and power pack. Recent work suggests that the addition of a precharger to a conventional precipitator is an economical way to convert it from single- to two-stage operation when handling high-resistivity ash (Offen and Altman 1991).

9.3 ELECTROSTATIC PRECIPITATION THEORY

Your objectives in studying this section are to

1. Develop the grade-efficiency equation for an ESP.
2. Estimate the average operating electric field strength for an ESP.
3. Estimate the total charge on a particle by diffusion and field charging.
4. Estimate grade efficiencies of an ESP with Feldman's model.

The theory of ESP operation requires many scientific disciplines to describe it thoroughly. The ESP is basically an electrical machine. The principal actions are the charging of particles and forcing them to the collector plates. The transport of the particles is affected by the level of turbulence in the gas. The particle properties also have a major effect on the operation of the ESP.

9.3.1 Particle collection

The electric field in the collecting zone produces a force on a particle proportional to the magnitude of the field and to the particle charge:

$$F_e = q_p E_c \tag{9.1}$$

where

F_e = force due to the electric field, newtons (N)
q_p = charge on the particle, coulombs (C)
E_c = strength of the electric field in the collecting zone, volt/meter (V/m)

The motion of the particles under the influence of the electric field is opposed by the viscous drag force of the gas. When the drag force exactly balances the electrostatic force, the particle attains its terminal velocity, u_t. Assuming that the particle obeys Stokes's law,

$$u_t = \frac{q_p E_c C_c}{3\pi\mu D_p} \tag{9.2}$$

where

C_c = Cunningham correction factor
D_p = particle diameter

Example 9.1 Terminal Velocity Of Charged Particle In Electric Field

Dust particles of 1.0-μm diameter with an electric charge of 3×10^{-16} C and a density of 1,000 kg/m^3 come under the influence of an electric field with a strength of 100,000 V/m. The particles are suspended in air at 298 K and 1 atm.
(a) Estimate the terminal velocity of the particles.
(b) Calculate the ratio of the electrostatic force to the force of gravity on the particle.

Solution

(a) For 1.0-μm diameter particles in air at 298 K and 1 atm, the Cunningham correction factor is $C_c = 1.168$. From Eq. (9.2), $u_t = (3 \times 10^{-16})\,(10^5)\,(1.168)/[(3\pi)(1.84 \times 10^{-5})(10^{-6})] =$ 0.202 m/s

(b) The ratio of electrostatic to gravity force acting on the particle is:

$$\frac{F_e}{F_g} = \frac{6 q_p E_c}{\pi \rho_p g D_p^3} = \frac{(6)\left(3 \times 10^{-16}\right)\left(10^5\right)}{\pi\,(1000)(9.8)\left(10^{-18}\right)} = 5{,}847$$

Equation (9.2) gives the particle velocity with respect to still air. In the ESP, the flow is usually very turbulent, with instantaneous gas velocities of the same magnitude as the particles' terminal velocity, but in random directions. The eddying motion owing to turbulence causes a uniform distribution of particles through most of the flow passage. The capture zone is limited to the laminar boundary layer in the vicinity of the collecting electrode. The physical situation is similar to that of turbulent-flow settling chambers (see Problem 7.13). The resulting grade efficiency equation is

$$\eta(D_p) = 1 - \exp\left[-\frac{A_c u_t}{Q}\right] \tag{9.3}$$

where

A_c = total collecting electrode area
Q = gas volumetric flow rate

Example 9.2 Collection Efficiency of Tubular ESP

Assume that the conditions of Example 9.1 correspond to a tubular ESP with a diameter of the collecting electrode of 2.9 m and a length of 5.0 m. If the gas flow rate is 2.0 m^3/s, estimate the collection efficiency for 1.0-μm diameter particles. Corroborate the assumptions of turbulent flow and Stokes's law applicability.

Solution

Figure 9.2 shows schematically the tubular precipitator arrangement. Assuming turbulent flow, Eq. (9.3) gives the grade efficiency. The collection area for a tubular precipitator is the area of the curved surface, $A_c = \pi DL = \pi(2.9)(5) = 45.55$ m^2. Therefore, $\eta(1\ \mu m) = 1 - \exp[-(45.55)(0.202)/(2.0)] = 0.99$ (99%).

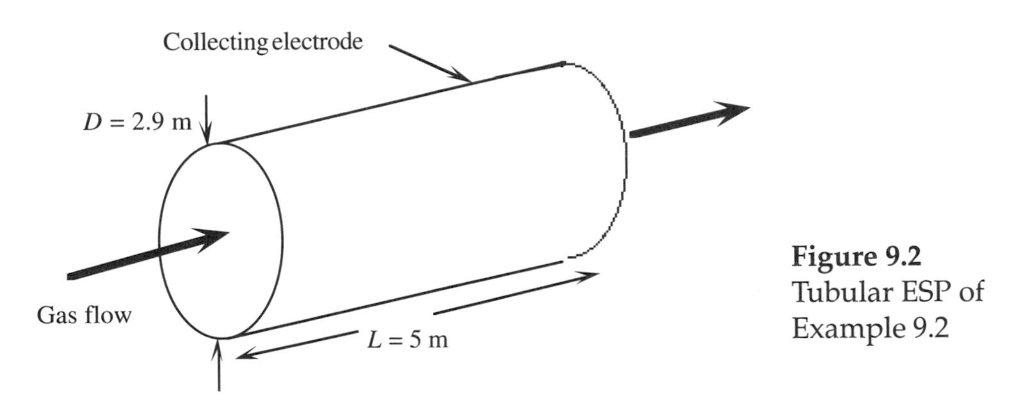

Figure 9.2 Tubular ESP of Example 9.2

To corroborate the assumption of turbulent flow, calculate the Reynolds number for the gas flow through the circular conduit, $Re = \rho vD/\mu$, where v is the average gas velocity. For the given conditions, $v = (4)(2)/[\pi(2.9)^2] = 0.3$ m/s. The Reynolds number is $Re = (1.185)(0.303)(2.9)/(1.84 \times 10^{-5}) = 56{,}600$. Therefore, the gas flow is fully turbulent.

To corroborate the applicability of Stokes's law, calculate the terminal Reynolds number for the particles, $Re_p = \rho u_t D_p/\mu = (1.185)(0.202)(10^{-6})/(1.84 \times 10^{-5}) = 0.013$. Therefore, Stokes's law applies.

9.3.2 Electrical operating point

The electrical operating point of an ESP section is thc value of voltage and current at which the section operates. The best collection occurs when the highest electric field is present, which roughly corresponds to the highest voltage on the electrodes. The term section represents one set of plates and electrodes having a common power source.

The lowest acceptable voltage is the voltage required for the formation of a corona. The negative corona is formed when an occasional free electron near the high-voltage electrode, produced by a cosmic ray, gains enough energy from the electric field to ionize the gas and produce more free electrons. The electric field for which this process is self-sustained has been determined experimentally. For round wires, the field at the surface of the wire is given by White (1963) as

$$E_0 = 3 \times 10^6 f\left[s.g. + 0.03\sqrt{s.g./r_w}\right] \tag{9.4}$$

where

E_0 = corona onset field at the wire surface, V/m

$s.g.$= specific gravity of the gas, relative to air at 293 K and 1 atm

r_w = radius of the wire, m

f = roughness factor

For a clean smooth wire, $f = 1.0$; for practical applications, $f = 0.6$ is a reasonable value (Flagan and Seinfeld 1988).

The voltage that must be applied to the wire to obtain this value of field is found by integrating the electric field from the wire to the collecting electrode. In cylindrical geometry, the field is inversely proportional to the radial distance. This leads to a logarithmic dependence of voltage on electrode dimensions. In the plate-wire geometry, the field spatial variation is more complex, but the voltage still exhibits the logarithmic dependence. The corona onset voltage, V_0, is given by

$$V_0 = E_0 r_w \ln\left(\frac{d}{r_w}\right) \tag{9.5}$$

where

d = outer cylinder radius in a tubular ESP

d = $(4/\pi)W$ for plate-wire ESP

W = wire-plate separation

Example 9.3 Corona Onset Voltage in Plate-Wire ESP

A plate-wire ESP handles air at 400 K and 110 kPa. The plate spacing is 300 mm, and the diameter of the discharge wires is 4 mm. Estimate the corona onset voltage.

Solution

Equation (9.4) gives the corona onset field at the wire surface. The specific gravity of the gas is $s.g. = PT_{ref}/P_{ref}T = (110)(293)/[(400)(101.3)] = 0.795$. Therefore, $E_0 = (3 \times 10^6)(0.6)$ $[0.795 + 0.03\ (0.795/0.002)0.5] = 2.51 \times 10^6$ V/m. Calculate $d = 4W/\pi = 4(150)/\pi = 191$ mm. Equation (9.5) gives the corona onset voltage, $V_0 = (2.51 \times 10^6)(0.002)\ \ln\ (191/2) =$ 22,900 V.

The electric field is strongest along the line from wire to plate and is approximated very well, except near the wire, by

$$E_{max} = \frac{V}{W} \tag{9.6}$$

where V = applied voltage. The electric field is not uniform along the direction of gas flow. Turner et al. (1988 a) suggest that the average field for an ESP section is given by

$$E_{av} = \frac{E_{max}}{K} \tag{9.7}$$

where K is a constant which depends on the ESP configuration and the presence of back corona. Table 9.1 gives values of K for some common situations.

Table 9.1 Ratio of Maximum to Average Eectric Fields in ESPs

Configuration	Back corona	K
Plate-wire	No[a]	1.75
Plate-wire	Severe	2.50
Flat plate	No[a]	1.26
Flat plate	Severe	1.80

[a] Resistivity of the collected dust less than 2×10^{11} ohm-cm.
Source: Turner et al. (1988 a).

When the electric field throughout the gap between the wire and the collecting electrode becomes strong enough, a spark occurs. The voltage cannot be increased beyond this point without severe sparking occurring. A reasonable estimate of the sparking field strength, E_s, is given by (Turner et al. 1988 a):

$$E_s = 6.3 \times 10^5 \left(\frac{273P}{T}\right)^{0.8} \tag{9.8}$$

where

E_s = sparking field strength, V/m
T = absolute temperature, K
P = gas pressure, atm
The ESP generally operates near this point. E_{max} is equal to or less than E_s.

Example 9.4 Average Electric Field in a Plate-Wire ESP

Estimate the average electric field and the operating voltage for the ESP of Example 9.3. The resistivity of the collected dust is 10^{11} ohm-cm.

Solution

Equation (9.8) gives E_s = 630,000 $[(273/400)(110/101.3)]^{0.8}$ = 496,000 V/m. = E_{max}.
Because the resistivity of the collected dust is less than 2×10^{11} ohm-cm, there is no back corona. From Table 9.1, K = 1.75, therefore, E_{av} = 496,000/ 1.75 = 283,400 V/m. The operating voltage is $V = WE_{max}$ = (0.15)(496,000) = 74.4 kV.

9.3.3 Particle Charging

Charging of particles takes place when ions bombard the surface of a particle. Once an ion is close to a particle, it is tightly bound because of the image charge within the particle.The image charge is a representation of the charge distortion that occurs when a real charge approaches a conducting surface. The distortion is equivalent to a charge of opposite magnitude to the real one, located as far below the surface as the real charge is above it. The motion of the fictitious charge is similar to the motion of an image in a mirror, hence the name. As more ions accumulate on a particle, the total charge tends to prevent further ionic bombardment.

There are two principal charging mechanisms: diffusion charging and field charging. Diffusion charging results from the thermal kinetic energy of the ions overcoming the repulsion of the ions already on the particle. Field charging results when ions follow electric field lines until they terminate on a particle. In general, both mechanisms operate for all sizes of particles. Field charging is the dominant mechanism for particles greater than about 2 μm, whereas diffusion charging dominates for particles smaller than about 0.5 μm.

Diffusion charging produces a logarithmically increasing level of charge on particles given by (White 1963)

$$q_d(t) = \frac{2\pi\varepsilon_0 k T D_p}{e} \ln(1 + t/\tau_d) \tag{9.9}$$

where

q_d = particle charge by diffusion

ε_0 = free space permittivity (8.845×10^{-12} $C^2/N\text{-}m^2$)

k = Boltzmann's constant (1.38×10^{-23} J/K)

e = electron charge (1.67×10^{-19} C)

t = exposure time

τ_d = characteristic time for diffusion charging given by

$$\tau_d = \frac{\varepsilon_0 \sqrt{8mk\pi T}}{e^2 N D_p}$$

N = number of ions per unit volume

m = mass of an ion

For typical operating conditions, $N = 2 \times 10^{15}$ ions/m^3, $m = 5.3 \times 10^{-26}$ kg (Crawford 1976, Licht 1980).

Diffusion charging never reaches a limit, but it becomes very slow after three characteristic times. For fixed exposure times, the charge on the particle is proportional to its diameter.

Example 9.5 Diffusion Charging

Calculate the charge by diffusion on a 1.0-μm diameter particle as a function of time. The gas is air at 400 K and 1 atm. The ion density is 2×10^{15} ions/m^3, and the mass of a typical ion is 5.3×10^{-26} kg.

Solution

Calculate the characteristic time for diffusion charging, $\tau_d = (8.85 \times 10^{-12})\ [(8)(5.3 \times 10^{-26})(1.38 \times 10^{-23})(\pi)(400)]^{0.5}/[(1.67 \times 10^{-19})^2(2 \times 10^{15})(10^{-6})] = 1.36 \times 10^{-5}$ s. From Eq. (8.9), $q_d = 3.068 \times 10^{-18} \ln [1 + t/(1.36 \times 10^{-5})]$ C. Figure 9.3 shows the variation with time of the charge acquired by the particle by diffusion.

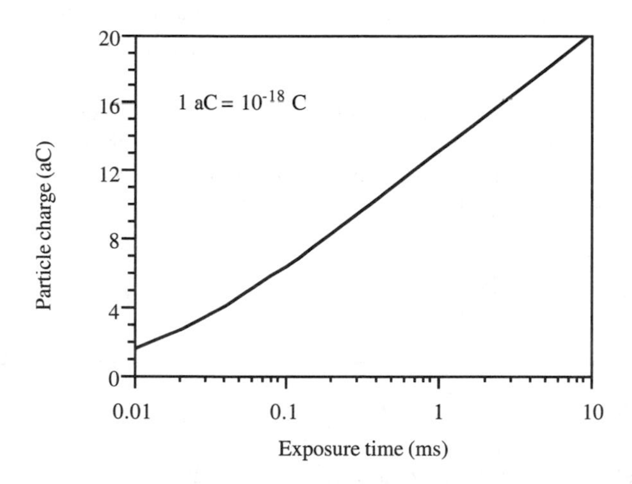

Figure 9.3 Charging by diffusion of a 1.0-μm diameter particle at 400 K.

Field charging also exhibits a characteristic time-dependence, given by

$$q_f(t) = \frac{q_s t}{t + \tau_f} \tag{9.10}$$

where

q_s = saturation charge (charge at infinite time)

τ_f = field charging time constant

The saturation charge is given by

$$q_s = \left(\frac{3\kappa}{\kappa + 2}\right)\pi \varepsilon_0 E D_p^2 \tag{9.11}$$

where

κ = dielectric constant of the particle

E = external electric field applied to the particle

The saturation charge is proportional to the square of the particle size, which explains why field charging is the dominant mechanism for larger particles. The field charging time constant is given by

$$\tau_f = \frac{4\varepsilon_0}{NeB} \tag{9.12}$$

where B = ion mobility, usually of the order of 10^{-4} m²/V-s. For typical conditions, $\tau_f = (4)(8.85 \times 10^{-12})/[(2 \times 10^{15})(1.61 \times 10^{-19})(10^{-4})] = 0.0011$ s. For practical purposes, the saturation charge may be taken as attained at $t = 100\ \tau_f. = 0.1$ s. Because the residence time of the particles in the precipitator generally exceeds a few seconds, it is usually assumed that the particles attain the saturation charge soon after entering the device.

Strictly speaking, both diffusion and field charging mechanisms operate at all times on all the particles, and neither mechanism is sufficient to explain the charges measured on the particles. It has been found empirically that a very good approximation to the measured charge is given by the sum of the charges predicted by Eqs. (9.9) and (9.10) independently of one another.

$$q_p(t) = q_d(t) + q_f(t) \tag{9.13}$$

Example 9.6 Total Particle Charge

For the following set of conditions, calculate the ratio q_p/q_s as a function of particle size for $t = 0.1$, 1.0, and 10.0 s.

$\kappa = 5.0$ $\quad E = 3 \times 10^5$ V/m $\quad T = 300$ K

$B = 2.2 \times 10^{-4}$ m^2/s-V $\quad N = 2 \times 10^{15}$ ions/m^3

$m = 5.3 \times 10^{-26}$ kg

Solution

The field charging time constant is from Eq. (9.12), $\tau_f = 0.000503$ s. Thus, assume that for $t > 0.05$ s the field charging mechanism is saturated. Equation (9.13) becomes

$$q_p = 1.787 \times 10^{-5} D_p^2 + 1.438 \times 10^{-12} D_p \ln\left(1 + 7.79 \times 10^{10} t D_p\right)$$

The first term in this equation is the saturation field charge. Therefore,

$$\frac{q_p}{q_s} = 1 + \frac{8.047 \times 10^{-8}}{D_p} \ln\left(1 + 7.79 \times 10^{10} t D_p\right)$$

where D_p is in meters and t in seconds. The following table presents the ratio q_p/q_s as a function of particle size and exposure time.

D_p (μm)	$t = 0.1$ s	q_p/q_s $t = 1.0$ s	$t = 10.0$ s
0.01	36.15	54.59	73.11
0.10	6.36	8.21	10.06
1.00	1.72	1.91	2.09
10.0	1.09	1.11	1.13

Feldman (1975) proposed another representation of the total charge on a particle to account for the behavior of the submicron particles. According to Cochet (1961) the combined effect of field charging and diffusion charging leads to:

$$q_p(t) = \left[\left(1 + \frac{2\lambda}{D_p}\right)^2 + \frac{2(\kappa - 1)}{\left(1 + \frac{2\lambda}{D_p}\right)(\kappa + 2)}\right] \pi \varepsilon_0 E D_p^2 \frac{t}{t + \tau_f} \tag{9.14}$$

where λ = mean-free path of the gas. For $t >> \tau_f$, Feldman incorporated Eqs. (9.14) and (9.2) into (9.3) to obtain

$$\eta(D_p) = 1 - \exp\left[-\frac{A_c \varepsilon_0 E_c E}{3\mu Q}(C_c F D_p)\right] \tag{9.15}$$

where

$$F = \left[\left(1 + \frac{2\lambda}{D_p}\right)^2 + \frac{2(\kappa - 1)}{\left(1 + \frac{2\lambda}{D_p}\right)(\kappa + 2)}\right] \tag{9.16}$$

In a single-stage precipitator, the electric field in the vicinity of the collecting electrode, E_c, and the charging electric field, E, are assumed equal to the average field, E_{av}. Equation (9.15) becomes

$$\eta(D_p) = 1 - \exp\left[-\frac{A_c \varepsilon_0 E_{av}^2}{3\mu Q}(C_c F D_p)\right] \tag{9.17}$$

Example 9.7 Grade-Efficiency Curve for Single-Stage ESP

The average electric field in a single-stage, plate-wire ESP is 300,000 V/m. The gas flowing through it is air at 298 K and 1 atm. The dielectric constant of the particles is 5.0. The *specific collection area* (SCA = A_c/Q) is 78.8 m²/(m³/s). Show that the grade efficiency for particulate removal goes through a minimum for particle diameters between 0.1 and 1.0 μm.

Solution

Equation (9.17) gives

$$\eta(D_p) = 1 - \exp\{-(78.8)(8.85 \times 10^{-12})(3 \times 10^5)^2\, C_c F D_p / [3(1.84 \times 10^{-5})]\}$$
$$= 1 - \exp\{-1.137 \times 10^6\, C_c F D_p\}.$$

For air at 298 K and 1 atm, $\lambda = 0.0667$ μm. For $\kappa = 5.0$, Eq. (9.16) becomes

$$F = \left(1 + \frac{0.1334}{D_p'}\right)^2 + \frac{8}{7\left(1 + \frac{0.1334}{D_p'}\right)}$$

where D_p' is in microns. The following table summarizes the calculations to estimate the ESP's grade efficiency for various particle sizes.

D_p (μm)	C_c	F	η
0.01	22.68	206.0	1.000
0.05	5.06	13.77	0.981
0.1	2.91	5.94	0.860
0.2	1.89	3.46	0.774
0.4	1.42	2.64	0.819
0.6	1.28	2.43	0.880
0.8	1.21	2.34	0.923
1.0	1.17	2.30	0.953
2.0	1.08	2.21	0.996

Figure 9.4 is a graphical representation of these results. It shows a "window" for particle sizes between 0.1 and 1.0 μm where the penetration increases significantly. This behavior was first documented experimentally by Abbott and Drehmel (1976) and has been observed by many others since.

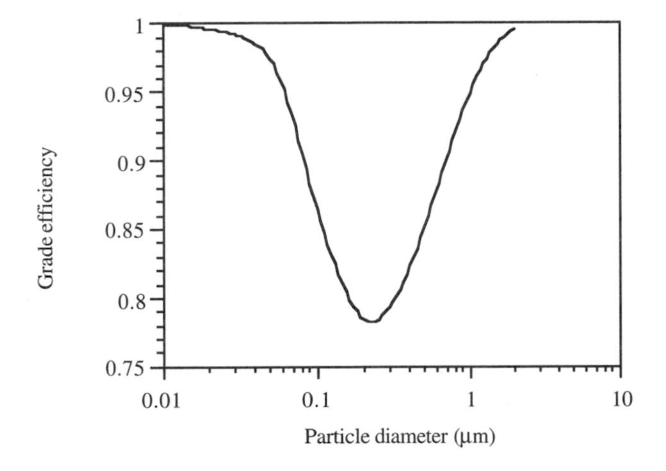

Figure 9.4 Grade efficiency versus particle size for an ESP

9.4 OVERALL EFFICIENCY

Your objectives in studying this section are to

1. Estimate the overall efficiency of an ESP with Deutsch-Anderson equation.
2. Estimate the overall efficiency through Gaussian quadrature formulas.

Equation (9.3) as it stands, without refinements, is known as the Deutsch-Anderson equation and has been used as the basis for much work on precipitators. Although it is a grade-efficiency equation, it has been the practice to use it for overall efficiency calculations. This is done by replacing u_t with a so-called *effective migration velocity*, w_e. This quantity is taken to represent the collection behavior of an entire dust of a certain kind and under a certain set of operating conditions. It is back calculated from experimental data and so is really a performance parameter. Equation (9.3) becomes:

$$\text{SCA} = \frac{A_c}{Q} = -\frac{\ln(1-\eta_M)}{w_e} \tag{9.18}$$

Turner et al. (1988 a) calculated migration velocities for three main precipitator types: plate-wire, flat plate, and wet wall ESPs of the plate-wire type. The following three tables, keyed to overall design efficiency, summarize the migration velocities under various conditions. In Table 9.2, the migration velocities for plate-wire ESPs are given for conditions of no back corona and severe back corona. In Table 9.3, they are given for a wet wall precipitator assuming no back corona and no rapping reentrainment. In Table 9.4, the flat plate ESP migration velocities are given only for no back corona conditions because flat plate ESPs appear to be less affected by high resistivity dusts than the plate-wire type.

It is generally expected from experience that the migration velocity will decrease with increasing efficiency. However, in Tables 9.2 through 9.4 the migration velocities show some fluctuations. This is because the number of sections must increase as the efficiency increases, and the changing sectionalization affects the overall migration velocity. This effect is particularly noticeable, for example, in Table 9.4 for glass plants.

Example 9.8 Design of Flat Plate ESP Based on Migration Velocity

Calculate the total collector plate area for a flat plate ESP to control fly-ash emissions from a coal-fired boiler burning bituminous coal. The flue gas stream is 24 m^3/s at 436 K. Analysis of the ash shows a resistivity of 1.2×10^{11} ohm-cm. The overall efficiency required is 99.9%.

Solution

Calculate the SCA from Eq. (9.18). The fly-ash migration velocity is 0.160 m/s (see Table 9.4). Then, SCA = –ln (1 – 0.999)/0.16 = 43.2 $m^2/(m^3/s)$. The total collector plate area is then Ac = (43.2)(24) = 1,037 m2.

A more general method to calculate overall efficiencies is based on the grade-efficiency equation for the ESP and the size distribution function of the aerosol population entering the device. If the aerosol is log-normally distributed, the calculation of overall efficiency can be done numerically through Gaussian quadrature formulas as shown in Chapter 7.

Table 9.2 Plate-Wire ESP Migration Velocities (m/s)

Particle source	Efficiency (%) 95	99	99.5	99.9
Bituminous coal fly-ash[a]				
(no BC)	0.126	0.101	0.093	0.082
(BC)	0.031	0.025	0.024	0.021
Other coal[a]				
(no BC)	0.097	0.079	0.079	0.072
(BC)	0.029	0.022	0.021	0.019
Cement kiln[b]				
(no BC)	0.015	0.015	0.018	0.018
(BC)	0.006	0.006	0.005	0.005
Iron/steel sinter plant dust with mechanical collector[a]				
(no BC)	0.068	0.062	0.066	0.063
(BC)	0.022	0.018	0.018	0.017
Incinerator fly-ash[c]				
(no BC)	0.153	0.114	0.106	0.094

[a] At 420 K. BC = back corona.
[b] At 590 K.
[c] At 395 K.
Source: From Turner, J.H., et al. *JAPCA*, **38**:458 (1988). Reprinted with permission from *JAPCA*.

Example 9.9 Design of ESP Based on Grade-Efficiency Equation

Estimate the overall efficiency of the ESP of Example 9.8 integrating the grade efficiency equation. The particles entering the device are log-normally distributed with MMD = 16 mm and σ_g =3.0. The dielectric constant of the particles is 5.0.

Solution

At a temperature of 436 K, Eq. (9.8) gives the maximum electric field strength, $E_{max} = E_s =$ 433,000 V/m. For a flat-plate configuration with no back corona, Table 9.1 gives $K = 1.26$. Then, $E_{av} = 433{,}000/1.26 = 344{,}000$ V/m. From Example 9.8, SCA = 43.2 s/m. The sub

Table 9.3 Wet Wall Plate-Wire ESP Migration Velocities (m/s)

	Efficiency (%)			
Particle source[a]	95	99	99.5	99.9
Bituminous coal fly-ash	0.314	0.330	0.338	0.249
Other coal	0.400	0.427	0.441	0.314
Cement kiln	0.064	0.056	0.050	0.057
Iron/steel sinter plant dust with mechanical collector	0.140	0.137	0.133	0.116

[a] All sources at 370 K, no back corona.

Source: From Turner, J. H. et al. *JAPCA*, **38**:458 (1988). Reprinted with permission from *JAPCA*.

Table 9.4 Flat plate ESP migration velocities (m/s).

	Efficiency (%)			
Particle source	95	99	99.5	99.9
Bituminous coal fly-ash[a]	0.132	0.151	0.186	0.160
Other coal[a]	0.155	0.112	0.151	0.135
Cement kiln[b]	0.024	0.023	0.032	0.031
Glass plant[c]	0.018	0.019	0.026	0.026
Iron/steel sinter plant dust with mechanical collector[a]	0.134	0.121	0.131	0.124
Incinerator fly-ash[d]	0.252	0.169	0.211	0.183

[a] At 420 K, no back corona.

[b] At 590 K, no back corona.

[c] At 530 K, no back corona.

[d] At 395 K, no back corona.

Source: From Turner, J. H. et al. *JAPCA*, **38**:458 (1988). Reprinted with permission from *JAPCA*.

program FUNCTION HETA(D) calculates the grade efficiency from Eqs. (9.16) and (9.17). Empirical correlations to estimate the viscosity and the mean-free path for air at atmospheric pressure as a function of temperature are part of the subprogram. The grade efficiency equation is integrated numerically with a 16-point Gauss-Hermite quadrature formula to give an overall efficiency of 99.6%.

```
FUNCTION HETA(D)
   REAL LAMBDA, DIEL, EL, SCA, VIS, E0, T, D
   PARAMETER (E0 = 8.85E-12)
   DATA T, DIEL, EL, SCA /436.0, 5., 3.44E5, 43.2/
   CUN(D)=1.+(2.*LAMBDA/D)*(1.257+.4*EXP(-.55*D/LAMBDA))
   AFAC(D)=(1.+2.*LAMBDA/D)**2+2.*((DIEL-1.)/(DIEL+2.))/
   *     (1.+2.*LAMBDA/D)
   VIS = 1.72E-5*(T/273.0)**0.71
   LAMBDA = 6.71E-11*T**1.21
     IF (0.55*D/LAMBDA .LT. 80. ) THEN
        CUNING=CUN(D)
     ELSE
        CUNING= 1. + 2.*LAMBDA/D
     ENDIF
     ARGUM=SCA*E0*EL**2*CUNING*AFAC(D)*D/(3.*VIS)
     IF (ARGUM .GT. 80.0) THEN
        PEN=0.0
     ELSE
        PEN=EXP(-ARGUM)
     ENDIF
   HETA = 1. - PEN
   END
```

9.5 CORRECTIONS TO THE MODEL

Your objectives in studying this section are to

1. Identify those phenomena affecting the performance of the ESP not incorporated in the grade-efficiency model.
2. Define the quality factor of the flow distribution and include it in the efficiency calculations.
3. Define the loss factor owing to reentrainment and gas sneakage, and include it in the efficiency calculations.

The experimental values of the effective migration velocity do not behave in all respects as predicted by the theoretical model. There are various aspects of the performance of an ESP not incorporated in the model. These are sometimes referred to as non-Deutschian phenomena. Some of the most significant are *non-uniform gas velocity distribution; gas sneakage;* and *rapping reentrainment*. Possible corrections to the model for these items follow.

9.5.1 Non-Uniform Gas Velocity Distribution

The theoretical model developed assumes that the fluid velocity parallel to the collector surface is everywhere the same and equal to the volumetric flow rate divided by the total cross-sectional area. This condition is almost impossible to achieve in practice. Uniform, low-turbulence gas flow is very important for optimum precipitator performance. The detrimental effect of non-uniform gas flow is twofold. First, owing to the exponential nature of the collection mechanism uneven treatment of the gas lowers collection efficiency in the high-velocity zones to an extent not compensated for in the low-velocity zones. Second, high-velocity regions near collection plates can sweep particles back into the main gas stream.

Although it is known that a poor gas velocity distribution results in reduced collection efficiency, it is difficult to formulate a mathematical description for gas flow quality. The following development, presented by McDonald and Dean (1982), demonstrates the general considerations to be made in accounting for the effects of a non uniform gas velocity distribution on collection efficiency. Eq. (9.18) can be written as

$$Pt^{id} = \exp\left(-\frac{A_c w_e}{A_1 v_a}\right) = \exp\left(-\frac{k}{v_a}\right) \tag{9.19}$$

where

Pt^{id} = ideal overall penetration
A_1 = inlet cross sectional area
v_a = average gas inlet velocity
k = $A_c w_e / A_1$

The precipitator can be divided into a number of imaginary channels corresponding to pitot tube traverse points. The penetration for all the channels can be summed and averaged to obtain the mean penetration with an actual velocity distribution instead of an assumed uniform velocity.

$$Pt = \frac{1}{N v_a} \sum_{j=1}^{N} v_j Pt_j^{id} = \frac{1}{N v_a} \sum_{j=1}^{N} v_j e^{-\frac{k}{v_j}} \tag{9.20}$$

where

N = number of points for velocity traverse

v_j = point values of gas velocity

For any practical velocity distribution and efficiency, the mean penetration calculated in this manner will be higher than that calculated based on an average uniform velocity. This is equivalent to a reduced "apparent" migration velocity. The ratio of the original migration velocity to the "apparent" migration velocity is called the *quality factor* ϕ, of the velocity distribution. Equation (9.20) can, then, be written as:

$$Pt = \exp\left(-\frac{k}{\phi v_a}\right) = \frac{1}{N v_a} \sum_{j=1}^{N} v_j e^{-\frac{k}{v_j}} \tag{9.21}$$

Solving Eq. (9.21) for the quality factor,

$$\phi = -\frac{k}{v_a \ln\left[\frac{1}{N v_a} \sum_{j=1}^{N} v_j e^{-\frac{k}{v_j}}\right]} = \frac{\ln\left(Pt^{id}\right)}{\ln\left(Pt\right)} \tag{9.22}$$

The use of Eq. (9.22) with measured velocity traverses in a working precipitator may yield values of ϕ as high as 2 to 3 in a poorly regulated flow pattern (McDonald and Dean 1982). When care is taken, through scale model studies, values of ϕ as low as 1.1 or 1.2 should be achieved (Licht 1980).

Example 9.10 Flow Quality Factor Calculations

(a) The ideal overall penetration for an ESP is 1% assuming a uniform velocity distribution. There are actually three zones of equal flow area in which $v_1 = v_a/2$, $v_2 = v_a$, and $v_3 = 3v_a/2$. Calculate the flow quality factor and the actual overall penetration.

(b) The ideal overall penetration for an ESP is 1% assuming a uniform velocity distribution. The flow quality factor is 2.0. Calculate the actual overall penetration.

Solution

(a) From Eq. (9.19), $k/v_a = -(\ln Pt^{id}) = -\ln(0.01) = 4.605$. The following table summarizes the calculations for the quality factor and actual penetration:

j	v_j/v_a	$\exp(-k/v_j)$	$v_j/v_a \exp(-k/v_j)$
1	0.5	0.0001	0.00005
2	1.0	0.01	0.01
3	1.5	0.0447	0.06705
			$\Sigma = 0.07710$

From Eq.(9.21), $Pt = \Sigma/N = 0.0771/3 = 0.0257$. Equation (9.22) yields $\phi = \ln(0.01)/\ln(0.0257) = 1.258$.

(b) From Eq. (9.22), $Pt = (Pt^{id})^{1/\phi} = (0.01)^{0.5} = 0.10$.

Comments

These examples emphasize the importance of a uniform velocity distribution in an ESP's performance. Part b shows a 10-fold increase in penetration for a flow quality factor of 2.0! However, Hein (1989) developed a computer model that shows the possibility of improving the performance of an ESP by using a controlled non-uniform gas distribution at the inlet and outlet faces of the precipitator when there are significant rapping reentrainment losses.

9.5.2 Sneakage and Rapping Reentrainment

Sneakage and rapping reentrainment losses are best considered on the basis of the sections within an ESP. Consider first the effect of sneakage. Assuming that the gas is well mixed between sections, the penetration for each section can be expressed as:

$$Pt_s = S_N + (1 - S_N)Pt_c(Q') \qquad (9.23)$$

where

Pt_s = section's overall penetration

S_N = fraction of the gas bypassing the section (sneakage)

Q' = gas volume flow in the collection zone = $Q(1 - S_N)$

$Pt_c\,(Q')$ = overall penetration in the collection zone

The penetration for the entire ESP is the product of the section penetrations. The sneakage sets a lower limit on the collection efficiency through each section.

The collected dust accumulates on the plates until they are rapped, when most of the material falls into the dust hopper. A fraction of it is reentrained by the gas flow and leaves the section. The average penetration for a section, including sneakage and reentrainment is

$$Pt_s = S_N + (1 - S_N)Pt_c(Q') + RR\{(1 - S_N)[1 - Pt_c(Q')]\} \tag{9.24}$$

where RR = fraction reentrained. Equation (9.24) can be written as

$$Pt_s = LF + (1 - LF)Pt_c(Q') \tag{9.25}$$

where LF (loss factor) is defined as

$$LF = S_N + RR - (S_N)(RR) \tag{9.26}$$

Fly-ash precipitators analyzed in this way have average values of $S_N = 0.07$, $RR = 0.12$, and $LF = 0.182$ (Turner et al. 1988 a). To achieve a given overall penetration, choose the minimum number of sections, N_s, such that:

$$N_s > \frac{\ln(Pt)}{\ln(LF)} \tag{9.27}$$

Example 9.11 Sneakage and Reentrainment Losses

Estimate the minimum number of sections for a fly-ash ESP if the overall efficiency must be 99.9%. Assume that the loss factor owing to sneakage and reentrainment is 0.182. Assuming that all the sections have the same overall penetration, estimate the penetration in the collection zone, Pt_c, of each section.

Solution

Equation (9.27) gives $N_s > \ln(1 - 0.999)/\ln(0.182) = 4.06$. Therefore, the ESP should have 5 sections. Assuming that all the sections have the same overall penetration,

$$Pt_c = \frac{Pt^{(1/N_s)} - LF}{1 - LF} \tag{9.28}$$

Substituting in Eq. (9.28), $Pt_c = [(0.001)0.2 - 0.182]/(1 - 0.182) = 0.0846$.

Example 9.12 Corrections to Penetration Predicted by the ESP Model

The collection area of the ESP of Example 9.9 is equally divided into six sections in the direction of the gas flow. Values for sneakage and rapping reentrainment are 10% and 12%, respectively; the flow quality factor is 1.2. Estimate the actual overall collection efficiency for the ESP.

Solution

The penetration predicted by Feldman's model, Pt^{id}, must be corrected for non-uniform gas flow distribution, number of sections, and reduced gas flow—owing to sneakage—through the collection zone. The corrected value corresponds to Pt_c, and is given by

$$Pt_c = \left(Pt^{id}\right)^{1/\left[\phi N_s(1 - S_N)\right]} \tag{9.29}$$

Substituting in Eq. (9.29), $Pt_c = (0.004)^{1/[(1.2)(6)(0.9)]} = 0.426$. The loss factor is $LF = 0.10 + 0.12 - (0.1)(0.12) = 0.208$. From Eq. (9.25), $Pt_s = 0.545$, then $Pt = (0.545)^6 = 0.026$, and $\eta_M = 97.4\%$.

Example 9.13 Design of ESP Including Corrections to Model

Consider the conditions of Examples 9.9 and 9.12. Redesign the ESP to maintain an actual overall collection efficiency of 99.6%.

Solution

Equation (9.27) gives the optimum number of sections the ESP should have: $N_s > \ln(0.004)/\ln(0.208) = 3.52$. Therefore, $N_s = 4$ sections. Equation (9.28) gives the actual overall penetration in the collection zone of each of the ESP's sections: $Pt_c = [(0.004)^{0.25} - 0.208]/(1 - 0.208) = 0.055$. Eq. (9.29), applied to a single section, gives $Pt^{id} = Pt_c^{\phi(1 - SN)} = (0.055)^{1.056} = 0.0468$. The computer program ESP calculates the collection area per section for a given value of Pt^{id}, as predicted by Feldman's model. The answer is 324.3 m^2/section. For four sections, the total collection area is 1,297.2 m^2.

```
      PROGRAM ESP
C
C  CALCULATES THE COLLECTION AREA REQUIRED FOR A GIVEN IDEAL
C  PENETRATION, PENID, AS PREDICTED BY FELDMAN'S MODEL
C
      REAL PENID, MMD, NMD, SIGG, Q, LAMBDA, VIS, X1, X2, SL
      REAL RTBIS, XACC, EL, CUN, AFAC, PEN, DIEL
      INTEGER N
      LOGICAL SUCCES
      COMMON /BLOCK1/ PENID, MMD, SIGG, Q, LAMBDA, VIS, EL, DIEL, E0
      EXTERNAL FUNCP
      CUN(D)=1.+(2.*LAMBDA/D)*(1.257+.4*EXP(-.55*D/LAMBDA))
      AFAC(D)=(1.+2.*LAMBDA/D)**2+2.*((DIEL-1.)/(DIEL+2.))/
     *      (1.+2.*LAMBDA/D)
      X1=-3.*VIS*Q*LOG(PENID)/(E0*EL**2*CUN(MMD)*AFAC(MMD)*MMD)
      X2=SIGG*X1
      CALL ZBRAC(FUNCP, X1, X2, SUCCES)
      IF (SUCCES) THEN
        XACC=ABS((X1-X2)/10000.)
        SL=RTBIS(FUNCP, X1, X2, XACC)
        PRINT *, ' COLLECTION AREA PER SECTION  ', SL, ' SQ METERS'
        GO TO 10
      ENDIF
      PRINT *, 'FAILURE IN BRACKETING SL '
10    STOP
      END
C
      BLOCK DATA
      REAL PENID, MMD, SIGG, Q, LAMBDA, VIS, EL, DIEL, E0
      COMMON /BLOCK1/ PENID, MMD, SIGG, Q, LAMBDA, VIS, EL, DIEL, E0
      DATA PENID/0.0468/, MMD/16.0/, SIGG/3.0/, Q/24.00/, LAMBDA/
     *  0.1050/, VIS/2.4E-5/, EL/3.44E5/, DIEL/5./, E0/8.85E-18/
      END
C
      FUNCTION FUNCP(SL)
      REAL PENID, MMD, SIGG, Q, LAMBDA, VIS, EL, DIEL, E0, PEN, X, A
      REAL NMD, EPS, PENC, FUNCP, SL
      INTEGER N
      COMMON /BLOCK1/ PENID, MMD, SIGG, Q, LAMBDA, VIS, EL, DIEL, E0
      PARAMETER (N=10, EPS=1.0E-06)
      DIMENSION X(N), A(N)
```

```
      CUN(D)=1.+(2.*LAMBDA/D)*(1.257+.4*EXP(-.55*D/LAMBDA))
      AFAC(D)=(1.+2.*LAMBDA/D)**2+2.*((DIEL-1.)/(DIEL+2.))/
     *     (1.+2.*LAMBDA/D)
      CALL HERMIT(N, X, A, EPS)
      NMD=EXP(LOG(MMD)-3*(LOG(SIGG))**2)
      SUMA=0.0
      DO 10 I=1, N
        D=EXP(SQRT(2.)*LOG(SIGG)*X(I)+LOG(NMD))
        IF (0.55*D/LAMBDA .LT. 80. ) THEN
           CUNING=CUN(D)
        ELSE
           CUNING= 1. + 2.*LAMBDA/D
        ENDIF
        ARGUM=SL*E0*EL**2*CUNING*AFAC(D)*D/(3.*Q*VIS)
        IF (ARGUM .GT. 80.0) THEN
          PEN=0.0
        ELSE
          PEN=EXP(-ARGUM)
        ENDIF
        SUMA=SUMA+A(I)*PEN*EXP(3.*SQRT(2.)*LOG(SIGG)*X(I))
 10   CONTINUE
      PENC=SUMA/(SQRT(3.1416)*EXP(4.5*(LOG(SIGG))**2))
      FUNCP=PENC-PENID
      END
```

9.6 PRACTICAL DESIGN CONSIDERATIONS

Your objectives in studying this section are to

1. Estimate the dimensions of an ESP, given the collection area.
2. Estimate the total power consumption of an ESP.
3. Understand the importance of flue gas conditioning to maintain proper resistivity of the collected dust.

The complete design of an ESP includes sizing and determining the configuration of the collecting and discharge electrodes, calculating the power consumption, and specifying rapping, dust removal, and flue gas conditioning systems.

9.6.1 Sizing the electrodes

The discharge electrode system is designed in conjunction with the collection electrode system to maximize the electric current and field strength. The shape of the discharge electrode may be in the form of cylindrical or square wires, barbed wire, or stamped from formed strips of metal of various shapes. Most ESPs in the U.S. use cylindrical wires of about 2.5 mm diameter kept taut by weights, whereas European designs favor rigid, mast-type supports for the wires. Recent designs in both continents use rigid discharge electrodes (Cooper and Alley 1986).

The number and size of the collection plates depend on the total collection area specified, the gas flow velocity, the plate spacing, and the aspect ratio (length-height). An ESP collecting a dry particulate material runs a risk of nonrapping, continuous reentrainment if the gas velocity becomes too large. For fly-ash applications, the maximum acceptable velocity is about 1.5 m/s for plate-wire ESPs and about 1 m/s for flat-plate ESPs.

Most U.S. utility precipitators have 230-mm plate spacing, although systems purchased in the last few years generally have 300-mm spacing which allows the use of rigid discharge electrodes. A few U.S. utilities are currently installing ESPs with the new European plate spacing standard of 400 mm (Offen and Altman 1991).

The Deutsch equation relates ESP performance to specific collecting surface area with no regard to the length-height aspect ratio. However, because of gas sneakage and rapping reentrainment losses, the actual precipitator's performance is dependent on how the surface is arranged. For collection efficiencies grater than 99%, the aspect ratio should be kept above 1.0.

The number of ducts for gas flow, Nd, is given by

$$N_d = \frac{Q}{2 W v_a H} \tag{9.30}$$

where

W = wire-plate spacing
H = plate height

Since each plate has two collecting surfaces, the total collection area is given by

$$A_c = 2 N_d R H^2 \tag{9.31}$$

where R is the aspect ratio. Combining Eqs. (9.30) and (9.31), the plate height is

$$H = \frac{(\mathrm{SCA}) v_a W}{R} \tag{9.32}$$

Example 9.14 Dimensions of Plate-Wire ESP

Estimate the dimensions of a plate-wire ESP treating 333 m^3/s of gas with a total plate area of 14,000 m^2 for 99% collection of fly-ash. Plate spacing must be 300 mm to allow the use of rigid discharge electrodes.

Solution

Choose typical values of gas velocity and aspect ratio: $v_a = 1.5$ m/s, $R = 1.0$. Calculate SCA $= A_c/Q = 14{,}000/333 = 42.0$ s/m. For a wire-plate spacing of 150 mm, Eq. (9.32) gives $H =$ 9.45 m. Since the aspect ratio is unity, $L = H = 9.45$ m. Equation (9.30) gives $N_d = 78.3$; choose $N_d = 79$ ducts. The width of the ESP is (79)(0.30) = 23.7 m.

Because the number of ducts was rounded to the next integer, the actual collection area is higher than that specified based on collection efficiency (14,110 m^2 versus 14,000 m^2), and the gas velocity is lower than the value initially chosen (1.49 m/s versus 1.5 m/s). Both factors improve the performance of the ESP. Figure 9.5 is a schematic diagram of the resulting design.

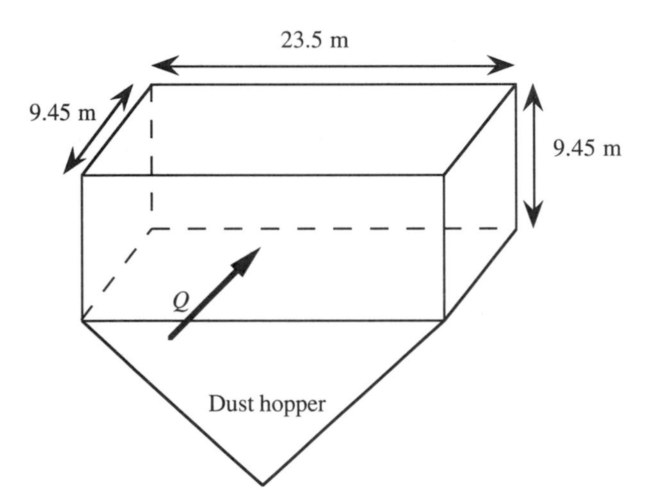

Figure 9.5 Dimensions of the ESP of Example 9.14

9.6.2 Power consumption

The two main sources of power consumption of an ESP are from corona power and gas pressure drop. Total pressure drop for a precipitator and associated ductwork is usually of the order of 1.0 kPa (Turner et al. 1988 a). Corona power is a strong function of overall penetration.

The following correlation is based on actual operational data presented by White (1984):

$$\dot{W}_c = Q\left(115.8 + \frac{1.17}{Pt}\right) \tag{9.33}$$

where $\dot{W}_c$ is the corona power in watts.

Example 9.15 Power Consumption of ESP

Estimate the total power consumption of the ESP of Example 9.14. Assume that the motor-fan efficiency is 60%.

Solution

Assume a pressure drop of 1.0 kPa. The fan power is (333)(1.0)/(0.6) = 555 kW. Equation (9.33) gives the corona power, $\dot{W}_c$ = 333(115.8 + 1.17/0.01) = 77.5 kW. The total power is 632.5 kW.

9.6.3 Flue Gas Conditioning Systems

The major factors affecting fly-ash resistivity are temperature and chemical composition (of the fly-ash and of the combustion gases). For a given chemical composition, fly-ash resistivity always exhibits a distinct maximum at temperatures about 395 to 450 K. The temperature of the maximum resistivity is unfortunate for power boiler operators. ESP operating temperatures cannot be much below 395 K without risking condensation of sulfuric acid on some of the colder surfaces. Conversely, operating at temperatures much higher than 450 K results in unnecessary loss of heat out the stack, reducing the thermal efficiency of the plant (Cooper and Alley 1986).

Resistivity decreases with increased fuel sulfur content. because of increased adsorption of conductive gases, such as SO_3, by the fly-ash. The combustion of high-sulfur coal (>2% sulfur) produces sufficient SO_3 to condition the fly-ash to a low resistivity, as shown on Figure 9.6. The combustion of low-sulfur coals (<1% sulfur) with highly alkaline ash produces only small amounts of SO_3, resulting in very high ash resistivities. In the intermediate case, the existing data for medium-sulfur coal exhibit considerable scatter in the relationship between the SO_2 and SO_3 in the gas entering the ESP (Harrison et al. 1988). The Clean Air Act Amendments of 1990 require electric utilities to reduce SO_2 emissions by approximately 10 million ton/yr by 2000. Many existing plants are switching to low-sulfur coal, with negative impacts on the electrical operation of ESPs.

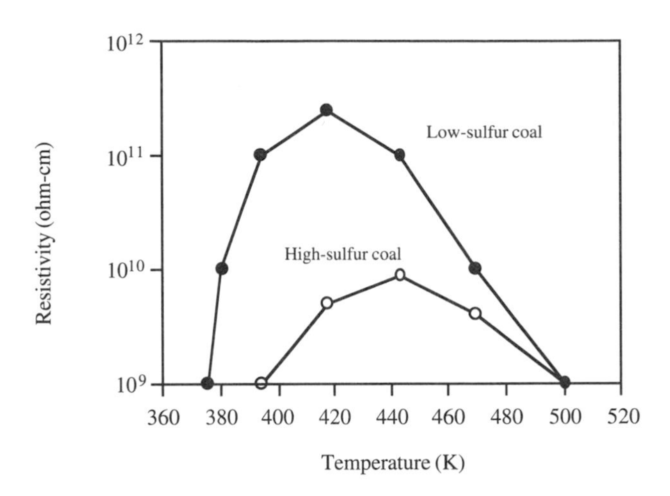

Figure 9.6 The effect of temperature and fuel sulfur content on fly ash resistivity

Other SO_2 "dry" reduction strategies, such as spray dryers, furnace sorbent injection, and conversion to fluidized-bed combustion, have a significant impact of ESPs located downstream. These techniques increase the particulate loading and change the resistivity of the collected material, generally making it more difficult to remove. Durham et al. (1990) reported resistivity values of sorbent–fly-ash mixtures of the order of 10^7 to 10^9 ohm-cm at spray dryer conditions. Although the injected material had a high resistivity at 420 K, the surface conditioning provided by the increased moisture and cooling was sufficient to reduce it dramatically. At such low resistivities, the electrostatic force holding the particles onto the collector plates is reduced and the particles are easily reentrained. They found that reentrainment had to account for 60% of the observed penetration per section to correctly predict ESP performance at spray dryer conditions (see Problem 9.19).

The cohesive characteristics of the sorbent–fly-ash mixture improve by using additives such as ammonia. The interaction of ammonia and SO_3 results in the formation of ammonium bisulfate and sulfate, which condense on the surface of the particles. These surface deposits are viscous and cohesive, which reduce rapping and nonrapping reentrainment.

The *atmospheric fluidized-bed combustor* (AFBC) is an emerging technology for the control of SO_x and NO_x emissions from coal-fired power plants. Crushed coal is fed into a bed of inert ash mixed with limestone or dolomite. The bed is fluidized by injecting air through the bottom at a controlled rate. The coal burns within the bed and the SO_x formed reacts with the limestone to form a dry calcium sulfate which is removed with the fly-ash in a downstream particulate collection device. The system can remove up to 95% of the SO_2 and up to 80% of the NO_x emissions (U.S. Congress, 1991). However, Altman (1988) reported severe ESP operational problems at a full-scale coal-fired power plant converted to AFBC operation due to the high resistivity of the collected dust. Flue gas conditioning to reduce the particulate resistivity was been considered.

Gas conditioning equipment is frequently used to upgrade existing ESPs. Conditioning agents used include SO_3, H_2SO_4, sodium compounds, and ammonia. A typical dose rate of the gaseous agents is 10 to 30 ppm by volume. There are several vendors and systems commercially available. Gas conditioning usually gives good results with reasonably small expense.

9.7 COSTING CONSIDERATIONS

Your objectives in studying this section are to

1. Estimate the TCI for various types of ESPs.
2. Estimate the TAC for electrostatic precipitation.

This section presents methods to estimate the total capital investment and total annual costs for an ESP system. This material is taken primarily from Turner et al. (1988b).

9.7.1 TCI

The TCI includes costs for the ESP structure, the internals, rappers, power supply, auxiliary equipment, and the usual direct and indirect installation costs. Land, working capital, and off-site facilities are not normally required.

ESP equipment costs are almost always correlated with the collecting area. Turner et al. (1988 b) obtained cost quotes from precipitator vendors and regressed them against their collecting areas. These costs, updated to June 1990, are given by

$$EC = aA_c^b \tag{9.34}$$

where

A_c = collecting area, m^2

a, b = regression parameters

Table 9.5 lists the values of a and b for different ranges of collecting area.

Table 9.5 Regression Parameters for ESP Equipment Cost Equation[a]

Plate area (m^2)	a	b
930 to 4,600	4,551	0.6276
4,600 to 93,000	715	0.8431

Source: From Turner, J. H. et al. *JAPCA*, **38**:715(1988). Reprinted with permission from *JAPCA*. [a]Updated to June 1990.

These costs apply to all ESP configurations, except the two-stage. They include the following: ESP casing, pyramidal hoppers, rigid electrodes and internal collecting plates, transformer-rectifier (TR) sets and microprocessor controls, rappers, inlet and outlet nozzles and diffuser plates, weather enclosure and stair access, structural supports, and insulation (7.6 cm of fiberglass encased in a metal skin).

Two-stage ESPs are usually limited to small sizes and are sold as modular units. To be consistent with industry practice, the equipment costs (in $ of June 1990) are given as a function of flow rate through the unit (Turner et al. 1988b),

$$EC = 27{,}200 + 41{,}500 \ln Q \qquad 1.0 < Q, \frac{\text{m}^3}{\text{s}} < 6.0 \tag{9.35}$$

where Q is the actual volume flow rate through the unit. This cost is for modular units fully assembled mechanically and electrically, and mounted on a steel structural skid.

The purchased equipment cost, B, is the sum of the equipment cost, auxiliary equipment, instruments and controls, taxes, and freight. TCI is estimated from a series of factors applied to the purchased equipment cost to obtain direct and indirect costs for installation. Table 9.6 summarizes the most important factors for average installation conditions. For two-stage ESPs, purchased as packaged systems, installation costs are greatly reduced. Turner et al. (1988b) suggest that, in this case, TCI = 1.25B.

9.7.2 TAC

Direct annual costs include operating and supervisory labor, maintenance (labor and materials), utilities, dust disposal, and waste water treatment for wet ESPs. Typical operating labor requirements are 2 h/shift. Supervisory labor is taken as 15% of operating labor. Maintenance labor is estimated as 660 h/yr; annual maintenance materials are approximately 1% of the purchased equipment cost.

If the collected dust cannot be recycled or sold, it must be disposed of. Disposal costs for nonhazardous wastes are typically $30/metric ton. Disposal of hazardous wastes may cost 10 times as much. If the dust collected can be reused or sold, a recovery credit should be taken.

For capital recovery calculations, ESP systems are assumed to have average lifetimes of 20 yr. Overhead is calculated as 60% of total labor plus maintenance materials. Property taxes, insurance, and administrative charges are estimated as 4% of the TCI.

Example 9.16 Capital Investment and Annual Cost of ESP

Assume a plate-wire ESP is required for the control of fly-ash emissions from a coal-fired boiler burning bituminous coal. The flue gas stream is 26.3 m^3/s at 435 K and has an inlet

Table 9.6 Capital Cost Factors for ESPs

Costs	Factor
Direct costs	
ESP and auxiliary equipment costs	A
Instruments and controls	$0.10A$
Taxes	$0.03A$
Freight	$0.05A$
Purchased equipment cost	$B = 1.18A$
Installation direct costs	$0.67B$
Total direct costs	$1.67B$ + SP[a] + Bldg[b]
Indirect costs	$0.57B$
TCI	$2.24B$ + SP + Bldg

Source: Vatavuk and Neveril (1980). [a]Site Preparation. [b]Buildings.

ash loading of 0.01 kg/m^3. The design SCA for 99.9% overall efficiency is 62.4 s/m. Estimate the TCI and TAC for this application. Assume that the unit operates for 8,640 h/yr. Operating labor rate is \$15/h, maintenance labor rate is \$20/h. The cost of electricity is \$0.06/kW-h, dust final disposal costs are \$30/metric ton. The minimum attractive rate of the return is 12%/yr. Assume that no buildings or special site preparation is needed. Auxiliary equipment needed—ductwork, fan, motor, and stack—cost \$65,000.

Solution

The total collection area is (62.4)(23.6) = 1,473 m^2. From Eq. (9.34) and Table 9.5, EC = 4,551 $(1473)^{0.6276}$ = \$443,100. The sum of the ESP and auxiliary equipment costs is A = 443,100 + 65,000 = \$508,100. From Table 9.6, the purchased equipment cost is B = 1.18 (508,100) = \$600,000. TCI = 2.24 (\$600,000) = \$1,344,000 (\$ of June 1990).

Assuming a total pressure drop of 1.0 kPa and a mechanical efficiency of 60 percent for the fan, the total fan and corona power requirements are given by [(1.0)(23.6)/0.6] + 23.6 [0.1158 + (0.00117/0.001)] = 70 kW. The total mass of dust collected yearly is (26.3)(0.01) (0.999)(3,600)(8,640)/1,000 = 7,330 metric ton. For a useful life of 20 yr and a yearly rate of return of 12%, the capital recovery factor is 0.1339/yr. The following table presents annual costs for this project.

Item	Annual cost ($/yr)
Direct annual costs	
Operating labor: (3)(360)(15) =	16,200
Supervisory: (0.15)(16,200) =	2,400
Maintenance labor: (660)(20) =	13,200
Maintenance materials: (0.01)(600,000) =	6,000
Electricity: (70)(8,640)(0.06) =	40,000
Waste disposal: (7,330)(30) =	219,900
Total direct annual costs	297,700
Indirect annual costs	
Overhead : (0.6)(37,800) =	22,680
Tax, insurance, administration: (0.04)(1,344,000) =	53,760
Capital recovery cost: (0.1339)(1,334,000) =	178,600
Total indirect annual costs	255,040
TAC	$552,740

9.8 CONCLUSION

The selection of particulate control equipment in the near future will depend largely on regulatory trends and the technical and commercial success of ongoing particulate control research and development efforts. Well-designed modern ESPs can meet emission targets lower than the current NSPS. Requirements for greater particulate reductions or stringent emission limits based on respirable particulates may limit the use of conventional ESPs. Several options for significantly enhancing precipitators collection efficiency presently under active consideration are wet ESPs; pulse energization tuned specifically for fine particle control; and the addition of a very compact pulse jet baghouse (to be described in detail in Chapter 10) as a polishing step following an ESP.

ESPs have been the workhorse of particulate control equipment for almost a century. However, their future is uncertain and will depend on the ability of researchers and manufacturers to retain the reliability and simplicity of current designs while improving their tolerance to variations in particulate characteristics.

REFERENCES

Abbott, J. H., and Drehmel, D. C. *Chem. Eng. Prog.*,**72**:47 (1976).

Altman R. F. *JAPCA*, **38**:1455 (1988).

Cochet, R. *Colloq. Intern. Centre Natl.Rech. Sci.*[Paris], **102:**331 (1961).

Cooper, C. D., and Alley, F. C. *Air Pollution Control: A Design Approach,* PWS Engineering, Boston, MA (1986).

Crawford, M. *Air Pollution Control Theory*, McGraw-Hill, New York (1976).

Durham, M. D., Rugg, D. E., Rhudy, R. G., and Puschaver, E. J. *J. Air Waste Manage. Assoc.*, **40**:112 (1990).

Feldman, P. L. Paper No. 75-02.3, 68th Annual Meeting, *Air Pollution Control Association*, Boston, MA (1975).

Flagan, R. C., and Seinfeld, J. H. *Fundamentals of Air Pollution Engineering*, Prentice Hall, Englewood Cliffs, NJ (1988).

Harrison, W. A., Nicholson, J. K., DuBard, J. L., Carlton, J. D., and Sparks, L. E. *JAPCA,* **38**:209 (1988).

Hein, A. G. *JAPCA*, **39**:766 (1989).

Licht, W. *Air Pollution Control Engineering: Basic Calculations for Particulate Collection*, Marcel Dekker, New York (1980).

McDonald, J. R., and Dean, A. H. *Electrostatic Precipitator Manual,* Noyes Data Corporation, Park Ridge, NJ (1982).

Offen, G. R., and Altman R. F. J. *Air Waste Manage.Assoc.,*.**41**:222 (1991).

Turner, J. H., Lawless, P. A., Yamamoto, T., Coy, D. W., Greiner, G. P., McKenna, J. D., and Vatavuk, W. M. *JAPCA*, **38:**458 (1988a).

Turner, J. H., Lawless, P. A., Yamamoto, T., Coy, D. W., Greiner, G. P., McKenna, J. D., and Vatavuk, W. M. *JAPCA,* **38**:715 (1988b).

U.S. Congress, Office of Technology Assessment, *Energy Technology Choices: Shaping Our Future,* OTA-E-493, Washington, DC (1991).

Vatavuk, W. M., and Neveril, R. B. *Chemical Engineering,* pp. 157-162 (November 3, 1980).

White, H. J. "Control of Particulates by Electrostatic Precipitation," Chap.12 in *Handbook of Air Pollution Technology*, S. Calvert and H. M. Englund (Eds.), Wiley, NewYork (1984).

White, H. J. *Industrial Electrostatic Precipitation* Addison-Wesley, Reading, MA (1963).

PROBLEMS

The problems at the end of each chapter have been grouped into four classes (designated by a superscript after the problem number)

Class a: Illustrates direct numerical application of the formulas in the text.
Class b: Requires elementary analysis of physical situations, based on the subject material in the chapter.
Class c: Requires somewhat more mature analysis.
Class d: Requires computer solution.

9.1[a]. Terminal velocity of a charged particle in an electric field

Repeat Example 9.1 for 2.0-μm diameter particles. Assume that the particle charge is proportional to its surface area. Estimate the terminal velocity of the particles.

Answer: 0.375 m/s

9.2[a]. Dimensions of a tubular ESP

A small tubular ESP is to produce a collection efficiency of 99% when handling 0.25 m^3/s of air at 773 K and 1 atm containing 2.5-μm particles. The strength of the electric field is 200,000 V/m and the particle charge is 10^{-15} C. Determine the dimensions of the ESP. The velocity of the gases through the precipitator must not exceed 0.318 m/s.

Answer: L = *1.28 m*

9.3[b]. ESP overall efficiency to satisfy NSPS

Consider the coal-fired power plant of Example 5.1. Assuming that 25% of the ash drops out of the furnace as slag, calculate the efficiency of an ESP to remove fly-ash if the plant is to meet the 1980 federal NSPS for particulates.

Answer: 99.61%

9.4[a]. Corona onset value in plate-wire ESP

A plate-wire ESP handles air at 700 K and 101.3 kPa. The plate spacing is 460 mm and the diameter of the discharge wires is 5 mm. Estimate the corona onset voltage.

Answer: 17,300 V

9.5[a]. Average electric field in a plate-wire ESP

Estimate the average electric field and the operating voltage for the ESP of Problem 9.4. The resistivity of the collected dust is 10^{12} ohm-cm.

Answer: 119 kV/m

9.6[a]. Average electric field in a flat plate ESP

Repeat Problem 9.5 but for a flat-plate ESP configuration.

Answer: 165 kV/m

9.7[a]. Particle saturation charge

Determine the time constant and saturation charge for field charging of 5.0-μm particles having a dielectric constant of 4.0 if the ion concentration is 10^{16} ions/m^3 and the field strength is 150 kV/m. Assume that the ion mobility is 2.2×10^{-4} m^2/V-s.

Answer: 0.21 fC

9.8[a]. Total particle charge according to Feldman

Estimate the total charge on the particles of Problem 9.7 according to Eq. (9.14) for a charging time of 0.1 s. Assume that the gas mean-free path is 0.1 μm.

Answer: 0.215 fC

9.9[b]. Effect of gas flow rate on ESP grade efficiency

Consider the ESP of Example 9.7. If the gas flow rate increases by 20 percent, estimate the percent increase in penetration for 1.0-μm particles.

Answer: 66.5%

9.10[b]. Effect of average electric field strength on ESP performance

Consider the ESP of Example 9.7. If the average electric field strength decreases by 20%, estimate the percent increase in penetration for 1.0-μm particles.

Answer: 200%

9.11[a]. Effective migration velocity

Estimate the effective migration velocity for the fly-ash of Example 9.9.

Answer: 0.128 m/s

9.12[a]. ESP for the control of emissions from a cement kiln

The gases from a cement kiln flow at a volumetric rate of 380 m^3/min at 590 K and 1 atm. A plate-wire ESP will remove 99.9% of the particulate matter carried by the gases. Assuming no back corona, estimate the total collection area based on the effective migration velocities of Table 9.2.

Answer: 2,430 m^2

9.13[b]. Effect of sulfur content of the fuel on ESP performance

Selzler and Watson (*JAPCA*, **24**:115, 1974) proposed the following empirical Deutsch-type equation for the overall penetration of an ESP for the control of fly-ash emissions from a pulverized coal-fired furnace:

$$Pt_M = \exp\left[-0.0456(\text{SCA})^{1.4}\left(\frac{KW}{Q}\right)^{0.6}\left(\frac{S}{AH}\right)^{0.22}\right]$$

where

SCA = specific collection area, s/m

Q = gas volumetric flow rate, m^3/s

KW = power input to the discharge electrode, kW

S = sulfur content of the fuel, weight percent

AH = ash content of the fuel, weight percent

(a) A precipitator with an overall efficiency of 97.7% operates on a coal with a sulfur-to-ash ratio of 1.5 : 12.3. To help meet SO_2 emissions standards, a coal is substituted with a sulfur-to-ash ratio of 0.8 : 12.5. If all the other operating variables are the same, estimate the new collection efficiency.

Answer: 96.2%

(b) What percent change in corona power is necessary if, for the new sulfur-to-ash ratio, it is desired to restore the collection efficiency to its original value?

Answer: 26.7% increase

9.14[a]. Corona power for a fly-ash ESP

Problem 9.13 gives Selzler and Watson's empirical equation for the overall collection efficiency of an ESP. A precipitator treats 1,322 m^3/s of flue gas to remove fly-ash particles from a pulverized coal-fired power plant. The design overall removal efficiency is 99.5%. The total collection area is 79,320 m^2; the coal contains 1.0% S and 12% ash. Estimate the corona power for these conditions.

Answer: 650 kW

9.15[d]. Effect of back corona on plate-wire ESP performance

Consider the conditions described in Example 9.9. Assuming that a plate-wire precipitator is used with the same SCA, integrate the grade-efficiency equation to calculate the resulting overall penetration

(a) With no back corona

Answer: 1.9%

b) With severe back corona

Answer: 6.5%

9.16[d]. Multiple cyclone precleaner for an ESP

The gas flow rate through the ESP of Example 9.9 increases to 30 m^3/s because of changes in the capacity of the boiler. Instead of increasing the collection area, the ESP is fitted with a multiple cyclone precleaner with an overall removal efficiency of 80%. The aerosol population leaving the cyclones is log-normally distributed with MMD = 7.0 μm and $\sigma_g = 2.5$. Estimate the overall efficiency of the combined system at the new gas flow rate.

Answer: 99.6%

9.17[a]. Effect of non-uniform gas velocity on ESP performance

Example 9.9 calculates the overall efficiency of an ESP assuming a uniform gas velocity distribution. If the distribution is characterized by a quality factor of 1.2, estimate the actual efficiency assuming no gas sneakage or reentrainment losses.

Answer: 99%

9.18[b]. Flow quality factor and statistical measures of velocity nonuniformity

McDonald and Dean (1982) presented the following empirical relationship based on a pilot plant study between ϕ, the normalized standard deviation of the gas velocity distribution (σ_f), and the ideal penetration predicted:

$$\phi = 1 + 0.766\left(1 - Pt^{id}\right)\sigma_f^{1.786} + 0.0755\, \sigma_f \ln Pt^{id}$$

where

$$\sigma_f = \frac{\sqrt{\frac{1}{N}\sum_{j=1}^{N}\left(v_a - v_j\right)^2}}{v_a}$$

(a) Estimate σ_f for the flow distribution conditions of Example 9.10. Estimate the flow quality factor predicted under those conditions by McDonald and Dean's correlation.

Answer: $\phi = 1.224$

(b) The ideal overall collection efficiency of an ESP is 99.9% assuming a uniform velocity distribution. If the normalized standard deviation of the actual distribution is 50%, estimate the resulting overall efficiency assuming no losses by gas sneakage or reentrainment.

Answer: 99%

9.19[c,d]. ESP performance in SO_2 dry scrubbing applications

Dry scrubbing processes offer potential cost-effective retrofit FGD technologies for existing coal-fired electric generating stations faced with acid rain legislation. However, the existing particulate control equipment must be capable of collecting both the flyash and the unreacted injected sorbent. The following data were obtained at the TVA 10 MW Spray Dryer/ESP Pilot Plant (Durham et al. 1990):

Parameter	Baseline Conditions	Spray Dryer Conditions
Flow rate, m^3/s	16.35	15.5
Temperature, K	430	336
Plate area per section, m^2	314.4	314.4
Number of sections	4	4
Plate spacing, mm	254	254
Average electric field, kV/m	300	370
Inlet size distribution		
MMD, μm	8.0	10.0
σ_g	2.5	2.3
Inlet dust concentration, g/m^3	2.634	21.131
Outlet dust concentration, g/m^3	0.00453	0.0355

Gas sneakage and the flow quality factor were estimated as 0.05, and 1.1 respectively for both sets of operating conditions. The dielectric constant of the particles for both conditions is about 8.0.

(a) Estimate the reentrainment fraction, *RR*, at baseline conditions.

Answer: 0.09

(b) Estimate *RR* at spray dryer conditions

Answer: 0.158

(c) Notice that at spray dryer conditions the outlet dust concentration increases significantly as compared to the baseline conditions. Design a new ESP that, at spray dryer conditions, achieves the outlet concentration that characterized baseline conditions. Specify the number of sections and the collection area per section.

9.20[b]. Dimensions of an ESP

Specify the dimensions of an ESP to process 50,000 m^3/min of a flue gas with 99.5% overall particulate removal efficiency. The effective migration velocity of the particles is 0.1 m/s. Assume that the plate spacing is 400 mm; the aspect ratio is 1.0; and the gas velocity is not to exceed 1.0 m/s. Specify the total collection area, the number of channels, the dimensions of the ESP, and the corona power requirements.

Answer: $A_c = 44{,}300\ m^2$

9.21[a]. Total capital investment for an ESP

Estimate the TCI and the annual capital recovery cost for the ESP of Problem 9.20. Assume that the cost of auxiliary equipment is $215,000. No special site preparation or buildings are needed. Assume a useful life of 20 yr and no salvage value. The minimum attractive rate of return is 15%/yr.

Answer: CRC = $2,586,400/yr

9.22[d]. Plate-wire ESP for a Portland cement kiln

In Problem 8.17, a multiple cyclone system was designed as a precleaner for the control of particulate emissions from a Portland cement kiln. For each of the alternatives considered in that problem, design a plate-wire ESP such that the combined system is capable of satisfying the particulate NSPS for the source. The plate spacing must be 400 mm. The dielectric constant of the particles is 6.14; their resistivity is below 10^{11} ohm-cm. From scale models, gas sneakage is expected to be about 0.10; rapping reentrainment about 0.10; and the flow quality factor 1.2. For each alternative, specify the following:

(a) Number of sections
(b) Total collection area
(c) TCI and TAC of the combined multiple cyclone-ESP system

Choose between the three alternatives suggested based on the EUAC measure of merit. Assume that the ESP has a useful life of 20 yr with no salvage value. Operating and maintenance labor rates are \$15/h and \$20/h, respectively. The dust collected in both devices can be recycled to the process at no additional cost.

10

Fabric Filters

10.1 INTRODUCTION

Fabric filtration is a widely accepted method for particulate emissions control. In fabric filtration, the particle-laden gas flows through a number of filter bags placed in parallel, leaving the dust retained by the fabric. Extended operation of a fabric filter, or baghouse, requires that the dust be periodically cleaned off the cloth surface and removed from the baghouse. After a new fabric goes through a few cycles of use and cleaning, it retains a residual cake of dust that becomes the filter medium. This phenomenon is responsible for the highly efficient filtering of small particles that characterizes baghouses.

The type of cloth fabric limits the temperature of operation of baghouses. Cotton fabric has the least resistance to high temperature (about 355 K), while fiberglass has the most (530 K). In many cases this requires that the waste gas stream be precooled before entering the baghouse. The cooling system then becomes an integral part of the design. Conversely, the temperature of the exhaust gas stream must be well above the dew point of any of its condensable constituents as liquid particles plug the fabric pores quickly.

Most of the energy requirements of the system are to overcome the gas pressure drop across the bags, dustcake, and associated ductwork. Typical values of pressure drop range from 1 to 5 kPa. The most important design parameter is the ratio of the gas volumetric flow rate to fabric area, known as the *gas-to-cloth ratio*. Other important process variables include particle characteristics (such as size distribution and stickiness), gas characteristics (temperature and corrosivity), and fabric properties.

Electric utilities have made significant progress in recent years in designing and operating baghouses for the collection of coal fly ash. Interest in baghouses can be expected to continue increasing as air emission standards become more stringent and concerns over inhalable particles and air toxic emissions increase. In many cases, the relative insensitivity of baghouses to changes in fly ash properties make them more attractive than ESPs

The oldest fly-ash baghouse was commissioned in 1973. By 1990, there were close to 100 baghouses in operation on utility boilers, representing more than 21,000 MW of generating capacity (Cushing et al. 1990).

10.2 TYPES OF FABRIC FILTERS

Your objective in studying this section is to

Describe the operational characteristics of the three most common baghouse designs: (1) shaker; (2) reverse air; and (3) pulse jet.

The most important distinction between fabric filters designs is the method used to clean the dust from the bags between filtration cycles. The distinguishing features of each cleaning method are described subsequently, following Turner et al (1987a).

10.2.1 Shaker Cleaning

For any type of cleaning, enough energy must be imparted to the fabric to overcome the adhesion forces holding the dust to the bags. In shaker cleaning, used with inside-to-outside gas flow, this is accomplished by suspending the bag from a motor-driven hook or framework that oscillates. The motion creates a sine wave along the fabric, which dislodges the previously collected dust. Chunks of agglomerated dust fall into a hopper below the compartment. The compartments operate in sequence so that one compartment at a time is cleaned. Figure 10.1 is a schematic diagram of a shaker baghouse.

Parameters that affect cleaning include the amplitude and frequency of the shaking motion and the tension of the mounted bag. Typical values of the first two parameters are 4 Hz for frequency and 50 to 75 mm for amplitude. The vigorous oscillations tend to stress the bags and require heavier and more durable fabrics. In the United States, woven fabrics are used almost exclusively for shaker cleaning, whereas in Europe felted fabrics are used.

10.2.2 Reverse-Air Cleaning

When fiberglass fabrics were introduced, a gentler means of cleaning the bags was needed to prevent premature bag failure. Reverse-air cleaning was developed as a less intensive way to impart energy to the bags. Gas flow to the bags is stopped in the compartment being cleaned, and a reverse flow of air is directed through the bags. This reversal of flow gently collapses the bags and the shear forces developed remove the dust from the surface of the bags. The reverse air for cleaning comes from a separate fan capable of supplying clean, dry air for one or two compartments at a gas-to-cloth ratio similar to that of the forward gas flow.

Many reverse-air baghouses are installing sonic horns to improve the performance of the cleaning step. Sonic energy using horns during the reverse flow concentrates more cleaning energy at the bag-dust cake interface. The resulting reduction in residual dust cake density reduces the pressure drop by 50% to 60% (Carr and Smith 1984b). Sonic horns operate on compressed air at

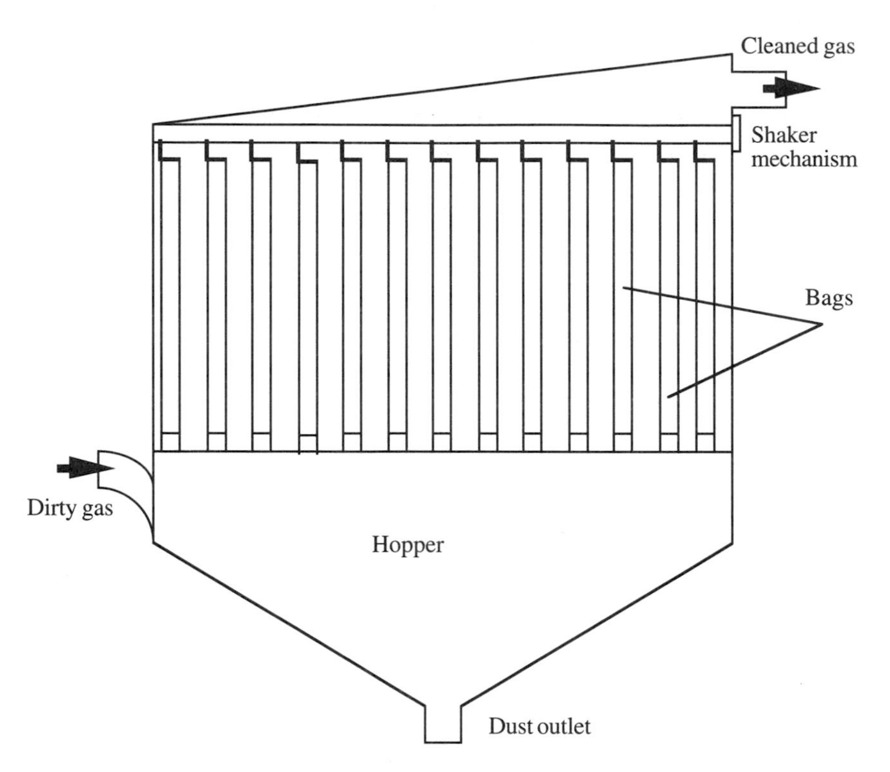

Figure 10.1 Schematic diagram of a shaker baghouse

pressures of about 400 to 900 kPa. It appears that reverse-air cleaning with sonic assistance is the cleaning method of choice for baghouses on pulverized coal-fired boilers (Cushing et al. 1990).

10.2.3 Pulse-Jet Cleaning

This form of cleaning forces a burst of compressed air down through the bag expanding it violently. As with shaker baghouses, the fabric reaches its extension limit and the dust separates from the bag. In pulse jets, however, filtering flows are opposite in direction when compared with shaker or reverse-air designs, with the outside-to-inside gas flow. Bags are mounted on wire cages to prevent collapse while the dusty gas flows through them. The top of the bag and cage assembly is attached to the baghouse structure, whereas the bottom end is loose and tends to move in the turbulent gas flow.

Most pulse-jet baghouses are not compartmented. Bags are cleaned by rows when a timer initiates the burst of cleaning air through a quick-opening valve. Usually 10% of the collector is pulsed at a time by zones. A pipe above each row of bags carries the compressed air. The pipe is pierced above each bag so that cleaning air exits directly downward into the bag. The pulse opposes and interrupts forward gas flow for only fractions of a second. However, the quick resumption of forward flow redeposits most of the dust back on the "clean" bag or on adjacent bags. Pulse jets normally operate at about twice the gas-to-cloth ratio of reverse-air baghouses.

10.3 FABRIC FILTRATION THEORY

Your objectives in studying this section are to

1. Describe, in qualitative terms, the mechanisms of particle penetration through a fabric filter.
2. Estimate the tubesheet pressure drop in baghouses.

The key to designing a baghouse is to determine the gas-to-cloth ratio that produces the optimum balance between pressure drop (operating cost) and baghouse size (capital cost). Although collection efficiency is another important measure of baghouse performance, properly designed and well-run baghouses are highly efficient

10.3.1 Penetration

Design overall efficiencies for fabric filters range from 98% to more than 99.9%. Most units currently in operation meet or exceed their design efficiencies (Cushing et al. 1990). Two basic mechanisms account for the particles that pass through the baghouse. Particles can escape collection through leaks in the ducting, tubesheet, or bag clamps, or through holes, tears, or improperly sewn seams in the bags themselves. The second mechanism for emissions is particle movement through the dustcake and the fabric, also known as *bleed-through*. Bleed-through is primarily a function of baghouse design and particle morphology. Smooth, spherical particles are less cohesive, and can seep through the dustcake and fabric easily, increasing emissions.

Well-developed equations exist in aerosol physics to calculate the fractional penetration of particles on various structures by filtration forces (Crawford 1976; Flagan and Seinfeld 1988). However, each requires detailed knowledge of the geometry of the collection medium. Because the structure of dust cakes is not defined—and, in fact, appears to be system dependent in ways that are now not understood—none can be used to predict fractional penetration.

Figure 10.2 shows actual baghouse penetration data reported by Ensor et al. (1981). Considering the dominant collection mechanisms, diffusion (which decreases with increasing particle size), and impaction and interception (which increase with particle size), the classic theoretical penetration curve exhibits a single maximum in the 0.1- to 0.3-μm range. Thus, additional mechanisms must act to create the bimodal curve of Figure 10.2. A reasonable explanation is to attribute the large-particle penetration mode to bleed-through (Carr and Smith 1984a).

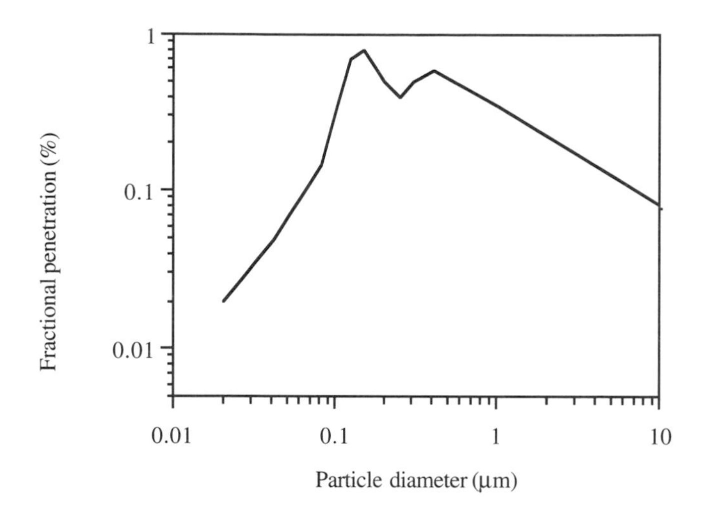

Figure 10.2 Measured baghouse fractional penetration (From Carr, R. C., and Smith, W. B. *JAPCA*, **34**:79, 1984; reprinted with permission from *JAPCA*)

Dennis and Klemm (1979) proposed the following semiempirical overall penetration equation based on the observations that penetration increased with increasing gas-to-cloth ratio and decreased as the dust cake on the fabric grew:

$$Pt = Pt_s + \left(Pt_0 - Pt_s\right)e^{-aW} + Pt_{bt} \tag{10.1}$$

where

Pt = overall penetration

Pt_s = low-level penetration due to pinholes in the cake-fabric structure; increases with the gas-to-cloth ratio

Pt_0 = penetration through a just-cleaned fabric area,

Pt_{bt} = penetration due to bleed-through; a function of particle morphology

W = dust mass per unit bag surface area, also known as areal density, kg/m^2

a = cake penetration decay rate

Viner et al. (1984) found that Eq. (10.1) overpredicts penetration just after a compartment has been cleaned; but, as the dust cake on the filter grows, the discrepancy between predicted and measured values diminishes. All the terms in Eq. (10.1) must be determined empirically for a given fabric-dust cake combination. The following example illustrates typical conditions in a reverse-air filter.

Example 10.1 Time Dependency of Penetration in Baghouse

The waste gases from a coal-fired boiler flow through a reverse-air baghouse for fly-ash removal. The following empirical equation for overall penetration as a function of operation time between cleaning cycles was developed:

$$Pt = 160V^{2.32} + \left[0.10 - 160V^{2.32}\right]\exp\left(-180C_i Vt\right) + \frac{5 \times 10^{-7}}{C_i}$$

where

V = gas-to-cloth ratio, m/s
C_i = inlet dust loading, kg/m3
t = operation time since last cleaning

The baghouse operates at an air-to-cloth ratio of 0.01 m/s and the inlet dust loading is 0.004 kg/m^3. Each compartment is cleaned every 20 min. Plot penetration as a function of time, and calculate the average overall penetration for the cycle.

Solution

Substituting the values of V and C_i in the penetration equation:

$$Pt = 0.0038 + 0.0963\exp\left(-0.432t'\right)$$

where t' is the time in minutes. Figure 10.3 shows the variation in penetration with time. Most of the particle emissions take place shortly after the bags are cleaned. To calculate the average penetration, integrate over a cycle and divide by the length of the cycle.

$$Pt_{av} = \frac{\int_0^{20} \left[0.0038 + 0.0963\exp\left(-0.432t'\right)\right] dt'}{20} = 0.0149$$

Vann Bush et al (1989) derived an expression to quantify the texture or surface morphology of bulk ash samples. Their morphology factor (M) is a dimensionless quantity based on volumetric size distribution data measured with a Coulter Counter (a device that measures particle size by a change in electrolytic resistivity), specific surface area (A), and the true density of the ash particles (ρ), and is defined as

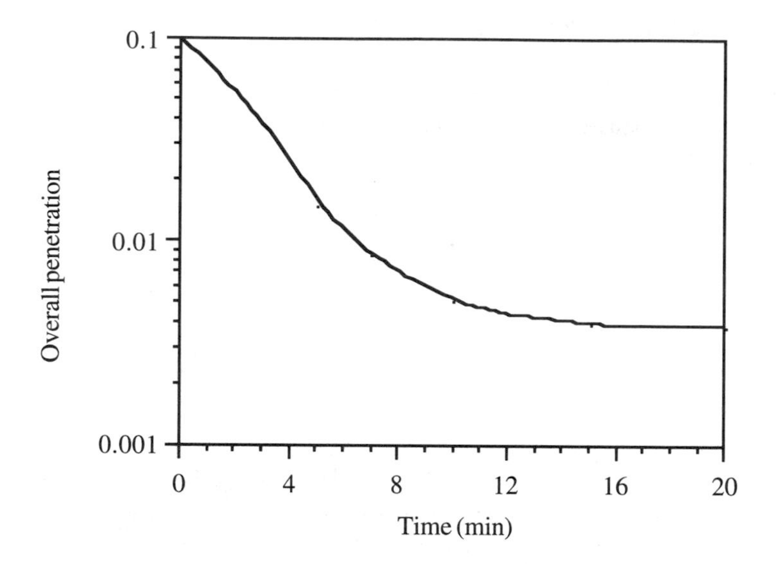

Figure 10.3 Penetration versus time in a reverse-air baghouse.

$$M = \frac{A\rho}{6} \exp\left[\frac{\sum_{i=1}^{N} n_i D_{pi}^3 \ln D_{pi}}{\sum_{i=1}^{N} n_i D_{pi}^3}\right] \tag{10.2}$$

where

n_i = the total number of particles counted in Coulter channel i

D_{pi} = the geometric mean diameter of Coulter channel i, μm

A low value of M indicates that the particles resemble smooth spheres, while a high value indicates that the particles have very irregular shapes and surfaces. For perfect spheres, M = 1.0. Field experience shows that ashes with low values of M (2.0 to 2.7) seep through the fabric, increasing emissions (Felix and Merritt 1986; Miller et al. 1985).

The mechanism of bleed-through is not well understood yet and there is no general theoretical model to predict its effect on particle penetration through baghouses. It is fortunate that fabric filters properly designed based on pressure drop considerations generally exhibit penetrations well below the applicable NSPS.

10.3.2 Pressure drop

There are several contributions to the total pressure drop across a baghouse including the pressure drop from the flow through the inlet and outlet ducts, from flow through the hopper regions, and from flow through the bags. The pressure drop across the bags (also called the *tubesheet pressure drop*) is a complex function of the physical properties of the dust and fabric and the manner in which the baghouse is designed and operated. The duct and hopper losses are constant and can be minimized effectively through proper design.

Fabric filtration is inherently a batch process that has been adapted to continuous operation through clever engineering. The dust collected on the bags must be removed periodically for continuous operation. Shaker and reverse-air designs are similar in the sense that they both normally use woven fabric bags run at relatively low gas-to-cloth ratios, and the filtration mechanism is cake filtration. The fabric merely serves as a support for the formation of a dust cake, which is the actual filtering medium. Pulse-jet baghouses generally use felt fabrics and run with a high gas-to-cloth ratio. The felt fabric plays a much more active role in the filtration process. The distinction between cake filtration and fabric filtration has important implications for calculating the pressure drop across the filter bags.

The design of a fabric filter begins with a set of specifications including average and maximum pressure drop, total gas flow, and other requirements. Based on these requirements, the designer specifies the maximum face velocity allowed. The standard way to relate tubesheet pressure drop to gas-to-cloth ratio is

$$\Delta P(t) = S(t)V \tag{10.3}$$

where

$\Delta P(t)$ = the pressure drop through the filter, a function of time, t
$S(t)$ = the drag through the fabric and cake, Pa-s/m
V = the average or design gas-to-cloth ratio, m/s

The drag across the fabric and cake is a function of the amount of dust collected on the surface of the bags. Assuming that the filter drag is a linear function of the dust load,

$$S(t) = S_e + K_2 W(t) \tag{10.4}$$

where

S_e = the drag of a dust-free (freshly cleaned) filter bag

K_2 = the dust cake flow resistance, s^{-1}

The dust mass as a function of time is $W = C_i Vt$, where C_i is the inlet dust concentration (kg/m^3). Substituting this expression in Eqs. (10.3) and (10.4):

$$\Delta P(t) = S_e V + K_2 C_i V^2 t \tag{10.5}$$

The constants S_e and K_2 depend on the fabric and the nature and size of the dust. The relationship between these constants and the dust and fabric properties are not understood well enough to permit accurate predictions and so must be determined empirically.

Example 10.2 Estimation of Parameters in Filter Drag Model

(a) Estimate the parameters in Eq. (10.4) based on the following data for a freshly cleaned fabric. The gas-to-cloth ratio is 0.0167 m/s and the inlet dust concentration is 0.005 kg/m^3.
(b) Estimate the pressure drop for the same conditions after 70 min of continuous operation.

Test Data

Time (min)	ΔP (Pa)
0	150
5	380
10	505
20	610
30	690
60	990

Solution

(a) Based on the test data, generate a plot of filter drag versus areal density. The following table illustrates the data to be plotted.

$S = \Delta P/V$ (kPa-s/m)	$W = C_i V t$ (kg/m^2)
9.00	0
22.80	0.025
30.30	0.050
36.60	0.100
41.40	0.150
59.40	0.030

Figure 10.4 shows an initial characteristic curvature followed by a linear dependency of filter drag on areal dust density. The nonlinear portion of the curve is due to a non-uniform initial flow through the fabric. The previous cleaning cycle usually dislodges the cake in

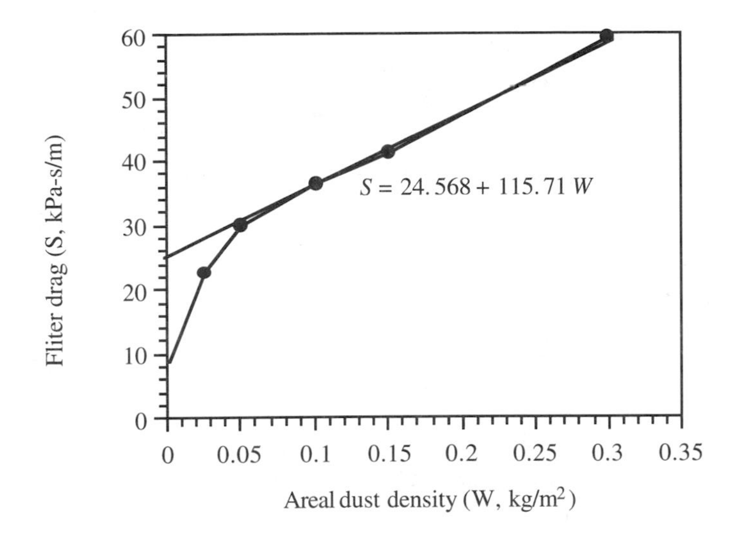

Figure 10.4 Filter drag versus dust areal density for Example 10.2

irregular chunks, leaving some parts of the bag very clean and others still quite dusty, resulting in spatial variations in the initial gas flow. A least-squares fit of the linear portion of the curve yields: $S_e = 24.57$ kPa-s/m, and $K_2 = 115.7$ kPa-s-m/kg $= 1.16 \times 10^5\ s^{-1}$.
(b) Equation (10.5) gives the pressure drop after 70 min of continuous operation: $\Delta P = 24{,}570(0.0167) + (1.16 \times 10^5)(0.005)(0.0167)^2(70)(60) = 1{,}090$ Pa.

Example 10.3 Cycle time for Reverse-Air Fabric Filter.

A reverse-air fabric filter has 1,000 m^2 of filtering area and treats 10 m^3/s of air carrying a dust concentration of 0.005 kg/m^3. Assume $S_e = 20.0$ kPa-s/m and $K_2 = 1.0 \times 10^5\ s^{-1}$. If the filter must be cleaned when the pressure drop reaches 2.0 kPa, after what period must the cleaning occur?

Solution

The gas-to-cloth ratio is $V = 10/1{,}000 = 0.01$ m/s. Solving Eq. (10.5) for the time of operation before the pressure drop exceeds 2,000 Pa, $t = [2{,}000 - (20{,}000)(0.01)]/[(10^5)(0.005)(0.01)^2] = 36{,}000$ s $= 10$ h.

Pulse-jet baghouses have been designed to operate in a variety of modes. Some remain on line at all times and are cleaned frequently. Others are taken off line for cleaning at relatively long intervals. A complete model of pulse-jet filtration therefore must account for the composite dust-fabric filtration occurring on a relatively clean bag, the cake filtration that result from prolonged periods on line, and the transition period between the two regimes.

If a compartment is taken off line for cleaning, the dust that is removed from the bags falls into the hopper before forward gas flow resumes. If a compartment is cleaned while on line, only a small fraction of the dust removed from the bag falls to the hopper. The remainder of the dislodged dust will be redeposited (i.e., "recycled") on the bag by the forward gas flow. The redeposited dust layer has different pressure drop characteristics than the freshly deposited dust. Dennis and Klemm (1979) proposed the following model of drag across a pulse-jet filter:

$$S = S_e + (K_2)_c W_c + K_2 W_0 \tag{10.6}$$

where

$(K_2)_c$ = specific dust resistance of the recycling dust
W_c = areal density of the recycling dust
K_2 = specific dust resistance of the freshly deposited dust
W_0 = areal density of the freshly deposited dust

This model can easily account for all three regimes of filtration in a pulse-jet baghouse.The pressure drop can thus be expressed as the sum of a relatively constant term and a term that increases with dust buildup.

$$\Delta P = (PE)_{\Delta w} + K_2 W_0 V \tag{10.7}$$

where

$$(PE)_{\Delta w} = [S_e + (K_2)_c W_c] V \tag{10.8}$$

The disadvantage of the model represented by Eqs. (10.7) and (10.8) is that the constants S_e, K_2, and W_c cannot be predicted at this time. Consequently, correlations of laboratory data must be used to determine the value of $(PE)_{\Delta w}$. For one fabric-dust combination of Dacron felt and coal fly ash, Dennis and Klemm (1980) developed the following correlation:

$$(PE)_{\Delta w} = 1{,}045\, V P_j^{-0.65} \tag{10.9}$$

where

$(PE)_{\Delta w}$ is in kPa
P_j = gauge pressure of the cleaning pulse, in kPa
V = gas-to-cloth ratio, in m/s

Example 10.4 Tubesheet Pressure Drop in Pulse-Jet Baghouse

A pulse-jet baghouse uses Dacron felt bags for the control of fly-ash emissions. The gas-to-cloth ratio is 0.024 m/s and the inlet dust loading is 0.01 kg/m^3. The bags are cleaned at 10-min intervals with pulses of air at 650 kPa gauge. If $K_2 = 1.5 \times 10^5$ s^{-1}, estimate the maximum tubesheet pressure drop.

Solution

The conditions are similar to those for which Eq. (10.9) was developed. Then, $(PE)_{\Delta w}$ = $(1{,}045)(0.024)(650)^{-0.65} = 0.372$ kPa. At the end of the 10-min cycle, $W_0 = (0.01)(0.024)(600) = 0.144$ kg/m^2. Equation (10.7) gives the tubesheet pressure drop: $\Delta P = 0.372 + (1.5 \times 10^5)(0.144)(0.024)/1{,}000 = 0.89$ kPa.

Example 10.5 Effect of Filter Inhomogeneities on Filter Drag Model (Cooper and Riff 1983)

Knowing the mean of a property, such as dust areal density, may be misleading for predicting the effect of that property. When estimating the specific dust resistance for a filter cake (K_2) from pressure drop data, Eq. (10.7) predicts that $K_2 \propto W^{-1}$. However, in a typical baghouse the areal density is not homogeneous, but changes from bag to bag. For example, Ellenbecker (1979) measured areal densities in a pulse-jet baghouse and found that the data could be fitted by a log-normal distribution with a median of 0.64 kg/m^2 and a geometric standard deviation, σ_g, of 1.15.

In the presence of inhomogeneities, the average value of W should be used for estimating K_2. The proper average to use is the *harmonic mean*, not the arithmetic mean. For a log-normal distribution, the ratio of the harmonic to the reciprocal of the arithmetic mean is

$$\frac{\overline{W^{-1}}}{\overline{W}^{-1}} = \exp\left(\ln^2 \sigma_g\right) \tag{10.10}$$

where the harmonic average for a log-normal distribution is:

$$\overline{W^{-1}} = \left[W_{50} \exp\left(-0.5 \ln^2 \sigma_g\right)\right]^{-1} \tag{10.11}$$

where W_{50} is the median.
(a) For Ellenbecker's data, estimate the error introduced in the estimation of K_2 if the arithmetic average of W is used instead of the harmonic average.
(b) Repeat part a for a geometric standard deviation of 2.0.

Solution

(a) Substituting $\sigma_g = 1.15$ in Eq. (10.10), $\overline{W^{-1}} / \overline{W}^{-1} = 1.02$. Therefore, using the arithmetic mean introduces an error of 2% in the estimation of K_2.

(b) For $\sigma_g = 2.0$, $\overline{W^{-1}} / \overline{W}^{-1} = 1.62$, therefore the error is 62%.

10.4 PRACTICAL DESIGN CONSIDERATIONS

Your objectives in studying this section are to

1. Choose a proper filter material for a given baghouse application.
2. Estimate a design gas-to-cloth ratio based on published data.
3. Estimate the net and gross cloth area required.
4. Appraise the potential of a waste heat boiler as a precooler.

10.4.1 Design gas-to-cloth ratio

The design gas-to-cloth ratio is difficult to estimate from theoretical principles. However, shortcut methods of varying complexity allow rapid estimation. One such method is to use gas-to-cloth ratio data available in the literature for similar applications. After a cleaning method and fabric type has been selected, the gas-to-cloth ratio can be estimated from Table 10.1.

Example 10.6 Net Cloth Area of Pulse-Jet Baghouse

A baghouse is required for controlling fly-ash emissions from a coal-fired boiler. The flue gas stream is 23.6 m3/s at 435 K. Estimate the net cloth area of a pulse-jet/Teflon felt baghouse for this application.

Table 10.1 Gas-to-Cloth Ratios[a] (cm/s)

Dust	Shaker/Woven Reverse-Air/Woven	Pulse-Jet/Felt
Alumina	1.27	4.07
Asbestos	1.52	5.08
Cocoa, chocolate	1.42	6.10
Cement	1.02	4.07
Coal	1.27	4.07
Enamel frit	1.27	4.57
Feeds, grain	1.78	7.11
Fertilizer	1.52	4.07
Flour	1.52	6.10
Fly ash	1.02	2.54
Graphite	1.02	2.54
Gypsum	1.02	5.08
Iron ore	1.52	5.59
Iron oxide	1.27	3.56
Iron sulfate	1.02	3.05
Leather dust	1.78	6.10
Lime	1.27	5.08
Limestone	1.37	4.07
Paint pigments	1.27	3.56
Paper	1.78	5.08
Rock dust	1.52	4.57
Sand	1.27	5.08
Sawdust	1.78	6.10
Silica	1.27	3.56
Soap, detergents	1.02	2.54
Starch	1.52	4.07
Sugar	1.02	3.56
Talc	1.27	5.08
Tobacco	1.78	6.61
Zinc oxide	1.02	2.54

a Generally safe design values. Source: From Turner J. H. et al. JAPCA, 37:749 (1987). Reprinted with permission from JAPCA.

Solution

Table 10.1 gives $V = Q/A_c = 2.54$ cm/s for filtration of fly-ash in pulse-jet filters. Then, the net cloth area is $A_c = (23.6)/(0.0254) = 930$ m2.

For continuously operated shaker and reverse-gas cleaned filters, the area must be increased to allow for shutting down of one or more compartments for cleaning and maintenance. A typical compartment uses bags 0.3 m in diameter and 10.7 m long, with the number of bags per compartment ranging from 40 to 648. Table 10.2 provides a guide for adjusting the net area to the gross area for these two types of baghouses. Because pulse-jet filters are cleaned on line, no additional filtration area is required, and the net and gross cloth areas are equal.

Table 10.2 Guide to Estimate Shaker And Reverse-Air Baghouse Gross Cloth Area

Net Cloth Area (m^2)	Multiply Net Area by
1–370	2
371–1,115	1.5
1,116–2,230	1.25
2,231–3,350	1.17
3,351–4,460	1.125
4,461–5,580	1.11
5,581–6690	1.10
6691–7,810	1.09
7,811–8,920	1.08
8921–10,040	1.07
10,041–12,270	1.06
12,271–16,730	1.05
> 16,730	1.04

Source: From Turner J. H. et al. *JAPCA*, **37**:749 (1987). Reprinted with permission from *JAPCA*.

Example 10.7 Gross Cloth Area of Reverse-Air Baghouse

Estimate the gross cloth area required for the conditions of Example 10.6 if a reverse-air baghouse is specified.

Solution

Table 10.1 gives $V = 1.02$ cm/s for fly-ash filtration in reverse-air baghouses. Then, the net cloth area required is $A_c = (23.6)/(0.0102) = 2{,}314$ m^2. From Table 10.2, the gross cloth area is $(2{,}314)(1.17) = 2{,}710$ m^2.

Comment

Notice that the reverse-air baghouse requires almost three times as much area as the pulse-jet to process the same flue gas flow rate.

10.4.2 Filter media

The type of filter material used in baghouses depends on the specific application in terms of chemical composition of the gas, operating temperature, dust loading, and the physical and chemical characteristics of the particulate. A variety of fabrics, either felted or woven (sometimes knit), is available. The selection of a specific material, weave, finish, or weight is based primarily on past experience. For some difficult applications, Gore-Tex, a polytetrafluoroethylene (PTFE) membrane laminated to a substrate fabric (felt or woven) may be used. Because of the violent action of mechanical shakers, spun or heavy weight staple yarn fabrics are commonly used with this type of cleaning. Lighter-weight filament yarn fabrics are used with reverse-air cleaning.

The type of material will limit the maximum operating gas temperature for the baghouse. Cotton fabric is among the least resistant to high temperatures (about 355 K) while fiberglass is the most resistant (about 530 K). Table 10.3 gives the maximum temperature limits of the leading fabric materials.

Example 10.8 Fabric Filter for Open-Hearth Steel Plant (Licht 1980).

The flue gases from an open-hearth steel plant flow at the rate of 110 m^3/s at 1,000 K and 101.3 kPa with an iron oxides particulate loading of 0.0026 kg/m^3. The water content of the gases is 8%. Design a fabric filter system to reduce the particulate loading of the flue gases to the corresponding NSPS.

Solution

The NSPS for particulate emissions from steel plants is 50 mg/dscm (see Table 1.3). To calculate the overall removal efficiency required, correct the actual particulate loading to dry, standard conditions: $(2{,}600)(1{,}000)/[(0.92)(273)] = 10{,}352$ mg/dscm. The overall efficiency

Table 10.3 Properties of Leading Baghouse Fabric Materials

Fabric	Temp.[a] (K)	Acid resistance	Base resistance
Cotton	355	Poor	Very good
Creslan[b]	395	Good	Very good
Dacron[c]	410	Good	Very good
Dynel[c]	345	Excellent	Fair
Fiberglas[d]	530	Fair-good	Fair-good
Filtron[e]	405	Excellent	Very good
Gore-Tex[f]	Depends on backing	Depends on backing	Depends on backing
Nomex[c]	465	Fair	Excellent
Nylon[c]	365	Fair	Excellent
Orlon[c]	400	Excellent	Fair-good
Polypropylene	365	Excellent	Excellent
Teflon[c]	505	Excellent	Excellent
Wool	365	Very good	Poor

[a] Maximum continuous operating temperature.
[b] American Cyanamid registered trademark.
[c] Du Pont registered trademark.
[d] Owens-Corning Fiberglass registered trademark.
[e] W. W. Criswell Division of Wheelabrator-Fry Inc.
[f] W. L. Gore and Co. registered trademark.
Source: From Turner J. H. et al. *JAPCA,* **37**:749 (1987). Reprinted with permission from *JAPCA*.

is $\eta_M = (10{,}352 - 50)/10{,}352 = 0.9952$ (99.52%). This efficiency is within the limits achievable by a properly designed fabric filter. However, the temperature of the gases is well above the operating limits of available fabric materials. A precooling system is required, but a fabric to operate as hot as possible should be sought.

Table 10.3 shows that fiberglass bags can operate continuously at temperatures up to 530 K. Previous experience shows that fiberglass filters can remove iron oxides dust with efficiencies higher than 99.6% in sonic-assisted, reverse-air baghouses at an air-to-cloth ratio of 1.27 cm/s (Licht 1980). Choose fiberglass bags at an operating temperature of 530 K.

There are, basically, three alternatives for cooling the gases to 530 K: (1) dilution with ambient air; (2) evaporation of a water spray into the gas ; and (3) cooling in a waste heat boiler. Dilution with ambient air will more than triple the filter area required, while evaporative cooling will increase it by about 30% (see Problem 10.10). They both waste the precious thermal energy contained in the gases. A waste heat boiler recovers close to 50% of that energy with no penalty in terms of filtering area required. Figure 10.5 is a schematic diagram of the proposed system.

Assume that the waste gases behave like air (ρ = 0.3524 kg/m^3, C_p = 1.08 kJ/kg-K) and calculate the enthalpy change they experience in going from 1,000 K to 530 K: ΔH_g = (110)(0.3524)(1.08)(530 – 1,000) = –19,677 kJ. The enthalpy change of 1 kg of liquid water at 298 K that is converted to saturated steam at 445 K (830 kPa) is 2,665.4 kJ. Assuming 10% energy losses, the amount of steam produced in the waste heat boiler is (0.9)(19,677) /(2,665.4) = 6.64 kg/s or 23,900 kg/h. To estimate the heat-transfer area, calculate the logarithmic mean temperature difference: LMTD = [(1,000 – 445) – (530 – 298)]/ln (555/232) = 370 K. Assume that the overall heat-transfer coefficient is 300 W/m^2-K. The heat-transfer area is (19,677)/[(0.3)(370)] = 177 m^2.

The gas flow rate entering the baghouse is (110)(530/1,000) = 58.3 m^3/s; the dust loading is (0.0026)(110/58.3) = 0.005 kg/m^3. For a gas-to-cloth ratio of 1.27 cm/s, the net cloth area is 58.3/0.0127 = 4,590 m^2. From Table 10.2, the gross cloth area is (1.11)(4,590) = 5,100 m^2. A reasonable design might use 10 compartments, with 50 bags per compartment each 0.3 m diameter and 10.7 m long.

The length of the filtration time depends on the maximum allowable pressure drop through the fabric-dustcake. Assume that, for this case, the maximum is 2.5 kPa For iron oxides dust on fiberglass, the parameters on the filter drag model— Eq. (10.5)—are S_e = 142 kPa-s/m and $K_2 = 1.21 \times 10^6$ s^{-1} (Licht 1980). Then, 2.5 = (142)(0.0127) + $(1.21 \times 10^6)(0.005)(0.0127)^2 t$ /1,000. Solving, t = 713 s = 12 minutes.

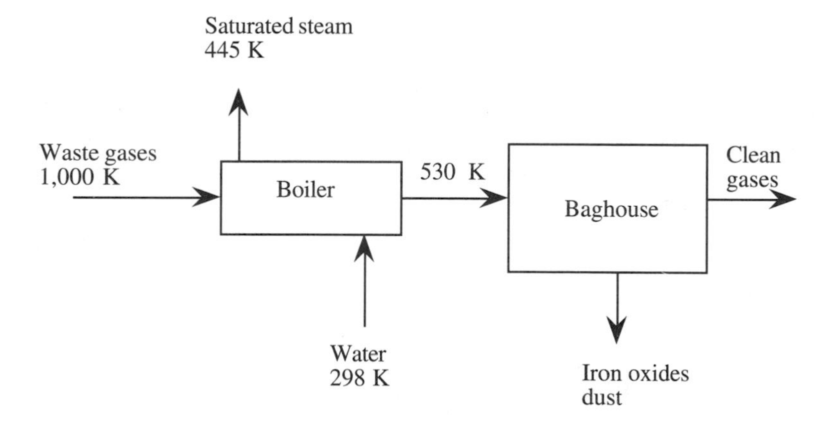

Figure 10.5 Waste-heat boiler of Example 10.8

10.5 COSTS OF FABRIC FILTRATION

Your objectives in studying this section are to

1. Estimate the equipment cost of the most common baghouse types.
2. Estimate the corresponding TCI.
3. Estimate the TAC associated with fabric filtration.

The annualized cost of owning and operating a baghouse can be very high. The capital cost for a fabric filter system can be estimated based on the gross cloth area. The cost also depends on the type of baghouse, whether it is made of stainless or mild steel, and whether or not it is insulated. Furthermore, standardized modular baghouses are less expensive than custom-built units.

10.5.1 Equipment Cost

Baghouse equipment costs consist of two components: the baghouse unit and the bags. Both costs are functions of the gross cloth area, Ac. The baghouse unit cost is, in turn, comprised of the basic baghouse cost and the costs of "add-ons" for stainless steel and insulation. The prices for the basic unit or add-ons are linear functions of the gross cloth area (Vatavuk 1990),

$$EC = a + bA_c \tag{10.12}$$

where a and b are regression constants listed in Table 10.4. Notice that the parameters in Table 10.4 apply only to certain gross cloth area ranges.

Table 10.5 gives selected bag prices (in $\$/m^2$) for pulse-jet, shaker, and reverse-air baghouses. The pulse-jet bag prices in Table 10.5 do not include the prices of protective cages. Mild steel cages will add approximately $\$13/m^2$ of filter area (June 1990 dollars). Stainless steel cages will cost about $\$32/m^2$.

Example 10.9 Equipment Cost of Reverse-Air Baghouse

Estimate the cost of the reverse-air baghouse of Example 10.8. The baghouse shell material can be mild steel. Insulation is required. The bags should be 0.3-m diameter with sewn-in snap rings.

Table 10.4 Parameters for Fabric Filter Equipment Costs[a]

Baghouse Type	Area Range m²	Component	*a*	*b*
Shaker—intermittent	370–1,500	Basic unit	4,120	84.6
		Stainless steel	14,000	42.9
		Insulation	2,200	5.7
Shaker—continuous	370–5,600	Basic unit	43,800	93.8
		Stainless steel	29,700	61.1
		Insulation	0	4.21
Pulse-jet—common-housing	370–1,500	Basic unit	11,280	69.8
		Stainless steel	12,700	59.1
		Insulation	1,670	11.7
Pulse-jet—modular	370–1,500	Basic unit	55,140	92.0
		Stainless steel	29,300	87.7
		Insulation	3,500	26.1
Reverse-air	930-7,500	Basic unit	34,200	88.0
		Stainless steel	16,500	68.5
		Insulation	1,320	10.0
Custom-built—all types	9,300-37,200	Basic unit	263,000	69.3
		Stainless steel	108,400	28.7
		Insulation	70,200	8.0

[a] Prices updated to June 1990. Source: From Turner J. H. et al. *JAPCA*, **37**:1105 (1987). Reprinted with permission from *JAPCA*.

Solution

For a reverse-air baghouse with $930 < A_c < 7{,}500$ m², Table 10.4 gives the following equipment costs:
Basic unit: 34,200 + (88)(5,100) = $483,000
Insulation add-on: 1,320 + (10)(5,100) = $52,320
For 0.3-m diameter Fiberglass bags with rings, Table 10.5 gives
Bags cost: (12.2)(5,100) = $62,220

Table 10.5 Selected Bag Prices ($ in June 1990/m^2)

	Pulse Jet		Shaker		Reverse Air	
Material[a]	TR[b]	BBR[c]	Strap[d]	Loop[e]	w/Rings[f]	w/o Rings
PE	5.4–7.3	4.0–4.6	5.6	5.4	5.7–5.8	4.0
PP	5.5–7.5	4.1–4.9	5.9	5.6	NA[g]	NA
NO	19.2–23.1	14.6–16.9	15.7	14.5	20.7–21.2	14.3–14.8
FG	13.3–15.9	11.7–15.3	NA	NA	9.3–12.2	6.5–8.5
TF	83.8–111	82.7–108	NA	NA	NA	NA
CO	NA	NA	5.5	4.8	NA	NA

[a] Materials: PE = polyester; PP = polypropylene; NO = Nomex; FG = Fiberglass with 10% Teflon; TF = Teflon felt; CO = cotton. [b,c] Bag removal method: TR = top bag removal; BBR = bottom bag removal. [d,e] Bag top design. [f] Prices are given for bags with and without sewn-in snap rings. [g] Not applicable. Ranges shown reflect different bag diameters. Pulse-jet bag diameters: 0.1–0.2 m; reverse air: 0.2–0.3 m; shaker: 0.13 m. Source: From Turner J. H. et al. *JAPCA*, **37**:1105 (1987). Reprinted with permission from *JAPCA*.

Example 10.9 Continuation

The equipment cost is EC = 483,000 + 52,320 + 62,220 = $598,000 in June 1990.

Example 10.10 Cost of Pulse-Jet Baghouse for Municipal Solid Waste (MSW) Incinerator

The air pollution control system of a MSW incinerator consists of a spray dryer followed by a pulse-jet fabric filter (Frame 1988). The flue gas entering the baghouse flows at the rate of 35 m^3/s at 500 K and 1 atm and contains significant amounts of HCl. A design gas-to-cloth ratio of 0.025 m/s is specified when using Teflon felt bags, which are extremely resistant to the harsh environment prevailing in the baghouse. Estimate the fabric filter equipment cost.

Solution

The net filtration area is (35)/(0.025) = 1,400 m^2. Assuming that the cleaning is done on-line, the gross filtration area is also 1,400 m^2 and a common housing design is appropriate. Because of the corrosive conditions and high temperature, stainless steel construction and insulation are required. The diameter of the bags is 0.1 m, and they are of the top removal type. Stainless steel cages are also required. The following table summarizes the disbursements that comprise the equipment cost for this system.

Item	Cost ($ in June 1990)
Basic unit	108,580
Stainless steel add-on	95,440
Insulation add-on	18,050
Teflon felt bags	117,320
Stainless steel cages	44,800
Equipment cost	384,190

10.5.2 TCI

The total capital investment is estimated from a series of factors applied to the purchased equipment cost to obtain direct and indirect costs of installation. The TCI is the sum of these three costs. Table 10.6 gives the required factors.

Example 10.11 TCI for Reverse-Air Baghouse

Estimate the total capital investment for the baghouse of Examples 10.8 and 10.9. Assume that no special site preparation or buildings are needed. The cost of auxiliary equipment, excluding the waste heat boiler, is $150,000.

Solution

From Example 10.9, the equipment cost for the baghouse—including the bags—is $598,000. The cost of the waste heat boiler to produce 23,900 kg/h of steam saturated at 445 K (830 kPa) must be estimated. Based on data published by Peters and Timmerhaus (1991), the cost of the boiler for steam pressures up to 1,800 kPa can be estimated from:

Table 10.6 Capital Cost Factors for Fabric Filters.

Costs	Factor
Direct costs	
Purchased equipment cost	
Fabric filter, plus bags, plus auxiliary equipment	A
Instruments and controls	0.10 A
Taxes and freight	0.08 A
Total purchased equipment cost (B)	1.18 A
Installation direct costs	0.72 B + SP[a] + Bldg[b]
Total direct costs	1.72 B + SP + Bldg
Indirect costs	0.45 B
Total capital investment	2.17 B + SP + Bldg

[a] SP = site preparation.. [b] Bldg = buildings. Source: From Turner J. H. et al. *JAPCA*, **37**:1105 (1987). Reprinted with permission from *JAPCA*.

$$EC = 40(\dot{m})^{0.84} \qquad 1{,}400 < \dot{m},\ kg/h < 180{,}000 \tag{10.13}$$

where

EC = cost of the boiler, in June 1990
$\dot{m}$ = rate of steam generation, kg/h

The cost calculated with Eq. (10.13) is for a complete package boiler plant, which includes feed-water deaerator, boiler feed pumps, chemical injection system, and stack. For a rate of steam production of 23,900 kg/h, Eq. (10.13) gives a boiler cost of \$190,500. From Table 10.6, A = 598,000 + 190,500 + 150,000 = \$938,500. The purchased equipment cost is B = 1.18 A = \$1,107,400. TCI = 2.17 B = \$2,403,000 (in June 1990).

10.5.3 TAC

Direct annual costs include operating and supervisory labor, replacement bags, maintenance labor and materials, utilities, and dust disposal. Indirect annual costs include capital recovery, property tax, insurance, administrative costs, and overhead.

Typical operating labor requirements are 2 to 4 h per shift for a wide range of filter sizes. Supervisory labor is taken as 15% of operating labor. Maintenance labor varies from 1 to 2 h per shift. Maintenance materials costs are assumed to be equal to maintenance labor costs (Vatavuk and Neveril 1980). The major replacement part items are filter bags, which have a typical operating life of 2 yr. The bag replacement labor is approximately $2.00/m^2 of bag area (Turner et al. 1987b).

Electric power is required to operate system fans and cleaning equipment. The pressure drop through the system is the sum of the pressure drop through the baghouse structure and ductwork and the tubesheet pressure drop. The contribution of the baghouse structure and ductwork is approximately 2 kPa. The tubesheet pressure drop is estimated with Eq. (10.3) when the filter drag coefficients are known. Cleaning energy for reverse-air systems can be calculated from the number of compartments to be cleaned at one time (usually one, sometimes two), and the reverse gas-to-cloth ratio (from one to two times the forward gas-to-cloth ratio). Reverse-air pressure drop is about 1.7 kPa. The reverse-air fan generally runs continuously. Typical energy consumption for a shaker cleaning system can be estimated from (Turner et al. 1984b):

$$W_s = 6.5 \times 10^{-5} A_c \tag{10.14}$$

where

W_s = power, in kW

A_c = gross cloth area, in m^2

Cooling process gases to acceptable temperatures by water evaporation or in a waste heat boiler require plant water. Pulse-jet filters use compressed air at pressures of about 500 to 800 kPa. Typical consumption is about 2 standard m^3/1,000 actual m^3. If the collected dust can not be recycled or sold, the costs for final disposal must be included. Disposal costs are site specific, but they are typically about $30 to $50/metric ton including transportation.

Capital recovery costs are based on the total depreciable investment less the cost of replacing the bags. The lifetime of the fabric filter system is typically 20 yr. Property taxes, insurance, and administrative charges are about 4% of the TCI. The overhead is calculated as 60% of the sum of all labor and the maintenance materials.

Example 10.12 TAC for Reverse-Air Baghouse

Estimate the TAC for the baghouse of Example 10.11. The system operates 8,640 h/yr. The operating labor rate is $12/h, the maintenance labor rate is $15/h. The cost of electricity is $0.06/kW-h, process water costs $0.26/metric ton. The dust recovered can be returned to the process at no additional cost. The recovery credit for the waste-heat steam is $2/metric ton. The pressure drop through the waste heat boiler is 2 kPa. The minimum attractive rate of return is 12%/yr; the system lifetime is 20 years. The mechanical efficiency of the fans is 65%.

Solution

Table 10.7 gives the direct and indirect annual costs as calculated from factors given previously. To calculate the annualized cost of replacing the bags, the total cost of replacement, including labor, taxes, and freight must be estimated. The cost of purchasing the bags is multiplied by a factor of 1.08 to account for taxes and freight. The cost of replacement labor is estimated at a rate of $\$2/m^2$ of filtering area. The capital recovery factor for a 2-yr lifetime and $i = 0.12$/yr is 0.592.

To calculate the electricity cost to operate the main fan, the total pressure drop through the system must be estimated. From Example 10.8, the maximum tubesheet pressure drop is 2.5 kPa. The total pressure drop including the waste heat boiler, is 6.5 kPa. Assuming that the fan is located at the entrance of the baghouse, the gas flow rate through the fan is 58.3 m^3/s. Therefore, the power to operate it is (58.3)(6.5)/(0.65) = 583 kW. A reverse-air fan will operate continuously to clean one (out of 10) compartment at a time. The reverse gas-to-cloth ratio is equal to the forward gas-to-cloth ratio: 0.0127 m/s. The reverse air flow is (5,100)(0.1)(0.0127) = 6.48 m^3/s. The corresponding pressure drop is 1.7 kPa, therefore, the power to operate the reverse-air fan is (6.48)(1.7)/(0.65) = 17 kW.

The capital recovery factor (0.134) for the depreciable investment corresponds to a lifetime of 20 yr and a minimum attractive rate of return of 12 %/yr. As usual when calculating the capital recovery cost, the total cost of replacing the bags is subtracted from the total depreciable investment to avoid double accounting. Because the recovered dust can be returned to the process at no additional cost, there is no cost associated with final disposition of the solid residue. There is a substantial recovery credit for the waste heat steam, which helps to partially offset the annual costs. The total annual cost for the system is \$504,800. The total amount of particulate removed is (0.0026)(110)(3,600) (8,640)/(1,000) = 8,900 metric ton/yr. Therefore, the cost effectiveness of the system is \$56.7/metric ton of dust collected.

10.6 CONCLUSION

Fabric filtration is rapidly becoming an accepted and frequently preferred particulate matter control technology. Interest in baghouses will continue increasing as air emissions standards become more stringent and concerns over fine particulates increase. They are gaining acceptance for use downstream of dry FGD systems and fluidized bed boilers. The baghouse provides additional SO_2 removal, and particulate collection is not affected by the high alkali content of the products of dry SO_2 removal. The relative insensitivity of fabric filters to variations in dust electric resistivity makes them more attractive than ESPs in many applications. *Electrically enhanced fabric filters,* an emerging technology, operate at higher gas-to-cloth ratios than conventional baghouses, and at greatly reduced pressure drops.

Table 10.7 Annual Costs for the Fabric Filter of Example 10.12

DAC	
Operating labor: 6 × 360 × \$12 =	\$25,900
Supervisory labor: 0.15 × 25,900 =	3,900
Maintenance	
Labor: 3 × 360 × \$15 =	16,200
Materials	16,200
Replacement parts, bags: (\$62,200 × 1.08 + \$2 × 5,100) × 0.592 =	45,810
Utilities	
Electricity: (583 + 17) × 8,640 × \$0.06 =	311,000
Process water for boiler: 23.9 × 8,640 × \$0.26 =	53,700
Total DAC	\$472,700
IAC	
Overhead: (25,900 + 3,900 + 2 × 16,200) × 0.60 =	37,300
Property tax, insurance, administration: 0.04 × \$2,403,000 =	96,120
Capital recovery cost	
(\$2,403,000 – \$2 × 5,100 – \$62,200 × 1.08) × 0.134 =	311,640
Total IAC	\$445,100
Recovery credit for waste heat steam: 23.9 × 8,640 × \$2 =	\$413,000
TAC	\$504,800

REFERENCES

Carr, R. C., and Smith, W. B. *JAPCA*, **34**:79 (1984a).

Carr, R. C., and Smith, W. B. *JAPCA*, **34**:584 (1984b).

Cooper, D. W., and Riff, M. *JAPCA*, **33**:770 (1983).

Crawford, M. *Air Pollution Control Theory*, McGraw-Hill, New York (1976).

Cushing, K. M., Merritt, R. L., and Chang, R. L. *J. Air Waste Manage. Assoc.*, **40:** 1051 (1990).

Dennis, R., and Klemm, H. A. *Fabric Filter Model Change: Vol. I, Detailed Technical Report*, EPA-600/7-79-043a, NTIS PB 293551 (February 1979).

Dennis, R., and Klemm, H. A. *JAPCA*, **30:**203 (1980).

Ellenbecker, M. J. "Pressure Drop in a Pulse-Jet Fabric Filter," Ph. D. thesis, Harvard University, Boston (1979).

Ensor, D. S. et al. "Kramer Station Fabric Filter Evaluation," RP1130-1, Final Report CS-1669, Electric Power Research Institute, Palo Alto, CA (1981).

Felix, L. G., and Merritt, R. L. *JAPCA*, **36:**1075 (1986).

Flagan, R. C., and Seinfeld, J. H. *Fundamentals of Air Pollution Engineering*, Prentice Hall, Englewood Cliffs, NJ (1988).

Frame, G. B. *JAPCA*, **38**:1081 (1988).

Licht, W. *Air Pollution Control Engineering: Basic Calculations for Particulate Collection*, Marcel Dekker, New York (1980).

Miller, R. L., Single, D. A., and Smit, W. R. "Successful baghouse operating experience at the Plains Electric Generation and Transmission Cooperative—Escalante Generating Station # 1," presented at the Third Conference on Fabric Filter Technology for Coal-Fired Power Plants (1985).

Peters, M. S., and Timmerhaus, K. D. *Plant Design and Economics for Chemical Engineers*, (4th ed.) McGraw-Hill, New York (1991).

Turner J. H., Viner, A. S., McKenna, J. D., Jenkins, R. E., and Vatavuk, W. M. *JAPCA,* **37**:749 (1987a).

Turner J. H., Viner, A. S., McKenna, J. D., Jenkins, R. E., and Vatavuk, W. M. *JAPCA,* **37**:1105 (1987b).

Vann Bush, P., Snyder, T. R., and Chang, R. L. *JAPCA*, **39:**228 (1989).

Vatavuk, W. M. and Neveril, R. B. *Chemical Engineering*, p. 157 (Nov. 3, 1980).

Vatavuk, W. M. *Estimating Costs of* Air *Pollution Control*, Lewis, Chelsea, MI (1990).

Viner A. S., Donovan, R. P., Ensor, D. S., and Hovis, L. S. *JAPCA,* **34**:872 (1984).

PROBLEMS

The problems at the end of each chapter have been grouped into four classes (designated by a superscript after the problem number)

Class a: Illustrates direct numerical application of the formulas in the text.
Class b: Requires elementary analysis of physical situations, based on the subject material in the chapter.
Class c: Requires somewhat more mature analysis.
Class d: Requires computer solution.

10.1[a]. Effect of the gas-to-cloth ratio on penetration

If the gas-to-cloth ratio of Example 10.1 increases to 0.015 m/s, calculate the average overall penetration for a 20-min cycle.

Answer: 1.65%

10.2[b]. Effect of operation cycle length on penetration

Determine the operation cycle length that reduces the average overall penetration of Example 10.1 to 1.0%.

Answer: 36 min

10.3[c]. Penetration and pressure drop through a fabric filter

The filter drag parameters for the fly-ash and fabric combination of Example 10.1 are $S_e = 100$ kPa-s/m and $K_2 = 10^6$ s^{-1}. Choose a gas-to-cloth ratio and operation time such that the average overall penetration for the cycle does not exceed 1.0% and the pressure drop does not exceed 2.5 kPa.

10.4^c. Alternative form of the morphology factor

(a)Show that an alternative expression for the morphology factor defined by Eq. (10.2) is

$$M = \frac{A\rho}{6} \prod_{i=1}^{N} D_{pi}^{x_i} = \frac{A\rho}{6} \text{ MGD}$$

where

x_i is the mass fraction collected in chanel i

MGD = mass-weighted geometric mean diameter

(b)The Monticello Generating Station is a coal-fired power plant operated by the Texas Utilities Generating Company (Felix et al. 1986). Part of the flue gases go to a 36-compartment shaker baghouse for particle removal. The baghouse experienced higher-than-expected pressure drops and large penetration spikes, even though it operated at reasonably low gas-to-cloth ratios. In an effort to find an explanation to these unexpected operational problems, samples of the fly ash were analyzed. The specific surface area was measured at 0.85 m^2/g and the true particle density at 2,420 kg/m^3 (Vann Bush et al. 1989). The following table gives a typical size distribution analysis.

Dpi (μm)	Mass Fraction	*Dpi* (μm)	Mass Fraction
0.02	5×10^{-5}	0.80	0.0503
0.05	2×10^{-4}	2.00	0.0600
0.08	1.6×10^{-3}	5.00	0.1500
0.10	0.0010	7.00	0.3540
0.20	0.0008	10.0	0.2640
0.30	0.0010	20.0	0.1020
0.40	0.0070	50.0	0.0010
0.60	0.0100	70.0	8×10^{-5}

Estimate the value of the morphology factor for the Monticello fly-ash and discuss whether the operational problems experienced can be explained in terms of the particle's morphology.

Answer: M = *2.22*

10.5[a]. Morphology of particles from atmospheric fluidized-bed combustor (AFBC)

The TVA atmospheric fluidized-bed combustor is a 20 MW power plant, which burns eastern bituminous coal. The flue gases flow through a reverse-air baghouse for particle removal. The fly-ash has the following properties: true density, 2,620 kg/m^3; specific surface area, 14.44 m^2/g; mass geometric mean diameter, 5.0 μm. Estimate the morphology factor for this fly-ash (see Problem 10.4). Do you expect penetration problems owing to bleed-through?

Answer: 31.5

10.6[b]. Filter drag model parameters estimation

Estimate the parameters in Eq. (10.4) based on the following test data.

Gas-to-cloth ratio = 0.013 m/s

Inlet dust loading = 0.005 kg/m^3

Time (min)	5	10	15	20	25	30
ΔP (kPa)	0.33	0.49	0.55	0.60	0.64	0.70

Answer: $K_2 = 1.96 \times 10^5\ s^{-1}$

10.7[a]. Maximum and average tubesheet pressure drops

The fabric filter of Problem 10.6 operates with an inlet dust loading of 0.01 kg/m^3. The filtration time is 1 h. Estimate the maximum and average tubesheet pressure drops during the filtration time.

Answer: $\Delta P_{max} = 1.58\ kPa$

10.8[a] Tubesheet pressure drop in a pulse-jet baghouse

A pulse-jet baghouse uses Dacron felt bags for the control of fly ash emissions. The gas-to-cloth ratio is 0.030 m/s and the inlet dust loading is 0.02 kg/m^3. The bags are cleaned at 10-min intervals with pulses of air at 690 kPa gauge. If $K_2 = 2.0 \times 10^5$ s^{-1}, estimate the maximum tubesheet pressure drop.

Answer: 2.61 kPa

10.9[b]. Areal density distribution in a baghouse

(a)The following areal density distribution data were measured in a baghouse:

Areal density (W, kg/m^2)	Cumulative percent less than W
0.56	1
0.77	5
1.10	20
1.60	50
2.40	80
4.10	98

Show that these data follow a log-normal distribution and estimate the median and geometric standard deviation

(b) Estimate the ratio of the harmonic mean to the reciprocal of the arithmetic mean.

Answer: 1.18

(c) Estimate the arithmetic mean of the distribution

Answer: 1.74 kg/m^2

10.10[b]. Precooling of waste gases upstream of fabric filters

(a) If the waste gases of Example 10.8 are precooled by dilution with ambient air at 298 K, estimate the volumetric flow rate of the gases at the baghouse inlet. Assume that the mean heat capacity of air in the range 298 to 530 K is 1.02 kJ/kg-K. Neglect the effect of heat losses.

Answer: 183 m^3/s

(b) Estimate the flow rate of the waste gases at the baghouse inlet if they are precooled by evaporation of water at 298 K.

Answer: 74.8 m^3/s

10.11[b]. Baghouse heat exchanger

The J. E. Baker Company of York, Pennsylvania processes dolomitic limestone into a variety of agricultural and industrial products (Krout, B., and Kilheffer, J. *JAPCA*,

31:293, 1984). They operate two rotary coal and coke-fired kilns. In 1978, they fitted Numer One kiln with a fiberglass reverse-air baghouse for particulate matter control. The system included a heat exchanger to lower the exhaust gas temperature from 756 K to 533 K before entering the baghouse. The cooling fluid is ambient air which enters the heat exchanger at 300 K and leaves at 590 K. The unit has a heat-transfer area of 1,374 m^2 for an inlet flow rate of 56.7 m^3/s. Estimate the overall heat-transfer coefficient for the unit. Assume that the exhaust gases behave like air.

Answer: 23 W/m^2-K

10.12[a]. Gross cloth area for a reverse-air fabric filter

Estimate the gross cloth area for the reverse-air baghouse of Problem 10.11. The particles coming out of the kiln are a mixture of lime dust and fly-ash.

Answer: 4,400 m^2

10.13[a]. Cloth area for a pulse-jet baghouse

The exhaust gases from a cement plant flow at the rate of 130 m^3/s at 500 K. Calculate the net cloth area for a pulse-jet baghouse to remove the cement dust. Assume on-line cleaning of the bags.

Answer: 3,190 m^2

10.14[a]. Reverse-air baghouse for a coal-fired power plant

Consider the coal-fired power plant of Example 5.1. The plant will consist of two, 150-MW boilers. Design a reverse-air fabric filter for each boiler to remove at least 99 percent of the fly ash in the flue gases. The bags are 0.3 m in diameter and 10.7 m long. There are 120 bags per compartment. Specify the gross cloth area and the number of compartments for each baghouse.

Answer: 22 compartments/unit

10.15[b]. Gas-to-cloth ratios for pulse-jet baghouses

A correlation was presented to estimate the gas-to-cloth ratios for process gas filtration with pulse-jet fabric filters as a function of the temperature, particle size, dust load, and material filtered (Turner et al. 1987a):

$$V = F\left(T - 256\right)^{-0.2335}\left(C_i\right)^{-0.06021}\left(0.7471 + 0.0853 \ln \text{MMD}\right)$$

where

V = gas-to-cloth ratio, m/s

T = temperature, K (between 285 and 410 K); for temperatures above 410 K, use $T = 410$ K

C_i = inlet mass loading, kg/m^3 (between 0.0001 and 0.23)

MMD = inlet aerosol mass median diameter, μm (between 3 and 100)

F = material factor given in the following table

		F		
0.1061	0.0849	0.0707	0.0636	0.0424
Flour	Asbestos	Cement	Fly-ash	Activated carbon
Tobacco	Lime (hydrated)	Coal	Metal oxides	Detergents
Grain	Talc	Limestone	Dyes	Soaps
Sawdust	Gypsum	Silica	Silicates	Milk powder

Estimate the gross cloth area for a pulse-jet filter to process the gases from the open-hearth steel plant of Example 10.8 The MMD of the iron oxides dust is 5 μm (Licht 1980).

Answer: 2440 m^2

10.16[b]. Profit from a baghouse heat exchanger

The cooling air of Problem 10.11 (leaving the heat exchanger at 590 K) is used for drying purposes in the plant, reducing the fuel required for the calcination process. The fuel savings are approximately $100,000/yr. The annual operation and maintenance expenses of the heat exchanger are $20,000/yr. Estimate the profitability index for the capital invested in the heat exchanger. Assume that the useful life of the unit is 15 yr with no salvage value. The following equation can be used to estimate the cost of heat exchangers (Vatavuk, W. M. and Neveril, R. B. *Chemical Engineering*, p.129, July 12, 1982):

$$EC = 14{,}814\, A^{-0.12} \exp\left[0.0672\,(\ln A)^2\right]$$

where

EC = cost in 1978 dollars

A = heat-transfer area, m^2

Assume that TCI = 2.34 EC.

10.17[a]. TCI for a fabric filter system

Estimate the total capital investment for the fabric filter system of Problem 10.14. Because the fabric filters will be located upstream of the FGD system, the flue gases are acidic at this point, which mandates stainless steel fabrication. At the temperaure of operation (530 K) insulation is required and the bags must be made of Fiberglass. The cost of auxiliary equipment for each baghouse is $300,000.

Answer: TCI = $19.9 millions in June 1990

10.18[b]. TAC of a reverse-gas baghouse

Estimate the total annual costs and the cost effectiveness ($/metric ton) for the baghouse system of Problem 10.17. The system operates 8,640 h/yr. The operating labor rate is $25/h, the maintenance labor rate is $28/h. The cost of electricity is $0.06/kW-h; the total pressure drop through the system is 5 kPa. The cost of final fly-ash disposal is $20/metric ton. The minimum attractive rate of return is 12%/yr; the system lifetime is 20 yr. The mechanical efficiency of the fans is 65 percent.

Answer: $77/metric ton

10.19[a]. TCI for a MSW incinerator pulse-jet baghouse

Estimate the total capital investment for the baghouse of Example 10.10. Assume that the cost of the auxiliary equipment is $90,000. No buildings or special site preparation is needed.

Answer: $1,214,000 in June 1990

10.20[c] Design of a pulse-jet filter for a Portland cement kiln

In Problem 8.17, a multiple cyclone system was designed as a precleaner for the control of particulate emissions from a Portland cement kiln. For each of the alternatives considered in that problem, design a pulse-jet fabric filter such that the combined system is capable of satisfying the particulate NSPS for the source. Use the correlation of Problem 10.15 to estimate the gas-to-cloth-ratio for each alternative. Assume that the tubesheet pressure drop can be calculated from Eqs. (10.7) to (10.9), with $K_2 = 1.5 \times 10^5$ s^{-1} and a cleaning interval of 10 min. The lifetimes of the bags and system are 2 yr and 20 yr, respectively, with no salvage value. Operating and maintenance labor rates are $15/h and $20/h, respectively. The dust collected in both devices can be recycled to the process at no additional cost. Compressed air (dried and filtered) is available at $7.00/1,000 standard m^3. For each alternative, specify the following:

(a) Total filtration area

(b) Filter media, number and dimensions of the bags

(c) TCI and TAC of the combined multiple cyclone–fabric filter system.

Choose between the three alternatives suggested based on the EUAC measure of merit. Compare your results to those of Problem 9.22.

Appendix A

Heat Capacity Equations

Heat Capacity Equations for Gases at Low Pressures

Compound	a	$b \times 100$	$c \times 10^5$	$d \times 10^9$	Range, °C
Acetaldehyde	51.06	12.15	-5.12	0.000	25-730
Acetone	71.96	20.10	-12.78	34.76	0-1,200
Acetylene	42.43	6.053	-5.033	18.20	0-1,200
Ammonia	35.15	2.954	0.4421	-6.686	0-1,200
Benzene	74.06	32.95	-25.20	77.57	0-1,200
Isobutane	89.46	30.13	-18.91	49.87	0-1,200
n-Butane	92.30	27.88	-15.47	34.98	0-1,200
Isobutene	82.88	25.64	-17.27	50.50	0-1,200
Cyclohexane	94.14	49.62	-31.90	80.63	0-1,200
Cyclopentane	73.39	39.28	-25.54	68.66	0-1,200
Ethane	49.37	13.92	-5.82	7.28	0-1,200
Ethyl alcohol	61.34	15.72	-8.75	19.83	0-1,200
Ethylene	40.75	11.47	-6.89	17.66	0-1,200
Formaldehyde	34.28	4.27	0.000	-8.69	0-1,200
n-Hexane	137.44	40.85	-23.92	57.66	0-1,200
n-Heptane	158.58	42.81	-16.20	0.000	25-1,230
Hydrogen sulfide	33.51	1.55	0.301	-3.292	0-1,500
Methane	34.31	5.47	0.366	-11.00	0-1,200
Methanol	42.93	8.30	-1.87	-8.03	0-700
Nitric oxide	29.50	0.819	-0.293	0.365	0-3,500
n-Pentane	114.8	34.09	-19.0	42.26	0-1,200
Propane	68.03	22.59	-13.11	31.71	0-1,200
Propylene	59.58	17.71	-10.17	24.6	0-1,200
Sulfur dioxide	38.91	3.90	-3.10	8.61	0-1,500
Toluene	94.18	38.00	-27.86	80.33	0-1,200

$$C_p = a + bT + cT^2 + dT^3$$

Units of C_p are J/(gmol)(K); units of T are °C.

Sources: Himmelblau, D. M. *Basic Principles and Calculations in Chemical Engineering,* 5th ed., Prentice Hall, Englewood Cliffs, NJ (1989). Smith, J. M., and Van Ness, H.C. *Introduction to Chemical Engineering Thermodynamics,* 3rd ed., McGraw-Hill, New York (1975).

Appendix B

Heating Values of Various Compounds

Lower Heating Value of Various Gases— Products are H2O (g) and CO2 (g) at 298 K.

Compound	Lower Heating Value (kJ/kg)
Acetone	29,130
Acetaldehyde	25,100
Acetylene	48,290
Benzene	40,580
Carbon monoxide	10,110
Isobutane	45,520
n-Butane	45,730
Isobutene	45,300
Cyclohexane	43,810
Cyclopentane	44,100
Ethane	47,440
Ethyl alcohol	25,960
Ethylene	47,080
Formaldehyde	17,230
n-Hexane	45,090
n-Heptane	44,830
Methane	50,150
Methanol	18,790
n-Octane	44,690
n-Pentane	45,320
Propane	46,530
Propylene	45,760
Toluene	40,950
o-, m-, and p-Xylene	41,130

Source: Developed from data presented in Perry, J. H. *Chemical Engineers Handbook,* 4th ed., McGraw-Hill, New York (1963).

Appendix C

Clean Air Act Amendments (CAAA) of 1990

On November 15, 1990, President George Bush signed the new amendments to the Clean Air Act. This is probably the most comprehensive environmental legislation of the decade, if not the century. Although the original act was less than 50 pages long, the 1990 amendments are nearly 800 pages long. The legal implications of this landmark legislation are such that some claim that, because of it, no lawyer will ever be unemployed again (Lee 1991).

The CAAA offer significant new approaches to air quality in several areas. The bill addresses acid rain emissions and will phase out production of chemicals contributing to depletion of the stratospheric ozone layer. A major new concept incorporated into the emissions limitations established by the law is a system of tradeable emissions credits. If a facility reduces emissions below the standard or ahead of the timetable set by the law, emissions credits are earned that can be applied to future emissions or sold to another facility. The "right" to emit SO_2 and NO_x will be traded as a free market item.

There are eleven sections, referred to as "titles," to the new law. Following is a brief description of the most important titles.

Title I—Attainment of Ambient Air Quality Standards

The law gives EPA new authority to categorize nonattainment areas—regions that exceed standards for CO, O_3, and PM_{10}. The ozone nonattainment areas will be divided into "severe," "serious," "moderate," and "marginal" classifications, as shown in Table C.1. Los Angeles, which has the nation's worst air quality, will be designated "extreme". Since the Clean Air Act failed to address small sources, the amendment defines as a major source in extreme nonattainment areas any facility emitting as little as 10 ton of VOCs and NO_x annually. In serious regions, major sources will be those emitting 50 ton a year. In moderate and marginal areas, the threshold will be 100 ton. The CO and PM_{10} nonattainment regions will be categorized as serious and moderate.

Table C.1 Ozone nonattainment areas designation

Area Designation	Ozone Level (ppm)	Deadline (yr)
Marginal	0.121–0.137	3
Moderate	0.138–0.159	6
Serious	0.160–0.179	10
Severe I	0.180–0.279	15
Severe II	0.190–0.279	17
Extreme	0.280+	20

Regions designated moderate or worse will be required to make steady progress toward air quality goals. States must reduce overall emissions from all sources 15% per year for the first 6 yr. NO_x sources will be required to install *Reasonably Available Control Technology* (RACT). EPA must identify RACT for 11 categories of hydrocarbon emissions sources. Any region with moderate or worse air quality will be required to install vapor recovery devices at gasoline stations and institute vehicle emissions inspection and maintenance programs.

Title II—Mobile Sources

The bill enacts stricter tailpipe emissions standards for vehicles. The Tier I standards set a NO_x emissions limit of 0.6 g per mile and a hydrocarbons (HC) standard of 0.4 g per mile. Both standards will be phased in over a 2-yr period, beginning with the 1994 model. Tier II standards would be set at 0.2 g/mi for NO_x and 0.125 g/mi for HC in the year 2003, if EPA finds that tougher emissions limits are needed. For CO, the bill sets a 10 g/mi standard in 1994, phased in over 3 yr. Should EPA find in 1996 that further controls are necessary, the Phase II CO limit will be 9.5 g/mi.

By 1995, reformulated gasolines will be required in the nine cities with the worst smog problems. The fuel will contain no more than 1% benzene, 25% aromatics, and 15% less air toxics and VOCs. By the end of the century, stricter standards will apply. In the 40 cities with CO nonattainment problems, the act requires the use of oxygenated fuels.

Title III—Air Toxics

The bill regulates 189 air toxics. Within a year, EPA must publish a list of source categories emitting 10 ton annually of any one toxic or 25 ton annually of a combination of toxic pollutants. The agency must then issue *Maximum Achievable Control Technology* (MACT) standards based on the best demonstrated control technology.

Eight years after the application of the MACT standards, health risk-based standards are slated to take effect if a facility's emissions present a cancer risk of more than one in 1 million.

Title IV—Acid Deposition

The law requires SO_2 emissions to be cut 10 million ton below the 17 million ton EPA estimates utilities emitted in 1980. To accomplish this, the agency will issue emissions allowances in two phases (Claussen 1991). The first phase begins in 1995 and affects 261 units in 110 coal-burning electric utility plants located in 21 eastern and midwestern states. The second phase begins in the year 2000 and sets restrictions on smaller cleaner plants fired by coal, oil, and gas. In both phases, affected utilities will be required to install systems that continuously monitor emissions in order to track progress and assure compliance.

The legislation also calls for a 2 million–ton reduction in NO_x emissions by the year 2000. A significant portion of this reduction will be achieved by utility boilers, which will be required to meet tough new emissions requirements. These requirements will also be implemented in two phases. EPA will establish emission limitations for two types of utility boilers (tangentially fired and dry bottom, wall-fired boilers) by mid-1992. Regulations for all other types of boilers will be issued by 1997.

To help bring about the mandated SO_2 emissions reduction in a cost-effective manner, EPA is implementing a market-based allowance-trading system that will provide power plants great flexibility in reducing emissions. EPA will allocate allowances to affected utilities each calendar year based on formulas provided in the bill. Each allowance permits a utility to emit one ton of sulfur dioxide. To be in compliance, utilities may not emit more SO_2 than they hold allowances for. This means that utilities have to either reduce emissions to the level of allowances they hold or obtain additional allowances to cover their emissions.

Utilities that reduce their emissions below the number of allowances they hold may elect to trade allowances within their systems, bank allowances for future use, or sell them to other utilities for profit. Allowance trading will be conducted nationwide, so that a utility in North Carolina, for example, will be able to trade with a utility in Puerto Rico. Anyone may hold allowances including affected utilities, brokers, environmental groups, and private citizens.

The allowance allocation for each unit affected by Phase I is listed in the bill. An individual's unit allocation is the product of a 2.5-pound SO_2 per million BTU emission rate multiplied by the unit's average fuel consumption for 1985-87. Although the lowest- emitting plants will be able to increase their emissions between 1990 and 2000 by about 20 percent, they may not thereafter exceed their year 2000 emission level. The legislation also establishes a permanent cap on the number of allowances EPA issues to utilities. Beginning in the year 2000, EPA will issue 8.95 million allowances annually. This should effectively limit emissions even if more plants are built and the combustion of fossil fuels increases.

Title V—Permits

EPA had until November 15, 1991, to issue a final rule on air pollution permits, creating a program similar to the *National Pollutant Discharge Elimination System* (NPDES) established under the Clean Water Act. The NPDES program has operated effectively for over 18 yr and applies to all point sources of discharge into the nation's waters. States will be required to implement the CAAA permit program within three years. EPA has review authority and can veto a state's permit program. Each major source will be required to have a renewable 5-yr permit outlining its compliance requirements.

Title V provides a framework on which EPA can construct a permit program that works, but a great deal of effort remains to be done to make this happen.

Title VI—Ozone Depleting Chemicals

The provisions to phase out ozone-depleting chemicals were the first agreements on the bill reached by congressional conferees. CFCs, halons, and carbon tetrachloride will be phased out by 2000; methyl chloroform by 2002. The law also requires the phaseout of hydrochlorofluorocarbons by 2030 and to cap their production by 2015.

Title VII—Enforcement

The bill presents serious civil and criminal liabilities for corporations and corporate officials. Criminal penalties for knowing violations, previously misdemeanors, are now felonies. A conviction of knowingly endangering the public could lead to 15 yr in prison. The bill allows private citizens and groups to seek penalties against violators. The penalties will go to a special fund EPA can use for compliance and enforcement efforts.

REFERENCES

Claussen, E. *EPA Journal,* **17**:21 (January-February 1991).

Lee, B. *J. Air Waste Manage. Assoc.*, **41**:16 (January 1991).

Appendix D

Gaussian Quadrature

Gaussian quadrature is a powerful method of numerical integration which employs unequally spaced base points to estimate the value of

$$\int_a^b f(x)\,dx$$

This method uses a Lagrange polynomial to approximate the function and then applies *orthogonal polynomials* to locate the loci of the base points (Carnahan et al. 1969).

The function to be integrated, $f(x)$, is replaced by a Lagrange polynomial of order n, $P_n(x)$, and its remainder, $R_n(x)$,

$$f(x) = P_n(x) + R_n(x) \tag{D.1}$$

$$f(x) = \sum_{i=0}^{n} L_i(x) f(x_i) + \prod_{i=0}^{n} (x - x_i) \frac{f^{(n+1)}(\xi)}{(n+1)!} \qquad a < \xi < b \tag{D.2}$$

where x_i are the base points, and

$$L_i(x) = \prod_{\substack{j=0 \\ j \neq i}}^{n} \frac{(x - x_j)}{(x_i - x_j)} \tag{D.3}$$

Substituting Eq. (D.1) in the integral:

$$\int_a^b f(x)\,dx = \int_a^b P_n(x)\,dx + \int_a^b R_n(x)\,dx \tag{D.4}$$

Without loss of generality, the interval $[a, b]$ is changed to $[-1, 1]$. The general transformation for converting between x in interval $[a, b]$ and z in interval $[-1, 1]$ is

$$z = \frac{2x - (a+b)}{b-a} \tag{D.5}$$

Define a new function $F(z)$ so that

$$F(z) = f(x) = f\left\{\frac{(b-a)z + (a+b)}{2}\right\} \tag{D.6}$$

Equation (D.2) becomes

$$F(z) = \sum_{i=0}^{n} L_i(z)F(z_i) + \prod_{i=0}^{n}(z - z_i)\frac{F^{(n+1)}(\xi)}{(n+1)!} \qquad -1 < \xi < 1 \tag{D.7}$$

If f(x) *is assumed to be a polynomial of degree* $2n + 1$, then the term $F^{(n+1)}(x)/(n+1)!$ must be a polynomial of degree n, which we will call $q_n(z)$. Then

$$F(z) = \sum_{i=0}^{n} L_i(z)F(z_i) + \prod_{i=0}^{n}(z - z_i)\,q_n(z) \qquad -1 < \xi < 1 \tag{D.8}$$

Integration of both sides of Eq. (D.8) between the limits -1 and 1 gives:

$$\int_{-1}^{1} F(z)\,dz = \int_{-1}^{1} \sum_{i=0}^{n} L_i(z)F(z_i)\,dz + \int_{-1}^{1} \prod_{i=0}^{n}(z - z_i)\,q_n(z)\,dz \tag{D.9}$$

Dropping the right-most integral and taking the summation operator outside of the integral sign in the remaining term , Eq. (D.9) becomes

$$\int_{-1}^{1} F(z)\,dz \approx \sum_{i=0}^{n} F(z_i) \int_{-1}^{1} L_i(z)\,dz \approx \sum_{i=0}^{n} w_i F(z_i) \tag{D.10}$$

where w_i = weight factor corresponding to the transformed base point z_i, given by

$$w_i = \int_{-1}^{1} L_i(z)\,dz = \int_{-1}^{1} \prod_{\substack{j=0 \\ j \neq i}}^{n} \left[\frac{(z - z_j)}{(z_i - z_j)}\right] dz \tag{D.11}$$

Equation (D.10) is a *quadrature integration formula.* The second integral on the right-hand side of Eq. (D.9),

$$\int_{-1}^{1} \prod_{i=0}^{n} (z - z_i)\, q_n(z)\,dz \tag{D.12}$$

is the error term for the quadrature formula. The object at this point is to select z_i in such a way that the error term vanishes. The *orthogonality property of Legendre polynomials* is used for that purpose (Wylie and Barrett 1982). The two polynomials in Eq. (D.12) are expanded in terms of Legendre polynomials. The values of z_i are chosen as the roots of the $(n + 1)$st-degree Legendre polynomial, which makes the error term vanish (Carnahan, et al.1969). The weight factors are calculated from Eq. (D.11).

The quadrature formula developed above is known as the *Gauss-Legendre quadrature* because of the use of the Legendre polynomials. Other orthogonal polynomials, such as Hermite, Chebyshev, or Laguerre may be used in a similar manner to develop a variety of Gauss quadrature formulas.

Returning to the integral in Eq. (D.4), and expressing it in terms of the original variables

$$\int_{a}^{b} f(x)\,dx \approx \frac{(b-a)}{2} \sum_{i=0}^{n} w_i f\left(\frac{z_i(b-a) + b + a}{2}\right) \tag{D.13}$$

Example D.1 Roots and Weight Factors for Three-Point Formula

Calculate the roots and weight factors for the three-point Gauss-Legendre quadrature formula.

Solution

The first step involves generating the 3rd-degree Legendre polynomial, $P_3(z)$. The following equation is a recurrence formula that applies to Legendre polynomials:

$$(n+1)\,P_{n+1}(z)-(2n+1)\,z\,P_n(z)+n\,P_{n-1}(z)=0 \tag{D.14}$$

The first two polynomials are: $P_0(z) = 1.0$; $P_1(z) = z$. Applying the recurrence formula twice, we obtain $P_3(z) = z(5z^2 - 3)/2$. The three roots are easily evaluated: $z_0 = -(3/5)^{0.5} = -0.7746$; $z_1 = 0.0$; $z_2 = 0.7746$.

Equation (D.11) gives the corresponding weight factors. Consider, for example, the evaluation of w_1 corresponding to $z_1 = 0.0$:

$$w_1 = \int_{-1}^{1} \left[\frac{z-0.7746}{0-0.7746}\right]\left[\frac{z+0.7746}{0+0.7746}\right] dz = 0.8889$$

The other two weight factors are calculated in an analogous manner; the answer is $w_0 = w_2 = 0.55556$.

Comments

This example illustrates why the Gauss-Legendre quadrature formulas have been little used in the past. The base points and weight factors are inconvenient numbers to evaluate and use in hand calculations. Fortunately, these values are tabulated in some mathematical tables (Abramowitz and Stegun 1972). Press et al. (1989) present a very efficient computer subroutine to estimate the roots and weight factors for any nth-degree Legendre polynomial. This dramatically simplifies the use of Gauss-Legendre formulas in computerized applications.

Example D.2 Gauss-Legendre Integration

Compute an estimate of ln 2 using a three-point Gauss-Legendre quadrature.

Solution

From the definition of the natural logarithms,

$$\ln 2 = \int_1^2 \frac{1}{x}\,dx$$

Comparing this expression to Eq. (D.4), $f(x) = 1/x$, $a = 1$, $b = 2$. From Eq. (D.13):

$$\ln 2 \approx \frac{1}{2}\sum_{i=0}^{2}(w_i)\left(\frac{2}{z_i+3}\right) \approx \sum_{i=0}^{2}\left(\frac{w_i}{z_i+3}\right)$$

$$\ln 2 = 0.5556 \times (3 - 0.7746)^{-1} + 0.8889 \times (3 + 0)^{-1} + 0.5556 \times (3 + 0.7746)^{-1}$$
$$= 0.69316 \text{ (accurate to four figures!)}$$

REFERENCES:

Abramowitz, M., and Stegun, I.A. *Handbook of Mathematical Functions with Formulas, Graphs, and Mathematical Tables*, 10th printing, Wiley, New York (1972).

Carnahan, B., et al. *Applied Numerical Methods,* Wiley, New York (1969).

Press, W. H., et al. *Numerical Recipes, The Art of Scientific Computing,* Cambridge University Press, New York (1989).

Wylie, C. R., and Barrett, L. C. *Advanced Engineering Mathematics*, 5th ed., McGraw Hill, New York (1982).

Index

A

B

C

D

E

F

G

H

I

L

M

N

O

P

R

S

T

U

V

W

Z